Computation of Three-Dimensional Complex Flows

Edited by Michel Deville,
Spyros Gavrilakis,
and Inge L. Ryhming

Notes on Numerical Fluid Mechanics (NNFM) Volume 53

Series Editors: Ernst Heinrich Hirschel, München (General Editor)
Kozo Fujii, Tokyo
Bram van Leer, Ann Arbor
Michael A. Leschziner, Manchester
Maurizio Pandolfi, Torino
Arthur Rizzi, Stockholm
Bernard Roux, Marseille

Volume 53 Computation of Three-Dimensional Complex Flows. Proceedings of the IMACS-COST Conference on Computational Fluid Dynamics, Lausanne, September 13–15, 1995 (M. Deville / S. Gavrilakis / I. L. Ryhming, Eds.)
Volume 52 Flow Simulation with High-Performance Computers II. DFG Priority Research Programme Results 1993–1995 (E. H. Hirschel, Ed.)
Volume 51 Numerical Treatment of Coupled Systems. Proceedings of the Eleventh GAMM-Seminar, Kiel, January 20–22, 1995 (W. Hackbusch / G. Wittum, Eds.)
Volume 50 Computational Fluid Dynamics on Parallel Systems. Proceedings of a CNRS-DFG Symposium in Stuttgart, December 9 and 10, 1993 (S. Wagner, Ed.)
Volume 49 Fast Solvers for Flow Problems. Proceedings of the Tenth GAMM-Seminar, Kiel, January 14–16, 1994 (W. Hackbusch / G. Wittum, Eds.)
Volume 48 Numerical Simulation in Science and Engineering. Proceedings of the FORTWIHR Symposium on High Performance Scientific Computing, München, June 17–18, 1993 (M. Griebel / Ch. Zenger, Eds.)
Volume 47 Numerical Methods for the Navier-Stokes Equations (F.-K. Hebeker, R. Rannacher, G. Wittum, Eds.)
Volume 46 Adaptive Methods – Algorithms, Theory, and Applications. Proceedings of the Ninth GAMM-Seminar, Kiel, January 22–24, 1993 (W. Hackbusch / G. Wittum, Eds.)
Volume 45 Numerical Methods for Advection – Diffusion Problems (C. B. Vreugdenhil / B. Koren, Eds.)
Volume 44 Multiblock Grid Generation – Results of the EC/BRITE-EURAM Project EUROMESH, 1990–1992 (N. P. Weatherill / M. J. Marchant / D. A. King, Eds.)
Volume 43 Nonlinear Hyperbolic Problems: Theoretical, Applied, and Computational Aspects Proceedings of the Fourth International Conference on Hyperbolic Problems, Taormina, Italy, April 3 to 8, 1992 (A. Donato / F. Oliveri, Eds.)
Volume 42 EUROVAL – A European Initiative on Validation of CFD Codes (W. Haase / F. Brandsma / E. Elsholz / M. Leschziner / D. Schwamborn, Eds.)
Volume 41 Incomplete Decompositions (ILU) – Algorithms, Theory, and Applications (W. Hackbusch / G. Wittum, Eds.)
Volume 40 Physics of Separated Flow – Numerical, Experimental, and Theoretical Aspects (K. Gersten, Ed.)
Volume 39 3-D Computation of Incompressible Internal Flows (G. Sottas / I. L. Ryhming, Eds.)
Volume 38 Flow Simulation on High-Performance Computers I (E. H. Hirschel, Ed.)
Volume 37 Supercomputers and Their Performance in Computational Fluid Mechanics (K. Fujii, Ed.)
Volume 36 Numerical Simulation of 3-D Incompressible Unsteady Viscous Laminar Flows (M. Deville / T.-H. Lê / Y. Morchoisne, Eds.)

Volumes 1 to 29, 45 are out of print.
The addresses of the Editors and further titles of the series are listed at the end of the book.

Computation of Three-Dimensional Complex Flows

Proceedings of the
IMACS-COST Conference on
Computational Fluid Dynamics
Lausanne, September 13–15, 1995

Edited by
Michel Deville, Spyros Gavrilakis,
and Inge L. Ryhming

All rights reserved
© Friedr. Vieweg & Sohn Verlagsgesellschaft mbH, Braunschweig/Wiesbaden, 1996

Vieweg ist a subsidiary company of Bertelsmann Professional Information.

No part of this publication may be reproduced, stored in a retrieval system or transmitted, mechanical, photocopying or otherwise, without prior permission of the copyright holder.

Produced by Langelüddecke, Braunschweig
Printed on acid-free paper
Printed in Germany

ISSN 0179-9614
ISBN 3-528-07653-4

This book is dedicated to the memory of
Prof. I.L. Ryhming,
who passed away, while it was prepared for publication.

FOREWORD

The IMACS-COST conference on "Computational Fluid Dynamics, Three-Dimensional Complex Flows" was held in Lausanne, Switzerland, September 13 − 15, 1995. The scientific sponsors of the conference were

- IMACS: International Association for Mathematics and Computers in Simulation,
- COST: European Cooperation in the field of Scientific and Technical Research,
- ERCOFTAC: European Research Community on Flow, Turbulence and Combustion.

The scientific interests of the IMACS and ERCOFTAC associations are closely related to computational fluid dynamics whereas the European Union programme COST covers a wider range of scientific subjects. The COST 'Action F1' launched in 1992 by Professor I.L. Ryhming deals with "Complex three-dimensional viscous flows: prediction, modelling, manipulation and control". It has several subtopics among which numerical methods and modelling issues are the main areas of research and development.

The meeting gathered together eighty-seven scientists, engineers and researchers from seventeen countries: Belgium, Finland, France, Germany, Greece, Hong Kong, Israel, Italy, Japan, the Netherlands, Norway, Russia, Spain, Sweden, Switzerland, United Kingdom, United States of America. All major numerical approximation methods were discussed: finite differences, finite volumes, finite elements, spectral methods. The topics covered by the sixty communications spanned the full spectrum of computational fluid dynamics: direct numerical simulation, large-eddy simulation, turbulence modelling, free surface flows, non Newtonian fluids, thermal convection, etc. The content of the reports ranged from subjects about fundamental aspects of fluid mechanics like the interaction of perturbed vortex rings or three-dimensional instabilities, to high-technology applications to high-speed trains, engines, combustion, hydraulic machines, etc. Some papers addressed the possibilities of large-scale computing offered by massively parallel computers.

Six invited speakers delivered keynote lectures :

- T. Gatski, NASA Langley Research Center, 'Prediction of complex flow fields using higher-order turbulent closures'
- P. Leca, ONERA, 'Trends in parallel computers and their use in CFD'
- A. T. Patera, MIT, 'Baeysian-validated computer-simulation surrogates for optimization and design'
- A. Quarteroni, Politechnico di Milano, 'Large scale simulation of flows of industrial interest'
- K. Morgan, University of Swansea, 'Unstructured mesh techniques for aerospace applications'
- P. Wesseling, T.U. Delft, 'Computation of turbulent flows in general domains'

The talk of K. Morgan was delivered by N. Weatherhill.

This volume presents the papers contributed to the conference. Contributions appear in alphabetical order according to the surname of the first author. Printed versions of the invited talks will be published at a later date together with extended versions of some papers appearing in this volume.

The organizers of the Conference would like to acknowledge the support of the Swiss Federal Institute of Technology Lausanne, COST, Fonds National Suisse de la Recherche Scientifique and Société de Banque Suisse.

Finally, we would like to express our gratitude to our colleagues and secretarial staff at the *Institut des Machines Hydrauliques et de Mécanique des Fluides* who helped us with the successful organization of the meeting.

Lausanne, December 20, 1995.   M. O. DEVILLE
                                          S. GAVRILAKIS
                                          I. L. RYHMING

IMACS-COST Conference EPFL 13-15.9.1995
Opening Address by Prof. H. Fernholz, TU Berlin

Ladies and Gentlemen,

Inge Ryhming should have welcomed you at this IMACS-COST Conference on Three-dimensional Complex Flows, the name of which is drawn from the first research cooperation in fluid dynamics under the auspices of the European Union, COST F1. This was suggested and started by Inge. He regrets very much not to be able to be here today but sends you his very best wishes for a successful meeting.

So let us welcome you all to EPFL on behalf of the sponsors IMACS, COST, ERCOFTAC and the members of the LOC who have prepared the conference. I am sure that the conference will run as smoothly as all the others which I have attended here.

It is a rare occasion for an experimentalist as myself to open a conference on «Computational Fluid Dynamics» and I could not resist the temptation to make a few remarks from the experimenter's point of view, especially since I started my career in fluid mechanics as a computor.

I remember vividly a discussion «Computation versus Experiment», mainly between my two senior colleagues Walz and Wille, 25 years ago. The apogee of the discussion was the statement of one speaker that in a few years time all important flows could be calculated and that experimenters would no longer be needed. Abrupt end of the discussion with the result that the main protagonists ceased to be on speaking terms for several weeks.

Time has taught us that «in a few years» must have been meant in an asymptotic sense, since similar remarks have been made in many public speeches in front of sponsors and young students who unfortunately were often naive enough to believe what they were told.

Experimenters had a hard time since, that is true, but they have not yet become extinct and the danger that they may have to be put on the list of «endangered species» may be less severe. There are two reasons for it, a local and a more general one:

> - the first is that EPFL has increased its experimental capacity considerably by recruiting Peter Monkewitz and ETH has announced specifically a professorship for «Experimental Fluid Mechanics»
> - the second is that a cooperation between computors and experimentors begins to emerge from the old battleground which gives me hope for the future.

I have noticed with pleasure, for example, that there are two presentations on your programme which have close connections to my own experiments and I am sure there are more cooperations between CFD and experiments that I can recognize at first sight.

For many reasons it is high time that this cooperation gets intensified. It is obvious that both sides can learn a lot from each other, that the combined intellects can achieve new results faster and often with a higher degree of confidence, and that all can get a better insight into the physics of the flows by making use of the shared wealth of data.

In Germany at least some pressure will be exerted on both parties in the future. This will mean that in all big projects research proposals must contain joint computation and experiments where ever possible.

So, computors begin now to look for a good cooperation with the few experimenters while they are still available.

I wish you a good conference and many fruitful discussions.

CONTENTS

	Page
S. AHMED, B.G. SHERLOCK, Q.G. RAYER: Numerical Prediction of Three-Dimensional Thermally Driven Flows in a Rotating Frame	1
R. ATMANI, R. ASKOVIC : Investigation of Separation of the Three-Dimensional Laminar Boundary Layer	8
J. D. BAUM, H. LUO, R. LÖHNER: Large-Scale Blast Simulations	15
M. BREUER, D. LAKEHAL, W. RODI: Flow around a Surface Mounted Cubical Obstacle: Comparison of LES and RANS-Results	22
O. DAUBERT, O. BONNIN, F. HOFMANN, M. HECKER: 3D Flow Computations under a Reactor Vessel Closure Head	31
H.-J. G. DIERSCH: Finite Element Analysis of Three-Dimensional Transient Free Convection Processes in Porous Media	38
P. DI MARTINO, A. TERLIZZI, G. CINQUE: Navier-Stokes 3D Computational Analysis of Incompressible Turbulent Flow in a Curved Rectangular Duct	45
D. DRIKAKIS, R. ZAHNER: Study of Incompressible Flows in Rectangular Channels Using High Order Schemes and Parallel Computing	52
P. DRTINA, M. KRAUSE: Numerical Prediction of Abrasion for Francis Turbine Guide Vanes	60
T.V. DURY, B.L. SMITH: Simulation of Turbulent Flow in a Cylindrical Drum with Multiple Outlets	69
J. DUŠEK, PH. FRAUNIÉ, C. DAUCHY, I. DANAILA: Secondary Instabilities and Transition to Turbulence in Wakes and Jets	78
M. FIEBIG, W. HAHNE, D. WEBER: Heat Transfer and Drag Augmentation of Multiple Rows of Winglet Vortex Generators in Transitional Channel Flow: A Comparison of Numerical and Experimental Methods	88
M.S. FOX-RABINOVITZ: Computational Dispersion Properties of 3-D Staggered Grids for Hydrostatic and Non-Hydrostatic Atmospheric Models	95
A. IVANČIĆ, A. OLIVA , C.D. PÉREZ SEGARRA , H. SCHWEIGER : Numerical Simulation of Oscillatory Regimes in Vertical Cylinders Heated from Below	102
Z. JAWORSKI, M.L. WYSZYNSKI, R.S. BADHAM, K.N. DYSTER, I.P.T. MOORE, N.G. OZCAN-TASKIN, A.W. NIENOW, J. McKEMMIE: Sliding Mesh Simulation of Transitional, Non-Newtonian Flow in a Baffled, Stirred Tank	109
S. KEGHIAN, M. GAZAIX: Computation of Three-Dimensional Laminar and Turbulent Shock Wave/Boundary Layer Interactions	116
K. KNOWLES, M. MYSZKO: Complex Three-Dimensional Jet Flows: Computation and Experimental Validation	123

CONTENTS (Continued)

	Page
V.P. KOROBEINIKOV, V.V. MARKOV: Numerical Modeling of Combustible Dusty Gas Mixture Flow	130
E. LE DEVEHAT, R. GREGOIRE, P. CRESPI, A. KESSLER: Computation of the 3D Turbulent Flow Surrounding a High-Speed Train Mock-Up Including the Inter-Car Gap and the Bogie and Comparisons with LDV Data	137
M.A. LESCHZINER, F.S. LIEN, N. INCE, C.A. LIN: Computational Modelling of Complex 3D Flows with Second-Moment Closure Coupled to Low-Re Near-Wall Models	144
R.C.K. LEUNG, N.W.M. KO: Interaction of Two Coaxial Weakly Perturbed Vortex Rings	154
V. MICHELASSI, F. MARTELLI, F. PIGARI: Aerodynamics of Combustion Chambers for Aeronautical Engines	162
G. MOMPEAN, M. O. DEVILLE: Time-Dependent 3D Numerical Simulation of Oldroyd-B Fluid Using Finite Volume Method	169
A. NAKKASYAN: A re-analysis of Stanford's 30 Degree Bend Experiment	177
C.W. NG, N.W.M. KO: Computation on Vortex Interactions behind Two Circular Cylinders	187
M. NORMANDIN, J.-R. CLERMONT: Three-Dimensional Study of Extrusion Processes by the Stream-Tube Method: Newtonian and Viscoelastic Results	195
R.R. NOURGALIEV, T.N. DINH, B.R. SEHGAL: Natural Convection in Volumetrically Heated and Side-Wall Heated Melt Pools: Three Dimensional Effects	202
A. ORELLANO, H. WENGLE: Numerical Simulation and Analysis of Manipulated Transitional Flow over a Fence	210
R.D. PARKER, J. BRUNS: Calculation of 3D Turbulent Boundary-Layer Flows in an S-Shaped Channel	217
F. PITTALUGA: Experimental Activities within COST-F1, Subgroup A-7, on Laminar-Turbulent Transition: A Summary Report	225
M. PROVANSAL, T. LEWEKE: The Modelling of the Wake of a Torus by the Ginzburg-Landau Equation	234
P. RAUTAHEIMO, T. SIIKONEN: Diagonalization of the Reynolds-Averaged Navier-Stokes Equations with the Reynolds-Stress Turbulence Model	240
E. RAZAFINDRAKOTO, J.L. GRANGE, L. FABRE, H. DELABRIÈRE, S. DELAROFF: 3 D Numerical Simulation of a Natural Draught Cooling Tower Flow	248
S. RHAM, C. ALBONE: An Evolutionary Far-Field Boundary Treatment for the Euler Equations	256

CONTENTS (Continued)

	Page
R.E. ROBINS, D.P. DELISI: 3-D Calculations Showing the Effects of Stratification on the Evolution of Trailing Vortices	264
P. SAGAUT, B. TROFF, T.H. LÊ, TA PHUOC LOC: Large Eddy Simulation of Turbulent Flow past a Backward Facing Step with a new Mixed Scale SGS Model	271
M. SALLABERGER, A. SEBESTYEN: Euler and Navier-Stokes Flow Analysis in a Francis-Runner	278
J. SALOM, J. CADAFALCH, A. OLIVA, M. COSTA: A Subdomain Method in Three-Dimensional Natural and Mixed Convection in Internal Flows	289
M.L. SAWLEY, O. BYRDE, J.-D. REYMOND, D. COBUT: Parallel Multi-Block Computations of Three-Dimensional Incompressible Flows	296
W. SPRÖSSIG, K. GÜRLEBECK: On the Treatment of Fluid Problems by Methods of Clifford Analysis	304
M. SUZUKI, T. MAEDA, N. ARAI: Numerical Simulation of Flow Around a Train	311
L.J.P. TIMMERMANS, F.T.M. NIEUWSTADT: Parallelization of a 3D Code for Simulation of Turbulent Channel Flow on the Cray T3D	318
R.P. TONG: A New Approach to Modelling an Unsteady Free Surface in Boundary Integral Methods with Application to Bubble-Structure Interactions	326
E. TRIC, G. LABROSSE, M. BETROUNI: Numerical Experimentation of the First Transition to Unsteadiness of Air Free Convection in a Differentially Heated Cubic Cavity with Non Active Adiabatic Walls	334
I. VALLET, G.A. GEROLYMOS, P. OTT, A. BÖLCS: Computation of Unsteady 3-D Transonic Flows due to Fluctuating Back-Pressure using k-ε Turbulence Closure	343
N. VANANDRUEL: Thermoconvection Simulation and Control in Melting Glass Furnaces	350
P.R. VOKE, Z. YANG: Computational Methods for Large-Eddy Simulation in 2D Complex Geometries	357
E. von LAVANTE, J. YAO: Three Dimensional Numerical Simulation of the Flow in a Four-Stroke I.C. Engine	365
K. YAKINTHOS, M. BALLAS, P. TAMAMIDIS, A. GOULAS: Numerical Simulation of Three-Dimensional Complex Flows Using a Pressure-Based non-Staggered Grid Method	372
S. ZALESKI, J. LI, R. SCARDOVELLI, G. ZANETTI: Flows with Interfaces: Dealing with Surface Tension and Reconnection	379
O. ZIKANOV: Numerical Simulation of Three-Dimensional Instabilities and Secondary Regimes in Spherical Couette Flow	386

Numerical prediction of three-dimensional thermally driven flows in a rotating frame

S. Ahmed[1], B. G. Sherlock[2] and Q. G. Rayer[3]

(1) Department of Computer Science, Parks College of Saint Louis University, 500 Falling Springs Road, Cahokia, Illinois, 62206, USA

(2) Department of Electrical Engineering, University of Path, Bath, UK

(3) Nuclear Electric Plc., Berkeley Technology Centre, Berkeley, Gloucestershire, GL13 9PB, UK

Abstract

An original approach has been used to solve the three-dimensional unsteady incompressible flow in a uniformly rotating frame. The governing equations of fluid motion are discretized using a control-volume procedure, which ensures the satisfaction of global conservation laws. Pressure and velocities are calculated iteratively at each time-level, yielding a semi-implicit scheme. Buoyancy effects are included by using the Boussinesq approximation. The current form of the energy equation neglects the presence of internal energy point sources and sinks, viscous dissipation of energy and work done by the body force. The model is tested by using it to simulate the three-dimensional incompressible flow in a fluid annulus rotating with uniform angular velocity, with differential heating provided by maintaining the side-walls of the convection chamber at separate temperatures. Fluid heat transports obtained from the model agree well with those obtained from experimental measurement.

1 Introduction

Geophysical fluid systems exhibit complex three-dimensional motions due to the interaction of differential heating and the earth's rotation. The understanding of these systems is assisted by the study of flows in a differentially heated rotating fluid annulus (Figure 1), which shares rotational and thermal forcing with them, but has simple well-defined boundary conditions [1]. Interest in rotating systems with barriers [2] is motivated by systems where zonal flow is obstructed by topographical features.

This paper presents a rigorous approach towards the numerical simulation of thermally driven three-dimensional flows. The incompressible Navier-Stokes

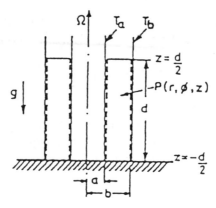

Figure 1 Diagram of fluid annulus. (r,ϕ,z) are cylindrical polar cordinates of a general point P, fixed in a frame rotating uniformly with the annulus at Ω rad. sec^{-1}. a=2.5cm, b=8.0cm, d=14cm.

equations in a rotating frame are solved, with buoyancy represented by the Boussinesq approximation. These governing equations are discretized using the control–volume approach [3, 4], thereby maintaining the statement of global conservation which is their essence. Preliminary results from this model have been presented [5], but the current paper extends them and gives a more detailed description of the model.

Previous computer models of the rotating fluid annulus have used grid–point finite–difference formulations which conserve mass, momentum and energy only in the limit as the simulation mesh becomes infinitely fine. They have been applied both to unobstructed flows [6, 7, 8] and to flows where the annulus is fully blocked by a thermally insulating radial barrier [2, 9]. The finite–difference model has so far failed to simulate several aspects of the flow in the annulus blocked by a radial barrier. Rayer [2] noticed defects in the velocity fields that may be similar to those mentioned by White [8]. Also the model did not calculate the correct value for the temperature drop observed across the thermally insulating barrier during experimental work. This may be due to an insufficiency of grid points adjacent to the barrier to adequately represent any boundary layer that may be present.

2 Computer Model

The new model uses the control–volume approach [3, 4]. The discretisation equation obtained by this method ensures that momentum and energy are exactly conserved over the computational domain. The physical variables of the system are expressed in non–dimensional terms for three–dimensional incompressible flow in a uniformly rotating frame. The reference velocity and characteristic time used for the non–dimensionalisation are

$$|\vec{\Omega}|\mathcal{L} = \omega\mathcal{L}, \text{ and } \frac{1}{|\vec{\Omega}|} = \frac{1}{\omega},$$

where $|\vec{\Omega}| = \omega$ and $\mathcal{L}$ is a characteristic length–scale. Then the non–dimensional variables become

$$\vec{x}^* = \frac{\vec{x}}{\mathcal{L}}, \quad \vec{u}^* = \frac{\vec{u}}{\omega\mathcal{L}}, \quad t^* = \omega t, \quad \mu^* = \frac{\mu}{\mu_0}, \quad \rho^* = \frac{\rho}{\rho_0}, \quad k^* = \frac{k}{k_0},$$

$$T^* = \frac{T - \mathcal{T}}{T_b - \mathcal{T}}, \quad \alpha^* = \alpha(T_b - \mathcal{T}), \quad p^* = \frac{p - \rho_0 g(d-z) - \frac{1}{2}\rho_0\omega^2 r^2}{\rho_0\omega^2\mathcal{L}^2}.$$

The parameters listed above are: position vector, $\vec{x} = (r, \phi, z)$; fluid veclcity, $\vec{u}^* = (u^*, v^*, w^*)$; reference fluid velocity, $\vec{\mathcal{U}} = (\mathcal{U}, \mathcal{V}, \mathcal{W})$; time, t; fluid viscosity, μ; fluid density, ρ; thermal conductivity, k; temperature, T; mean fluid temperature, $\mathcal{T}$; pressure, p; thermal expansion coefficient, α and gravity, g. T_b is the temperature of the hot side-wall. With these definitions the differential operators become,

$$\frac{\partial}{\partial t} = \omega\frac{\partial}{\partial t^*}, \quad \nabla = \frac{1}{\mathcal{L}}\nabla^*, \text{ where } \nabla^* = \hat{x}_i\frac{\partial}{\partial x_i^*}.$$

Thus using the Boussinesq approximation, the momentum conservation equations for three–dimensional incompressible baroclinic flow in a rotating frame can be written in non–dimensional form as

$$\frac{\partial \vec{u}^*}{\partial t^*} + \nabla^*.(\vec{u}^*\vec{u}^*) = -\nabla^* p^* + \text{Gr.Ro}^2 T^*\hat{z} - r^*\alpha^* T^*\hat{r} - 2u^*\hat{\phi} + 2v^*\hat{r} + \text{Ro}\nabla^{*2}\vec{u}^*, \tag{2.1}$$

where Gr is the Grashoff number

$$\text{Gr} = \frac{\rho_0^2 g\mathcal{L}^3\alpha(T_b - \mathcal{T})}{\mu_0^2},$$

and Ro is the Rossby number

$$\text{Ro} = \frac{\mu_0}{\rho_0\omega\mathcal{L}^2}.$$

The energy equation becomes

$$\frac{\partial T^*}{\partial t^*} + (\vec{u}^*.\nabla^*)T^* = \frac{\text{Ro}}{\text{Pr}}\nabla^*.(k^*\nabla^* T^*),$$

where Pr is the Prandtl number,

$$\text{Pr} = \frac{\mu_0 C_{P0}}{k_0}.$$

Equation (2.1) can be differenced in a control-volume manner with respect to the staggered grid and control-volume shown in Figure 2 to yield the discretised momentum conservation law, the r-component of which is

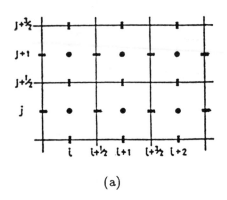
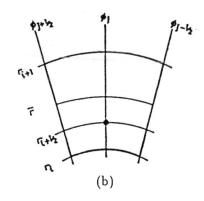

Figure 2 (a) Staggered grid used in control volume formulation. (b) Typical control-volume in (r,ϕ) plane.

$$\frac{u^{n+1}_{i+\frac{1}{2}}-u^n_{i+\frac{1}{2}}}{\Delta t} + \frac{[u_{i+1}u_{i+1}r_{i+1}-u_iu_ir_i]}{r_{i+\frac{1}{2}}\Delta r_{i+\frac{1}{2}}} + \frac{1}{r_{i+\frac{1}{2}}\Delta \phi_j}\left[(uv)_{i+\frac{1}{2},j+\frac{1}{2}}\cos(\phi_{j+\frac{1}{2}}-\phi_j)\right.$$

$$-(uv)_{i+\frac{1}{2},j-\frac{1}{2}}\cos(\phi_j-\phi_{j-\frac{1}{2}}) - v^2_{i+\frac{1}{2},j+\frac{1}{2}}\sin(\phi_{j+\frac{1}{2}}-\phi_j)$$

$$\left.- v^2_{i+\frac{1}{2},j-\frac{1}{2}}\sin(\phi_j-\phi_{j-\frac{1}{2}})\right] + \frac{1}{\Delta z_k}\left[(uw)_{i+\frac{1}{2},k+\frac{1}{2}}-(uw)_{i+\frac{1}{2},k-\frac{1}{2}}\right]=$$

$$\text{Ro}\left\{\frac{(u_{i+\frac{3}{2}}-u_{i+\frac{1}{2}})r_{i+1}}{\bar{r}\Delta r_{i+\frac{1}{2}}\Delta r_{i+1}} - \frac{(u_{i+\frac{1}{2}}-u_{i-\frac{1}{2}})r_i}{\bar{r}\Delta r_{i+\frac{1}{2}}\Delta r_i} + \frac{u_{i+\frac{1}{2},j+1}\cos(\phi_{j+1}-\phi_j)-u_{i+\frac{1}{2}}}{\bar{r}r_{i+\frac{1}{2}}\Delta\phi_j\Delta\phi_{j+\frac{1}{2}}}\right.$$

$$+\frac{u_{i+\frac{1}{2}}-u_{i+\frac{1}{2},j-1}\cos(\phi_j-\phi_{j-1})}{\bar{r}r_{i+\frac{1}{2}}\Delta\phi_j\Delta\phi_{j-\frac{1}{2}}} - \frac{u_{i+\frac{1}{2},j+1}\sin(\phi_{j+1}-\phi_j)}{\bar{r}r_{i+\frac{1}{2}}\Delta\phi_j\Delta\phi_{j+\frac{1}{2}}}$$

$$\left.+\frac{u_{i+\frac{1}{2},j-1}\sin(\phi_j-\phi_{j-1})}{\bar{r}r_{i+\frac{1}{2}}\Delta\phi_j\Delta\phi_{j-\frac{1}{2}}} + \frac{u_{i+\frac{1}{2},k+1}-u_{i+\frac{1}{2}}}{\Delta z_k\Delta z_{k+\frac{1}{2}}} - \frac{u_{i+\frac{1}{2},k}-u_{i+\frac{1}{2},k-1}}{\Delta z_k\Delta z_{k-\frac{1}{2}}}\right\}.$$

The fluxes at each cell face are calculated by upwind approximation. Discretization equations for the other flow variables are obtained similarly. The numerical scheme proceeds by first calculating u, v and w at the $n+1$ time-level explicitly from the momentum equations. However, this solution will not in general satisfy the continuity equation. Following the approach of [3] a Poisson equation for pressure correction is solved,

$$\nabla^2 p' = -\frac{1}{A}\nabla.\vec{u},$$

where A is a fictitious time–interval, followed by a velocity correction

$$\vec{u}' = A\nabla p'.$$

The corrected velocities and pressures are used to construct new fluxes at the cell faces. Upon iteration, this process yields velocities at the $n+1$ time–level which satisfy both the momentum and mass continuity equations. Because of the iterative update of the fluxes, the scheme is semi–implicit. The newly calculated velocity field is then used to solve the energy equation for the fluid temperature at time–level $n+1$ in an explicit manner. Marching in time continues until the changes in the flow variables are sufficiently small to indicate that the procedure has converged. This convergence indicates that the flow has reached a steady-state, although the scheme, as described, is also capable of modelling time–varying flows.

3 Results

Representative flow fields (Figure 3) have been simulated for an annulus blocked by a radial barrier, using $T_b - T_a = \Delta T = 4°C$ and $\Omega = 0.5 rad.sec^{-1}$. The plots are in the form of fluid velocity vectors plotted in the (r,z)–plane opposite the barrier (Figure 3a) and in the (r,ϕ)–plane at mid–depth in the fluid (Figure 3b). Figure 3(a) shows fluid rising by the hot outer cylinder and sinking by the cold inner cylinder, accompanied by radial inflow near the top of the convection chamber and radial outflow at the bottom. This aspect of the simulated flow is consistent with experimental observations, although the circulations in a horizontal plane (Figure 3b) differ somewhat from [2]. The results also show that vertical motions take place mainly in the side–wall boundary–layers, as would be expected by consideration of the Taylor–Proudman theorem [10], which predicts that vertical velocities should be inhibited in the body of the fluid.

Fluid heat transfer (Table 1) and temperature data (Table 2) have been calculated using $T_b - T_a = \Delta T = 10°C$ and $\Omega = 0.4 rad.sec^{-1}$, and $T_b - T_a = \Delta T = 4°C$ and $\Omega = 0.5 rad.sec^{-1}$. Table 1 shows values of the fluid heat transfer in terms of the Nusselt number

$$\mathrm{Nu} = \frac{H \ln(b/a)}{2\pi k \, \Delta T \, d},$$

where H is the total fluid heat transport calculated in the model as the heat conduction through the inner cylinder at $r = a$. The three columns give Nu from the model calculations; experimental measurements for nearly equivalent cases ($\Delta T = 9.98°C$, $\Omega = 0.402 rad.sec^{-1}$ and $\Delta T = 4.03°C$ and $\Omega = 0.401, 0.601 rad.sec^{-1}$) [2] and from a correlation for Nu [11] derived for a stationary annulus, $\mathrm{Nu} = (0.203 \pm 0.010)\mathrm{Ra}^{\frac{1}{4}}$ where Ra is the Rayleigh number,

$$\mathrm{Ra} = \frac{g\alpha \, \Delta T \, (b-a)^3}{\nu \kappa},$$

and ν and κ are the kinematic viscosity and thermometric conductivity for the fluid. The results show excellent agreement between the model and experimental measurements, and illustrate the result [12] that the barrier makes the fluid heat transport largely independent of Ω.

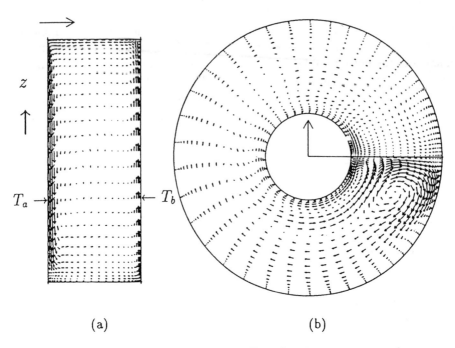

Figure 3 Fluid velocity vector plots, $\Delta T = 4°C$, $\Omega = 0.5$ rad. sec^{-1}, the scale arrow denotes a velocity of 1 cm. sec^{-1}. (a) Flow in the (r,z) plane opposite the radial barrier. (b) Flow in the (r,ϕ) plane at mid-depth in the annular chamber.

Table 1: Comparison of model and measurements of fluid heat transport.

ΔT °C	Ω rad.sec^{-1}	Nu model	Nu expt [7]	Nu($\Omega = 0$) corrln [8]
10	0.4	13.3	13.9 ± 0.1	13.6 ± 0.7
4	0.5	10.3	11.0 ± 0.2	10.8 ± 0.5

Table 2 shows the fluid temperature drop observed across the sides of the thermally insluating radial barrier at mid-height and mid-radius, ΔT_B. ΔT_B serves as a measure of the azimuthal temperature gradient in the fluid, and has been shown to be linked to the radial overturning cell observed in the blocked annulus [2]. The calculated values from the model have been compared with measurements [2]. It can be seen that although the calculated values of fluid heat transport agree quite closely with measurements, there is a considerable discrepancy between the calculated and measured values of ΔT_B. This indicates that the model is failing to properly reproduce the azimuthal temperature field in the annulus.

Table 2: Comparison of model and measurements of barrier temperature drop.

ΔT °C	Ω $rad.sec^{-1}$	ΔT_B model	ΔT_B expt [7]
10	0.4	2.0	0.15 ± 0.01
4	0.5	0.7	0.17 ± 0.01

Bibliography

[1] R Hide (1977), Experiments with rotating fluids, Quart. J. Roy. Met. Soc., 103, 1–28.

[2] Q G Rayer (1992), An experimental investigation of heat transfer by large-scale motions in rotating fluids, D.Phil. thesis, Oxford University, UK.

[3] S V Patankar (1980), Numerical heat transfer and fluid flow, Hemisphere, ISBN 0891169199.

[4] D A Anderson, J C Tannehill and R H Pletcher (1984), Computational fluid mechanics and heat transfer, Hemisphere Pub Corp, New York, ISBN 0-89116-471-5.

[5] Q G Rayer, B G Sherlock and S Ahmed (1995), Simulation of the flow in a differentially heated rotating fluid, Syst. Anal. Modelling Simulation, Vol. 18-19, 149-152.

[6] I N James, P R Jonas and L Farnell (1981), A combined laboratory and numerical study of fully developed steady baroclinic waves in a cylindrical annulus, Quart. J. Roy. Met. Soc., 107, 51–78.

[7] P Hignett, A A White, R D Carter, W N D Jackson and R M Small (1985), A comparison of laboratory measurements and numerical simulations of baroclinic wave flows in a rotating cylindrical annulus, Quart. J. Roy. Met. Soc., 111, 131–154.

[8] A A White (1988), The dynamics of rotating fluids: numerical modelling of annulus flows. Met. Mag., 117, 54–63.

[9] Q G Rayer (1994), A numerical investigation of the flow in a fully blocked differentially heated rotating fluid annulus, Int. J. Modern Phys. C, Vol. 5, No. 2, 203–206.

[10] J Proudman (1916), On the motion of solids in a liquid possessing vorticity, proc. Roy. Soc. London A, 92, 408–424.

[11] M Bowden (1961), An experimental investigation of heat transfer in rotating fluids, Ph.D. thesis, University of Durham, UK.

[12] M Bowden and H F Eden (1968), Effect of a radial barrier on the convective flow in a rotating fluid annulus, J. Geophys. Res., 73, 6887-6896.

[13] W W Fowlis and R Hide (1965), Thermal convection in a rotating annulus of liquid: effect of viscosity on the transition between axisymmetric and non-axisymmetric flow regimes, J. Atmos. Sci., 22, 541–558.

INVESTIGATION OF SEPARATION OF THE THREE-DIMENSIONAL LAMINAR BOUNDARY LAYER

R. ATMANI and R. ASKOVIC
Laboratoire de mécanique des fluides, Université de Valenciennes
et du Hainaut-Cambrésis, 59326 Aulnoy lez Valenciennes, France

SUMMARY

Three-dimensional boundary layer around a blunt body at low incidence, including the problem of boundary layer separation, is discussed. A universalisation of the boundary layer equations is first made in the sense that neither equation nor boundary conditions depend on particular problem data. The universality is achieved by transferring sets of parameters which express the influence of external velocity and shape of the body, characteristic for each particular problem, into the new variables. Subsequently, the solutions of the obtained universal equations are found in the form of series expansions in mentioned parameters. Then an application of the proposed method to calculate all boundary layer characteristics on a flatten ellipsoid (6:3:1) with the satisfactory results is done. Finally, the separation line on a prolate spheroid (6:1) at incidence of 6 degrees, by using the electochemical method, is experimentally determined.

1. INTRODUCTION

Three-dimensional boundary layers in the strict sense have been studied seriously since 1950. The long period (1904-1950) of mainly two-dimensional investigation of boundary layer is due to the inherent mathematical difficulties of really three-dimensional problems. It is to be noticed that bodies of revolution placed without incidence in an originally uniform airflow are sometimes referred to as "three-dimensional" configurations, but they lead in reality, as can be shown by Mangler's well-known transformation, to two-dimensional problems. Or configurations of flow around semi infinite or infinite cylindrical bodies with small angles of yaw or conical bodies with small incidence, as well as internal or external flows along edges of unlimited wedges, are three-dimensional, but degenerated in the sense that they do not correspond to physically realisable problems ("quasi two-dimensional" problems). These two-dimensional or quasi two-dimensional problems will not be treated afterwards. As a matter of fact, the word "three-dimensional" refers exclusively here to spatial dimensions.

2. BOUNDARY LAYER EQUATIONS

The simplest cases of steady three-dimensional boundary layers are the cases of laminar flow around a yawed body of revolution at low incidence or a triaxial ellipsoid with or without incidence or yaw. By using so-called streamline coordinates ([1],[2]), it may safely be assumed that the flow takes place in well defined "laminae" preserving their individuality with respect to n or with respect to z while proceeding downstream. We can, therefore, reasonably consider the boundary layer flow along an outer streamline as largely independent of the flow along neighbouring streamlines; this suggests an individual application of the condition w<<u ("principle of prevalence"[3]) to each lamina of boundary layer flow along a given outer

streamline -under the strict condition that integrations of the resulting simplified equations shall be executed only inside z-wise laminae, i.e. with respect to the downstream direction s and the distance n from the body surface only.

Or adopting the streamline coordinate system, the parametric curves s=constant and z=constant on the surface of the body are chosen to coincide with the projections of the external flow streamlines on the surface, and their orthogonal potential lines, respectively. In the external flow we then have $w_e = 0$, and the equations for the conservation of flow momentum in incompressible flow reduce simply to:

$$\frac{\partial p}{\partial s} = -\rho u_e \frac{\partial u_e}{\partial s}$$

$$\frac{\partial p}{\partial z} = \frac{\rho}{e_1} \frac{\partial e_1}{\partial z} u_e^2 \ .$$

The above relations show that the use of streamline coordinates enables us to relate the pressure terms in the boundary layer equations in a simple manner to the external flow velocity distribution.

Introduction of these relations into momentum boundary layer equations in the case of the small crossflow leads to the following equations in the streamwise and crossflow directions, plus the continuity equation:

$$\left. \begin{array}{l} u\dfrac{\partial u}{\partial s} + v\dfrac{\partial u}{\partial n} = u_e \dfrac{\partial u_e}{\partial s} + \gamma \dfrac{\partial^2 u}{\partial n^2} \\ \dfrac{\partial (e_2 u)}{\partial s} + e_2 \dfrac{\partial v}{\partial n} = 0 \end{array} \right\} \quad (1)$$

$$u\frac{\partial w}{\partial s} + v\frac{\partial w}{\partial n} - \frac{1}{e_1}\frac{\partial e_1}{\partial z}u^2 = \frac{1}{e_1}\frac{\partial e_1}{\partial z}u_e^2 + \gamma \frac{\partial^2 w}{\partial n^2} \quad (2)$$

with the conditions:

$$u = v = w = 0 \text{ for } n = 0; \quad u = u_e, w_e = 0 \text{ for } n \to \infty. \quad (3)$$

Here (s,z,n) denote above-mentioned streamline coordinates, $e_1(s,z)$, $e_2(s,z)$, $e_3 = 1$ -metric coefficients of the system of coordinates, (u,w,v) -velocity boundary layer components at (s,z,n), $u_e(s,z)$ -external velocity distribution and γ -fluid viscosity.

So with the assumption of small crossflow ("principle of prevalence"), the equations (1) do not contain terms relating to the crossflow, and will be recognized as being of the same form as the equations for axisymmetric flow. Accordingly, the streamwise flow can be calculated by any available method for axisymmetrical boundary layers, for instance by the method of universalisation in the sense of Loitsianski [4] already utilized for the three-dimensional boundary layer calculation [7]. Once the streamwise flow is calculated, i.e. once u and v are known, equation (2) can be solved retrospectively for w.

3. MATHEMATICAL ANALYSIS

Introducing the stream function $\psi(s,z,n)$ as follows:

$$u = \frac{1}{e_2}\frac{\partial \psi}{\partial n} \text{ and } v = -\frac{1}{e_2}\frac{\partial \psi}{\partial s} \quad (4)$$

the equation of continuity is identically satisfied and the equations in the streamwise and crossflow directions (1) and (2) will be reduced to:

$$\frac{\partial^3 \psi}{\partial n^3} + \frac{1}{\gamma e_2}\frac{\partial \psi}{\partial s}\frac{\partial^2 \psi}{\partial n^2} - \frac{1}{\gamma e_2}\frac{\partial \psi}{\partial n}\frac{\partial^2 \psi}{\partial s \partial n} + \frac{1}{\gamma e_2^2}\frac{\partial e_2}{\partial s}(\frac{\partial \psi}{\partial n})^2 = -\frac{e_2}{\gamma}u_e\frac{\partial u_e}{\partial s} \quad (5)$$

$$\frac{1}{e_2}\frac{\partial \psi}{\partial n}\frac{\partial w}{\partial s} - \frac{1}{e_2}\frac{\partial \psi}{\partial s}\frac{\partial w}{\partial n} - \gamma \frac{\partial^2 w}{\partial n^2} = \frac{1}{e_1}\frac{\partial e_1}{\partial z}\left[\frac{1}{e_2^2}(\frac{\partial \psi}{\partial n})^2 - u_e^2\right] \quad (6)$$

with the following conditions:

$$\left.\begin{array}{l}\dfrac{\partial \psi}{\partial n}=0,\ \dfrac{\partial \psi}{\partial s}=0,\ w=0;\quad for\quad n=0\\[2mm]\dfrac{\partial \psi}{\partial n}=e_2 u_e,\ w=0\quad for\quad n\to\infty\end{array}\right\} \quad (7)$$

The universality of these equations (5) and (6) can be achieved by introducing a new normal variable:

$$\eta = K\frac{n}{\delta_{ss}} \qquad (8)$$

and by transferring the following four sets of parameters:

$$\left.\begin{array}{l} f_k = u_e^{k-1}\dfrac{\partial^k u_e}{\partial s^k}\overset{*}{Z}{}^k,\ \alpha_k = u_e^k\dfrac{1}{e_2}\dfrac{\partial^k e_2}{\partial s^k}\overset{*}{Z}{}^k \\[3mm] \beta_k = u_e^k\dfrac{1}{e_1}\dfrac{\partial^k e_1}{\partial s^k}\overset{*}{Z}{}^k,\ \Delta_k = u_e^k\left[\dfrac{\partial^k}{\partial s^k}\ln(\dfrac{\partial e_1}{\partial z})\right]\overset{*}{Z}{}^k \end{array}\right\} k=1,2,3.. \qquad (9)$$

into the new variables instead of s and z. Here δ_{ss} denotes the boundary layer momentum thickness, δ_s -displacement thickness, $H=\dfrac{\delta_s}{\delta_{ss}}$, $\overset{*}{Z}=\dfrac{\delta_{ss}^2}{\gamma}$ and K= constant.

Now, if we are looking for the solutions of the equations (5) and (6) in the form:

$$\psi(s,z,n) = \frac{e_2\delta_{ss}}{K}u_e(s,z)F(\eta,\{f_k\},\{\alpha_k\}) \qquad (10)$$

$$w = \frac{1}{e_1}\frac{\partial e_1}{\partial z}u_e^2\overset{*}{Z}Q(\eta,\{f_k\},\{\alpha_k\},\{\beta_k\},\{\Delta_k\}). \qquad (11)$$

F and Q being two real continuous and infinitely derivable functions, then after replacing into the corresponding equations we obtain:

$$K^2\frac{\partial^3 F}{\partial \eta^3} + \frac{1}{2}EF\frac{\partial^2 F}{\partial \eta^2} + (f_1+\alpha_1)F\frac{\partial^2 F}{\partial \eta^2} + f_1\left[1-\left(\frac{\partial F}{\partial \eta}\right)^2\right] +$$
$$\sum_{k=1}^{\infty} a_k(\frac{\partial F}{\partial f_k}\frac{\partial^2 F}{\partial \eta^2} - \frac{\partial^2 F}{\partial \eta \partial f_k}\frac{\partial F}{\partial \eta}) + \sum_{k=1}^{\infty} b_k(\frac{\partial F}{\partial \alpha_k}\frac{\partial^2 F}{\partial \eta^2} - \frac{\partial^2 F}{\partial \eta \partial \alpha_k}\frac{\partial F}{\partial \eta}) = 0, \qquad (12)$$

$$K^2\frac{\partial^2 Q}{\partial \eta^2} + (\frac{1}{2}E+f_1+\alpha_1)F\frac{\partial Q}{\partial \eta} + \sum_{k=1}^{\infty}(a_k\frac{\partial F}{\partial f_k}+b_k\frac{\partial F}{\partial \alpha_k})\frac{\partial Q}{\partial \eta} - (E+2f_1-$$
$$\beta_1+\Delta_1)\frac{\partial F}{\partial \eta}Q - \sum_{k=1}^{\infty}(a_k\frac{\partial Q}{\partial f_k}+b_k\frac{\partial Q}{\partial \alpha_k}+p_k\frac{\partial Q}{\partial \beta_k}+p_k\frac{\partial Q}{\partial \Delta_k})\frac{\partial F}{\partial \eta}=1-\left(\frac{\partial F}{\partial \eta}\right)^2 \qquad (13)$$

where: $a_k = (k-1)f_k f_1 + f_{k+1} + kEf_k$
$b_k = (kf_1 - \alpha_1)\alpha_k + \alpha_{k+1} + kE\alpha_{k+1}$
$p_k = (kf_1 - \alpha_1)\alpha_k + \alpha_{k+1} + kE\alpha_k$
$q_k = kf_1\Delta_k + \Delta_{k+1} + kE\Delta_k$.

The boundary conditions are:

$$\left.\begin{array}{l} F = \dfrac{\partial F}{\partial \eta}=0\ for\ \eta=0;\ \dfrac{\partial F}{\partial \eta}\to 1\ for\ \eta\to\infty \\[2mm] Q=0\ for\ \eta=0;\ Q\to 0\ for\ \eta\to\infty. \end{array}\right\} \qquad (14)$$

As neither the equations (12) and (13) nor the boundary layer conditions (14) depend on the particular problem data, that means the equations (12) and (13) can be interpreted as universal in the sense of Loitsianski [4] and solved once for all.

Anyway, for determining the new variables (8) and (9), we should solve separately the equation:

$$u_e\frac{d\overset{*}{Z}}{ds} = E(f_k,\alpha_k) \qquad (15)$$

where:

$$E(f_k, \alpha_k) = 2K\left(\frac{\partial^2 F}{\partial \eta^2}\right)_{\eta=0} - 2f_1(H+2) - 2\alpha_1 \, . \qquad (16)$$

The universal parametric equations (12) and (13) as well as the particular equation (15) could be solved either by a direct numerical integration or by series expansions. For a moment, we found [8] these solutions in the form of series expansions in mentioned parameters by using an implicit Crank-Nicolson type of finite difference method, on a meshing of the outer flow following steamlines and potential lines (s,z) around the flatten ellipsoid at 6° of incidence (figure 1).

Figure1: Repartition of the steamlines on ellipsoid (6/3/1) at 6 ° of incidence.

This numerical study permitted us to calculate different boundary layer characteristics such as: profiles of longitudinal and crossflow velocities (figure 2), evolution of the displacement thickness (figure 3), evolution of longitudinal skin friction (figure 4), etc.

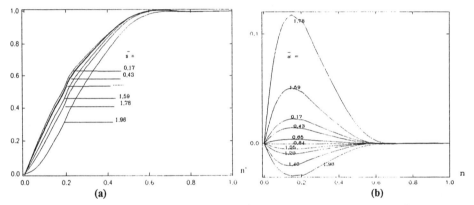

Figure 2 : Profile of velocities into the boundary layer for the different distances $\bar{s}$ according to the line J. (a) longitudinal velocity u/ue. (b) crossflow velocity w/ue.

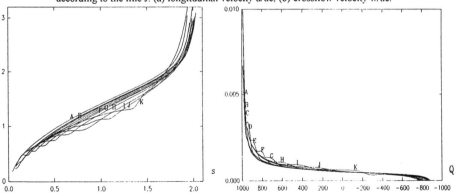

Figure 3: Evolution of the displacement thickness $\delta_s(mm)$ on all streamlines covering leeside of ellipsoïd.

Figure 4: Evolution of longitudinal skin friction C_f on leeside.

4. SEPARATION OF THREE-DIMENSIONAL BOUNDARY LAYER

We were particularly interested to the problem of boundary layer separation because of its considerable practical importance. Separation is the generic name given to a class of flow phenomena. One, and perhaps the most important, feature which characterizes this class is that the flow becomes detached from the body surface allowing a flow region of indefinite extent to develop between the body and the outer, quasi-inviscid flow. Separation provides a mechanism whereby vorticity, which in attached flow is confined within the boundary layer, can be transported into the interior of the fluid.

Prediction of the onset of separation is very essential. When significant areas of separation exist on a body, the pressure distribution ceases to conform closely to that corresponding to potential flow, and the body suffers a loss of lift, rapidly increasing drag, erratic changes of moments, and, often, buffeting. Under such conditions there is little doubt that separation has occurred somewhere on the body. On the other hand, the detection of limit regions of separation, or the assignment of a precise location to the onset of separation - even where it is of substantial extent, frequently present difficulty in three-dimensional flows. The difficulty arises because of the lack of an unambiguous definition of separation.

The numerical study of three-dimensional separation is based on results of the previous boundary layer calculation. We determine the separation line, as a line from which irregularities of numerical computing appear, like rush thickening of boundary layer and/or integral thicknesses or weakening from a certain order of the skin friction at the wall..

This analysis remains more or less unable to determine with precision the position of this line, and the pattern of separation which is concerned. We prefer taking those facts as consequences of separation and we opt for their utilization as criteria of verification of passage in the area of separated flow. Those effects are illustrated in figures 3 and 4 where we notice a rush leap of δ_s towards very large values, and a passage of the longitudinal skin friction C_f towards negative values and this following the streamline in question, in a well determined order.

We study the case of low incidences where the "principle of prevalence" seems to be satisfied (cross flow smaller than the longitudinal one,(Fig: 2-(a),(b)), that is why it looks much more precise to use the Wang criterion for the low incidences and calculate separation line depending on whether it is a closed type (bubble) or an open type (free vortex layer). At 6 degrees of incidence the separation is a closed type, it appears at the end of body and it is calculated according to longitudinal component of skin friction, because in that case, according to Wang[5], the vanishing of this component is added to the one of the transversal component, which means a reversal of the resultant flow. Computation of boundary layer can be continued without taking into account whether partial reversal occurs in the one or the other direction, but not both.

The application done on flatten ellipsoid shows a pace of separation line similar to the one obtained by Wang on an ellipsoid of revolution, in particular the existence of narrow zone where fluid takes off before the rest of obstacle surface. This zone seems to be more prolonged by the fact of flattening which is presented by the body in the y direction, and shows a peak situated at the intersection of lee side and windward parts of body (Fig 5).

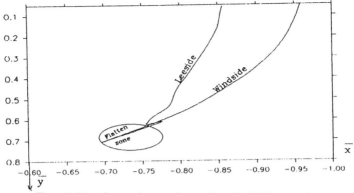

Figure 5: Line of separation on a flatten ellipsoïd (6/3/1). (x,y) plane.

5. EXPERIMENTAL MEASUREMENT

The experimental determination of the separation line is usually based on the measurement of skin friction. We measured [8] the wall velocity gradients $\left(\frac{\partial u}{\partial y}\right)_{y'=y'_{wall}}$ and/or $\left(\frac{\partial v}{\partial y}\right)_{y'=y'_{wall}}$ by using electrochemical method developed in our Laboratory of fluid mechanics [6]. This method allows to bind parietal velocity gradient to flux of diffusion of chemical substance used in the polarographic solution. During our measures we used for a moment, single electrochemical probes placed on a meridional line of a prolate spheroid (6:1), located in the middle of experimental vein of the hydraulic canal. Ellipsoid rotation around its longitudinal axis allows to bring back electrochemical probes in different azimuths θ and so to measure the mass transfer coefficient K (binded directly to the wall velocity gradient S by the relation K=0.807/S/$^{1/3}$) on the whole back of the body. The figure (6-a) shows the separation line obtained experimentally on the prolate spheroid in coordonatres(x,θ) for three flow velocities (0.5, 1, 1.5m/s).Then it is projected on the (x,y) plane (fig.6-b). The general shape of this curve confirms the one previously obtained by numerical computation, notably the existence of a narrow zone situated from 70° to 90° where the fluid detaches earlier than at the rest of the surface.

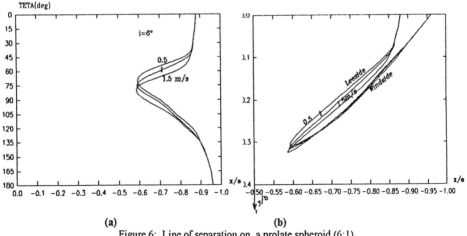

Figure 6: Line of separation on a prolate spheroid (6:1).
(a) in (x,θ) coordinates, (b) in (x,y) coordonates

6. CONCLUSION

In the case of three-dimensional flows, the separation can be defined in more than one way. Discussions of three-dimensional boundary layer separation in the literature have attempted to find a rational extension of the zero-skin-friction criterion. Attempts have been made to establish the identity of "separation lines ".

Among the definitions of these may be found: (a) envelopes of limiting streamlines; (b) lines dividing flow which has come from different regions; (c) lines of singularities (problems of topology); (d) lines on which some component of the skin friction vanishes. Each of these is valid under certain conditions, but none is universally valid.

In the present work, we use the definition (d), i.e. at low incidence of a blunt body the separation line is identified as the zero of the longitudinal skin friction component (K.C. Wang) So the separation line on a flatten spheroid (6:3:1) is calculated, as well as experimentally determined on a prolate spheroid (6:1) at incidence of 6 degrees by using the electrochemical method, a method permitting to follow the evolution of the wall velocity gradient.

ACKNOWLEDGEMENT. The authors thank Professor C. Tournier and Mr L. Labraga for their helpful suggestions during the course of the work.

7. REFERENCES

[1] PRANDTL L. :Uber Reibungschlichten bei dreidimensionalen Stromungen. British M.A.P. Volkenrode Rep. and Transl. N°64, 1946.

[2] HAYES W. D. : The Tridimensional Boundary Layer. NAVORD Report 1313, 1951.

[3] EICHELBRENNER E. A. et OUDART A. : Méthode de Calcul de la Couche Limite Tridimensionnelle. Application à un corps fuselé Incliné sur le vent. -O.N.E.R.A.Publication 76, 1955.

[4] LOITSIANSKII L. G. : Universal Equations and Parametric Approximations in the Boundary Layer theory (in Russian) - Prikl . Mech. Math. 29 (1), p.p. 70-87, 1965.

[5] WANG K. C. : Boundary Layers over a Blunt Body at Low Incidence with circumferential Reversed Flow. -J. of Fluid Mech., Vol 72, 1975, pp. 49-65.

[6] TOURNIER C. : Etude de l'Ecoulement Instationnaire autour des cylindres par Tribometrie Electrochimique -These d' Etat, Paris VI, 1976.

[7] ASKOVIC R. : Equations Universelles de la Couche Limite Laminaire Tridimensionnelle en Régime Instationnaire et leur traitement. Com. IUTAM. Symp. on the unsteady boundary layer, p.p. 637-670, Québec, 1971.

[8] ATMANI R. "Contribution à l'Etude de la Couche Limite Tridimensionnelle et de son Décollement autour des Corps Fuselés". Doctoral thesis . University of Valenciennes, 1995.

LARGE-SCALE BLAST SIMULATIONS

Joseph D. Baum, Hong Luo
Science Applications International Corporation
1710 Goodridge Drive, MS 2-3-1
McLean, VA 22102

and

Rainald Löhner
Institute for Computational Sciences and Informatics
George Mason University
Fairfax, VA 22030

ABSTRACT

This paper describes the application of FEFLO96, a three-dimensional, adaptive, finite element, edge-based, ALE shock capturing methodology on unstructured tetrahedral grids, to large-scale simulations of blast wave interaction with structures. The first simulation applied the CFD methodology to the numerical simulation of blast wave diffraction within the B-2 level of the World Trade Center garage. This simulation modeled blast wave diffraction about hundreds of rigidly-modeled structures, spread over a large area. The second simulation applied a new loose-coupling algorithm that combined FEFLO96 and DYNA3D, a state-of-the-art Computational Structural Dynamics (CSD) methodology, to the simulation of shock interaction with a structurally-responding truck.

INTRODUCTION

The numerical simulation of an explosion within a confined environment requires simulating blast initiation and evolution, shock wave diffraction, structural response to the blast, structural failure, blast transmission through open and/or breaking walls, floors, and ceilings, rigid body response to the shock (e.g., vehicle translation due to shock loading), and energy absorption by deforming structures. This list indicates that methodologies from widely disparate fields such as computational fluid dynamics (CFD), computational structural dynamics (CSD), and material modeling must be combined to completely assess structure vulnerability to internal blasts, an integration effort that was completed recently [1]. To quantify blast energy depletion due to structural deformation, we decided to perform two simulations: the first will treat all structures as rigid and non-responding; the second will model the structural coupling. The results of the first simulation are presented here. The coupled simulation will be conducted in the future.

The objective of developing a coupled CFD/CSD methodology was to enable cost-effective simulation of fluid-structure interactions, specifically, shock wave diffraction about complex-geometry structures, and the structural response to the shock loading. Since both CFD and CSD have reached a high degree of reliability, rather than develop a new methodology from scratch, we chose to couple two state-of-the-art methodologies. As the present research effort is directed toward nonlinear applications, we selected the FEFLO96 methodology [2-7], which solves the time-dependent, compressible Euler and Reynolds-averaged Navier-Stokes equations for the fluid; we chose the DYNA3D methodology [8] for CSD, which solves the large deformation, large strain formulation equations for the structure. Given that the geometrical complexity of the problems envisioned can be severe and the deformations considerable, automatic grid generation on unstructured grids was employed for both the fluid and the structure: tetrahedral elements for CFD, bricks for CSD. The basic CFD and CSD codes were altered as little as possible. The CSD code defines the structural surface location and velocity, while the CFD code defines the loads. The transfer of loads, displacements, and velocities is carried out via fast interpolation algorithms.

THE INTEGRATED NUMERICAL METHODOLOGY

The numerical methodology used here incorporates a CAD-like utility, a grid generator for both CFD and CSD, CFD and CSD solvers, and post-processing tools.

Our CAD tool is FECAD, a suite of efficient, user-friendly utilities that allows the user to quickly produce a FRGEN3D-compatible, error-free input for the generation of both the CFD and CSD meshes. In addition to basic CAD-like operations such as shrinking, translation, rotation and surface lofting, FECAD also eases the merging of several parts of the surface into one cohesive, well-defined input file, and allows import of data sets produced by other CAD programs. Most important, FECAD has a whole series of built-in diagnostics to avoid such undesirable features as doubly defined points, isolated points or lines, badly defined lines or surfaces, and lines or surfaces directed incorrectly.

The unstructured grid generator, FRGEN3D, is based on the advancing front method [9]. The CFD mesh is composed of triangular (surface) and tetrahedral (volume) elements, while the CSD mesh includes beams, quads (shells) and bricks or hexahedra (solids). The hexahedral elements are formed by filling the volume with tetrahedra, then splitting each into four hexahedra. Although the angles of a typical hex are less than perfect, extensive testing against perfect-angle bricks for both linear and nonlinear test problems produced almost identical results. This, nevertheless, necessitated the replacement of the Belytschko-Tsay hourglass control model (default model in DYNA3D), with the Flanagan-Belytschko hourglass control model (model no. 3 in DYNA3D), which incurred a 30% performance penalty.

The flow solver applied is FEFLO96, an extensively tested and validated three-dimensional, adaptive, finite element, edge-based [6], ALE shock capturing methodology on unstructured tetrahedral grids [2-7]. H-refinement is the preferred approach for grid adaption [4]. The high order scheme used is the consistent-mass Taylor-Galerkin algorithm. Combined with a modified second order Lapidus artificial viscosity scheme, the resulting scheme is second order accurate in space and fourth order accurate in phase. The spatio-temporal adaptation is based on local H-refinement, where the refinement/deletion criterion is a modified H2-seminorm [4]. Based on past experience with simulations of shock wave propagation processes in both 2-D and 3-D, density was chosen as the critical parameter for the refinement/deletion criteria.

The structural dynamics solver is DYNA3D [8]. This methodology is based on unstructured grids, a spatial discretization using finite element techniques, large deformation formulation for the solids in order to handle severe plastic deformation and failure, and explicit time integration for speed. DYNA3D incorporates many materials with different equations of state, as well as many kinematic options, such as slidelines and contacts. Furthermore, DYNA3D is a well-proven and benchmarked solver used extensively in the CSD community.

Post-processing is performed with the FEPOST3D and MOVIESUBS packages. FEPOST3D is based on FEPLOT4D, but performs all the CPU-intensive filtering operations on the Cray.

NUMERICAL APPLICATIONS

Blast in the World Trade Center

Solid Body Definition: In the absence of CAD data (the latest blueprints are from 1968), the complete B-2 level geometry was reconstructed from blueprints. All wall locations, ramps, exhaust plenum spaces, mechanical rooms, stairways, toll booths, curbs, A/C, and supply and elevator shafts were modeled as accurately as possible from the structural and architectural drawings (Fig 1a). Ten typical car groups were modeled here, including a small passenger car (Hyundai), a mid-size passenger car (Chevy Cavalier), a full-size sedan (Ford Taurus), a full-size station wagon (Ford Taurus wagon), a small sport utility vehicle (a basic Jeep), a small pick-up (Dodge Ram), a large sport utility vehicle (Ford Explorer), a large pick-up (Ford F-150), a mini-van (Dodge Caravan), and a full-size van (Ford). These cars were parked at the correct parking location.

Blast Wave Initiation and Diffraction: Blast initiation assumed a vehicle parked near the right curb of the exit lane, leading to the northbound exit to West Street (Ramp B). The simulation was initiated with a 2-D detonation code which modeled a charge of one thousand pounds of ANFO HE, placed on the floor, and integrated the solution until the leading shock was only a short distance from the south wall of tower A (the north tower). The solution was then overlayed on a 3-D mesh. As a consequence of the point-blast initiation, higher pressures on the ceiling than on the floor should be expected later on in the simulation, as the flow stagnates on the ceiling while propagating parallel to the floor.

The numerical simulation of shock diffraction within the B-2 floor of the World Trade Center is especially suited for an adaptive refinement methodology. The large dimensions of the computational domain (20,249 cm long, 17,743 cm wide, and 580 cm high) are in contrast to the requirement to model the shock as a sharp discontinuity and model the smallest possible physical features of interest (e.g., while modeling the complete domain, also model shock diffraction about a wall that is less than 20 cm thick, a ratio of 1:1000). These contrasting requirements can be fulfilled only by using an adaptive methodology. Still, some compromises on mesh size and total number of nodes had to be made. When the simulation was initiated, the largest memory supercomputer available to us was a Cray C-90, with 128 MWords static memory. In an effort to optimize flow resolution within the limited available memory, we initiated the simulation by modeling only the zone fairly close to blast point. In addition, we allowed only two levels of refinement. Thus, the initial mesh before refinement had only about 370K points and 1.95 million elements. At t=25.7 ms, when the leading shock was close to the boundary of the modeled domain, we remeshed, modeling a larger portion of the B-2 floor. We then interpolated the solution from the old mesh to the new mesh, adapted twice (two levels of refinement), and continued the computation. By this time we gained access to a Cray YMP/M-92 supercomputer with 256 MWords static memory, so our initial mesh size was about 780K points and 4.1M elements. Finally, at t=108.9 ms, we generated a third mesh that included the complete domain. The initial mesh included about 1.54M points and 8.0M elements. After solution adaptation, the final mesh included about 3.52M points and 18.1M elements. This part of the simulation was executed on a new Cray YMP/M-92 with 1 GWords static memory.

The complex shock evolution within the structure is described in great detail in Ref 5. Due to severe space limitations we will attempt to demonstrate the complexity of the shock diffraction process by analyzing the flow at two times. Better understanding of the complex diffraction processes can be gained by carefully analyzing the video data, which includes 1459 dumps.

Examination of the leading shock at t=12.51 ms (Fig 1b) shows on the right (east) the diffracted wave about the south-east corner room of tower A, and shock diffraction about a column. Closer to the center of the domain we observe a nicely detailed shock diffraction about a column, with almost symmetric rarefaction waves on both sides, and a reflected shock upstream of the column. A similar diffraction process is observed about a column toward the left of the expanding shock. Shock diffraction about the cars situated between these columns is more complex. The reflected pressure amplitude depends on the car profile (i.e., percentage of blocked area from floor to ceiling), car width, and shock incident angle. The impacting shocks were partially transmitted under the cars (the wheels were not modeled).

The dominant shocks on the west (left) side at t=95.81 ms (Fig 1c) are the shock reflections from the walls of the exit ramps (A and B), shock reflection from the second row of cars (moving in a northwest direction), and shock reflection from the wall of Ramp N (left side, leading upstairs). The primary shock was partially reflected from the side wall of Ramp N, and the row of cars north of Ramp T (center, leading downstairs), expanded around the wall surrounding Ramp N, and about the vehicles parked west of Ramp T. The front propagating south shows, from left to right; a large amplitude shock and a Mach stem near the west wall; a lower amplitude shock diffracting about the parked cars; the incident shock in the traffic lane, followed closely by the reflection from the side wall of Ramp N; the broad-side impact of this reflected shock on the row of cars north of Ramp T; shock diffraction about the southwest corner of Ramp T; and the impact of the shocks reflected from Ramps A and B on the central two rows of cars. Naturally, each shock impact produced a reflected shock, plus an assortment of shocks that reverberate between the cars.

At the center, shock stagnation at the north wall of Ramp T (and the resulting high pressure) explains the significant damage to the cars parked nearby, and the collapse of the wall. Since the cars to the left (west) and at the center of the wall are parked normal to the wall, the shock front is fairly uniform. On the right (east), however, the two vehicles parked diagonally perturbed the flow sufficiently to produce a curved (almost spherical) reflected shock. At the center-right we observe the sharp shock transmitted in the traffic lane between Ramp T and the west wall of tower B, the multiple reflections from rooms near the northwest corner of tower B, and from the broad-sided parked cars. At t=95.8 ms, the sharp, non-oscillatory captured primary shock diffracts about the south wall of Ramp T and the transfer shaft room near the west wall of tower B. The reflected shock from the north wall of Ramp T expands at this time near the ramp entrance.

In the Secret Service area, we observe the primary curved shock and the multiple diffractions about the columns. The shock interacted at t=95.81 ms with the walls of the pipe room attached to the north wall of tower B, to produce a strong reflection, an expansion, and a Mach stem. The weakened incident shock expanded to the north. The results at t=95.81 ms show near the north wall of tower B a reflected shock and a spherical reflection with a circular imprint on the floor. Shock wave reflection from a corner of the support beams produced this large amplitude spherical reflection.

The solution was integrated to t=287.73 ms, at which time the shocks have filled the complete computational domain.

Shock Impact on a Truck

The second simulation modeled shock impact on a standard five ton truck carrying a command and control center. The structure incorporates all the CSD elements typically used: beams, shells and solids. Since in this simulation we had to generate, for the first time, a CFD mesh about shells with zero thickness, special routines had to be developed. Using the available CAD data, we first defined the line/spline/surface definition for the CSD surface. We then generated the CSD surface and volume meshes. These were generated by forming triangles on the surfaces and tetrahedra in the volume, then splitting each triangle into three quads by connecting the mid-faces to the centroid, and each tetrahedron into four hexahedra. The CFD mesh was then generated using the CSD surface quads as a discrete front surface, generating the CFD surface triangulation (with finer resolution than the CSD surface mesh), and the volume tetrahedra. The initial CAD data has 6306 points, 3718 lines/splines, 1604 patches and included 22 different materials (rubber, steel, aluminum, etc.). The CSD mesh included 41,574 shells, 49,420 bricks, 904 beams, and 101,002 nodes. The initial CFD mesh included 36,970 boundary points, 175,228 nodes and 930,125 elements.

The truck was impacted on the back by a large-amplitude shock angled at 22.5 degrees off the plane of symmetry. Pressure contours on the truck at t=7.93 ms, and the CSD surfaces at t=11.273 ms, t=16.57 ms and t=26.76 ms, are shown in Figs 2a through 2d, respectively. While the incident shock propagates from the back to the front of the truck, shock reflections from the command and control box, wheels, Z and C beams and other protruding boxes are observed on the floor. Also noted is the buckling of the shelter and the increased pressure loading at areas inclined to the shock at higher angles. Large deformations are observed, especially on the command and control center, with reduced deformation wherever the shelter plates were reinforced by beams, and the wrapping of the back bumper around the two massive longitudinal support beams.

The top back portion of the truck (including the shelter) was lifted by about 60 cm at t=26.76 ms (Fig 2d). Simultaneously, the wheels were lifted off the ground by about 40 cm. This resulted from the damping action of the suspension system. The suspension system was defined to be made of rubber, with a steel support at the top (attached to the longitudinal beams and the shelter) and a steel support on the bottom (attached to the wheel assembly). A steel screw was clamped between the top and bottom steel supports, to limit the travel distance between the command and control shelter and the wheel assembly. Close examination of the results shows the suspension system stretching to the maximum distance allowed, and the bending of the supports, resulting in reduced lifting of the wheel assembly.

CONCLUSIONS

This paper described the application of a recently developed three-dimensional, adaptive, finite element, edge-based, ALE shock capturing methodology on unstructured tetrahedral grids (FEFLO96) to two large-scale blast diffraction simulations. The first modeled blast wave diffraction inside the B-2 level of the World Trade Center. The second employed a coupled CFD/CSD methodology to model blast wave diffraction about a responding truck.

The WTC blast simulation investigated long duration shock diffraction about hundreds of structures. Several interesting 3-D shock diffraction processes were identified. Among these are the formation of "hot spots" resulting from three-dimensional wave focusing, shock reverberation between cars, shock diffraction about cars and columns, and the formation of three-dimensional Mach stems.

The application of the coupled CFD/CSD algorithm to the simulation of blast wave interaction with a truck demonstrated the ability to model structural deformation concurrently with the blast propagation. The new methodology required the development of some unique computational tools such as flow modeling about zero-thickness structures, and CFD mesh generation from discrete CSD data.

ACKNOWLEDGEMENTS

This work was supported by the Defense Nuclear Agency. Dr. Michael E. Giltrud served as the project technical monitor.

REFERENCES

[1] R. Löhner, et al - Fluid-Structure Interaction Using a Loose Coupling Algorithm and Adaptive Unstructured Grids. AIAA-95-2259 (1995).

[2] R. Löhner, K. Morgan, J. Peraire and M. Vahdati - Finite Element Flux-Corrected Transport (FEM-FCT) for the Euler and Navier-Stokes Equations; ICASE Rep. 87-4, *Int. J. Num. Meth. Fluids* 7, 1093-1109 (1987).

[3] J.D. Baum and R. Löhner - Numerical Simulation of Shock-Box Interaction Using an Adaptive Finite Element Scheme; *AIAA J.* 32, 4, 682-692 (1994).

[4] R. Löhner and J.D. Baum - Adaptive H-Refinement on 3-D Unstructured Grids for Transient Problems; *Int. J. Num. Meth. Fluids* 14, 1407-1419 (1992).

[5] J. D. Baum, H. Luo and R. Löhner - Numerical Simulation of Blast in the World Trade Center; AIAA-95-0085 (1995).

[6] H. Luo, J.D. Baum and R. Löhner - Edge-Based Finite Element Scheme for the Euler Equations; *AIAA J.* 32, 6, 1183-1190 (1994).

[7] J. D. Baum, H. Luo and R. Löhner - A New ALE Adaptive Unstructured Methodology for the Simulation of Moving Bodies; AIAA-94-0414 (1994).

[8] R.G. Whirley and J.O. Hallquist - DYNA3D, A Nonlinear Explicit, Three-Dimensional Finite Element Code for Solid and Structural Mechanics - User Manual; UCRL-MA-107254 (1991), also *Comp. Meth. Appl. Mech. Eng.* 33, 725-757 (1982).

[9] R. Löhner and P. Parikh - Three-Dimensional Grid Generation by the Advancing Front Method; *Int. J. Num. Meth. Fluids* 8. 1135-1149 (1988).

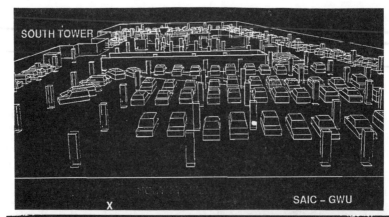
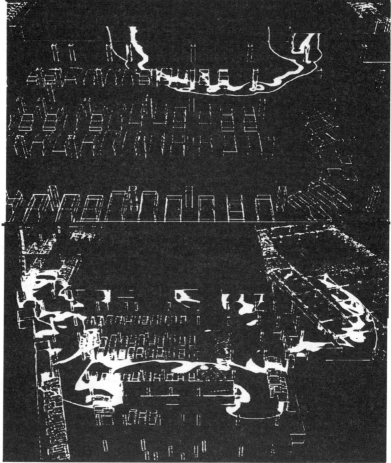

Fig 1. CFD Surface Mesh and Pressure Contours: a) CFD Surface Mesh of the Central Parking Area; Expanded view of Pressure Contours at t=12.51 ms; and c) Expanded View of Pressure Contours at t=95.81 ms.

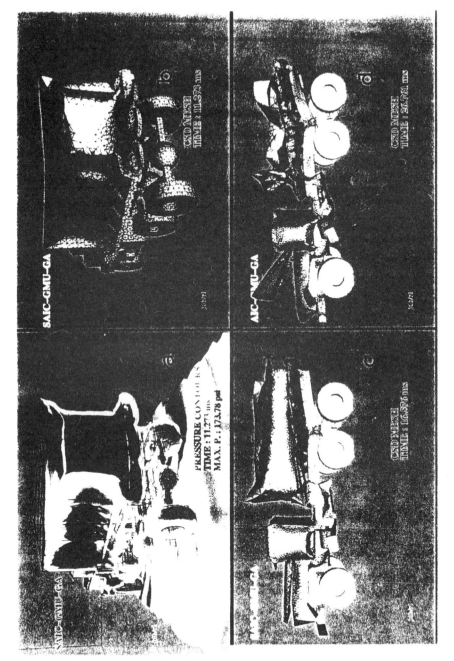

Fig 2. Pressure Contours on the Truck at t=11.27 ms, and CSD Surface Mesh at t=11.273 ms, t=16.57 ms, and t=26.76 ms.

FLOW AROUND A SURFACE MOUNTED CUBICAL OBSTACLE: COMPARISON OF LES AND RANS-RESULTS

M. Breuer [*], D. Lakehal, W. Rodi

Institut für Hydromechanik, Universität Karlsruhe (TH),
Kaiserstr. 12, D-76128 Karlsruhe, Germany

SUMMARY

The paper deals with a comparative study of LES and RANS ($k-\varepsilon$ model) results for a typical bluff–body flow, namely the flow around a surface mounted cubical obstacle placed in a plane channel. For this test case detailed experimental data (Re=40,000) are available [11]. Two slightly different numerical solution procedures based on a 3-D finite–volume method are used in this investigation. The Reynolds–averaged equations for incompressible flow are solved implicitly [10], whereas in the LES code [1, 2, 3, 4, 5] an explicit second order Adams–Bashforth scheme is applied. Different formulations of the $k - \varepsilon$ turbulence model are used in the RANS simulations, the standard version with wall functions, a RNG version, a modified version proposed by Kato and Launder [7], and a two–layer approach. For modeling the non-resolvable subgrid–scale motion in the LES two different models are applied, namely the well known Smagorinsky model [16] as well as the dynamic model originally proposed by Germano et al. [6]. The capability of the different methods is demonstrated by comparison with the measurements.

1. INTRODUCTION

Turbulent flows of practical interest are in general very complex including phenomena such as separation, reattachment and vortex shedding. An appropriate description by Reynolds–Averaged Navier–Stokes (RANS) equations combined with statistical turbulence models is difficult to achieve. This is due to the necessity of modeling the whole spectrum of turbulent scales. The method of direct numerical simulation requires no model assumptions but will not be applicable to engineering flows in the foreseeable future because of extremely high computing costs. The concept of large–eddy simulation (LES) seems to be a promising way of solving such flow problems. In LES the large eddies that depend strongly on the special flow configuration are resolved numerically whereas only the fine–scale turbulence has to be modeled by a subgrid–scale model.

The goal of the work reported here is the development of a large–eddy simulation technique for practically relevant flows and a comparative study of LES and RANS ($k - \varepsilon$ model) results. The flow around a surface mounted cubical obstacle inside a plane channel was chosen as a typical bluff-body flow. Detailed experimental data ($Re = U_B H/\nu = 40,000$, U_B = bulk velocity, H = obstacle height) have been provided by Martinuzzi et al. [11]. In the first section of the paper the governing equations, turbulence models, methods of solution, and boundary conditions for both RANS and LES approach are explained. The second part deals with a detailed comparison of the different computed results as well as with the experimental data.

2. GOVERNING EQUATIONS

Three–dimensional unsteady flows are described by the Navier–Stokes equations. These are the starting-point for both the RANS approach as well as the LES technique. The procedures to derive the governing equations are also very similar. For the RANS approach the Reynolds averaging procedure is introduced which separates all instantaneous quantities in a turbulent flow field into a time or ensemble averaged mean value and a fluctuating part. Then the Reynolds averaged equations describe the motion of the time or ensemble averaged mean flow. Due to the averaging of the non–linear convective terms the unknown Reynolds stress tensor τ_{ij} appears. In the LES approach a similar averaging procedure is applied to the governing equations. However, in contrast to the RANS approach the averaging is now accomplished with respect to space and not to time. In the context of LES this is called filtering rather than averaging. The goal is to separate different length scales in the turbulent flow field. All large scale structures which

---

[*] Present address: Lehrstuhl für Strömungsmechanik, Universität Erlangen-Nürnberg, Cauerstr. 4, D-91058 Erlangen

can be resolved by the numerical method applied should be separated from the small scale structures (subgrid–scales) which cannot be captured on a given grid. The governing equations for LES have the same form as for the RANS approach. However, the meaning of some variables has changed. In LES the equations describe the motion of the resolvable part (grid scale) of the flow field. The new stress tensor τ_{ij}, resulting from the filtering of the convective terms, is now called Reynolds **subgrid–scale** stress tensor. This expresses the main difference between RANS and LES. In a RANS simulation the Reynolds stress tensor has to be modeled, which describes the influence of the whole spectrum of turbulent scales on the time or ensemble averaged mean flow. In LES, however, a large part of the spectrum of turbulent motions is directly computed by the numerical scheme and only the subgrid–scales have to be modeled. It is well known that the smaller eddies in a turbulent flow are easier to model (more homogeneous and isotropic) than the whole spectrum of turbulent motions, but the price which has to be paid in LES for this advantage is the necessity to resolve the large scale structures. In both approaches the stress tensor τ_{ij} is modeled using the eddy viscosity concept [19]. In this, τ_{ij} is linearly related to the deformation tensor $S_{ij} = 1/2\,(U_{i,j} + U_{j,i})$ by $\tau_{ij} = 2/3\,\delta_{ij}k - 2\,\nu_t S_{ij}$. In the RANS approach the trace of the stress tensor $1/3\,\delta_{ij}\tau_{kk}$ is expressed by the turbulent kinetic energy $k = 1/2\,\overline{u_i u_i}$, whereas in LES this part is normally added to the pressure.

3. RANS: TWO–EQUATIONS TURBULENCE MODELS

The eddy viscosity is determined according to the algebraic expression [14] $\nu_t = C_\mu k^2/\varepsilon$. This relation requires values of the turbulent kinetic energy k as well as the dissipation rate ε.

3.1 Standard $k - \varepsilon$ model [14]

In the standard $k-\varepsilon$ model the turbulence parameters k and ε are obtained by solving transport equations for these quantities together with those describing the mean flow. In both transport equations the term $P_k = -\tau_{ij}\,U_{i,j}$ appears which represents the rate of production of turbulent kinetic energy resulting from the interaction of the turbulent stresses and the mean flow. The empirical constants appearing in the model are given the following standard values: $C_\mu = 0.09$; $C_1 = 1.44$; $C_2 = 1.92$; $\sigma_k = 1$. and $\sigma_\varepsilon = 1.3$.

3.2 RNG $k - \varepsilon$ model [20]

In [20], the renormalization group (RNG) theory was applied to derive the $k - \varepsilon$ turbulence model. In this theory, the small scale fluctuations are removed successively from the governing equations leading to averaged equations. The resulting $k - \varepsilon$ model has an extra term R_ε in the ε–equation, for which a model is introduced, which reads: $R_\varepsilon = C_\mu \eta^3 (1 - \eta/\eta_0)/(1 + \beta\eta^3)\varepsilon^2/k$ where $\eta = (2S_{ij}S_{ij})^{1/2}k/\varepsilon$ and $\eta_0 = 4.38$, $\beta = 0.015$. The model constants following from the RNG theory are: $C_\mu = 0.084$; $C_1 = 1.42$; $C_2 = 1.68$; and $\sigma_k = \sigma_\varepsilon = 0.7179$.

3.3 Kato–Launder $k - \varepsilon$ model [7]

One of the most important drawbacks of the standard $k-\varepsilon$ model is that it leads to an excessive production of k in stagnation regions, i.e. in impinging flows. This phenomenon is a consequence of the inability of eddy viscosity models to simulate correctly the difference in normal Reynolds stresses governing the production of k in such regions. Hence, Kato and Launder [7] suggested as an adhoc measure to replace the original production term $P_k = C_\mu \varepsilon S^2$ by $P_k = C_\mu \varepsilon S \Omega$, where $S = (1/2\,S_{ij}S_{ij})^{1/2}k/\varepsilon$ is the strain rate parameter and $\Omega = (1/2\,\Omega_{ij}\Omega_{ij})^{1/2}k/\varepsilon$ the rotation parameter ($\Omega_{ij} = 1/2\,(U_{i,j} - U_{j,i})$). In shear flows $\Omega \approx S$ so that the original production term is recovered while in stagnation regions $\Omega \approx 0$ so that the desired effect of suppressing the k–production is achieved.

3.4 Two–Layer Model [15]

The assumptions on which the wall functions used in the standard $k - \varepsilon$ model are based are not really valid. Hence, a two–layer model combining the $k - \varepsilon$ model in the outer region ($\nu_t/\nu > 36$) with a one–equation model in the viscosity-affected near-wall region ($\nu_t/\nu < 36$) was proposed [15]. In the one–equation model, the dissipation rate ε needed for both eddy viscosity formulation and k transport equation is not determined via a transport equation but from the following prescribed length–scale distribution:

$$\varepsilon = \frac{k^{3/2}}{L}\left(1 + \frac{\nu C_\epsilon}{k^{1/2} L}\right) \quad \text{with: } L = C_D \kappa y_n, \quad C_\epsilon = 13.2, \quad C_D = 6.41, \quad \kappa = 0.41. \tag{3.1}$$

L stands for the near–wall length–scale, and y_n the normal distance from the wall. However, the turbulent kinetic energy k is determined in the same way as in the $k - \varepsilon$ model. The eddy viscosity ν_t is obtained from:

$$\nu_t = f_\mu C'_\mu k^{1/2} L \quad \text{with: } f_\mu = 1 - exp(-0.0198 R_y), \quad R_y = k^{1/2} y_n / \nu, \quad C'_\mu = 0.084. \tag{3.2}$$

It should be noted that R_y involves $k^{1/2}$ as velocity scale and not U_τ which changes sign in separated flows.

4. LES: SUBGRID–SCALE MODELS

4.1 Smagorinsky Model [16]

In principle the LES concept leads to a similar closure problem as the RANS approach. Therefore a similar classification of turbulence models starting with zero–equation models and ending up with Reynolds stress models is possible. However, the non–resolvable small–scale turbulence in a LES is much less problem-dependent than the large–scale turbulence so that the subgrid–scale turbulence can be represented by relatively simple models, e.g. zero–equation eddy–viscosity models. The well known and mostly used Smagorinsky model [16] is based on the Boussinesq's approach. The eddy viscosity ν_t itself is a function of the strain rate tensor S_{ij}:

$$\nu_t = l^2 \mid S_{ij} \mid = (C_s \overline{\Delta})^2 \mid S_{ij} \mid \quad \text{with: } \mid S_{ij} \mid = \sqrt{2 S_{ij} S_{ij}} \quad \overline{\Delta} = (\Delta x \, \Delta y \, \Delta z)^{1/3} \tag{4.3}$$

C_s is called the Smagorinsky constant. Taking into account the reduction of the subgrid length l near solid walls, the length scale is usually multiplied by a Van Driest damping function. Here the main advantages of LES become clear. The difficult problem of RANS to determine a characteristic length scale of the turbulent flow does not appear in LES, because it is fixed by the size of the computational cell.

4.2 Dynamic Model [6]

One of the major drawbacks of the Smagorinsky model is that the Smagorinsky constant C_s was found to depend on the flow problem considered. Secondly, in an inhomogeneous flow, the optimum choice for C_s may be different for different points in the flow. Furthermore, the Smagorinsky model needs some additional assumptions to describe flows undergoing transition or near solid walls. The dynamic model, originally proposed by Germano et al. [6], eliminates some of these disadvantages by calculating a 'Smagorinsky constant' as a function of time and position from the smallest scales of the resolved motion. Based on the local equilibrium approach (production = dissipation), the eddy viscosity is again evaluated from eq.(4.3). However, in contrast to the Smagorinsky model C_s is no longer a constant but a local, time–dependent variable. The dynamic model itself represents a method for determining this unknown variable C_s from the information already contained in the resolved velocity field. A systematic procedure for computing turbulent flows by LES without the necessity of prior experience to properly adjust the Smagorinsky constant was derived by Germano et al. [6]. The basic formalism behind the method need not be repeated here. Following a suggestion of Lilly [9] the original model was slightly improved by using a least–squares approach to obtain values for C_s.

Tests with this formulation of the dynamic model have shown that in principle negative values of C_s are possible. However, in practice these are not able to represent the backscatter effect because negative eddy viscosities strongly destabilize the numerical algorithm. Furthermore, the value C_s is an instantaneous and local quantity varying very much in space and time. This also leads to numerical instabilities. Different possibilities have been tested to remove this problem. Depending on the flow considered, different kinds of averaging procedures can be applied. If the flow is homogeneous in a certain direction, averaging can be applied over this direction. For fully inhomogeneous flows, like the flow around a surface mounted cubical obstacle, only an averaging procedure in time is applicable. In order not to restrict the values of C_s to a fully time–independent function and to allow variations with low frequencies, a special form of time averaging (lowpass filtering) is chosen, which is well known as a recursive lowpass digital filter [1]. With an appropriate value for the parameter of this filter function all high frequency oscillations are damped out and only the low frequency variations remain. This seems to be better than fully freezing C_s. In addition, negative eddy viscosities are clipped.

5. METHODS OF SOLUTION

Because the LES code (**LESOCC** [1, 2, 3, 4, 5]) is a descendant of the RANS program (**FAST–3D** [10]) both have many features in common, which will be described first. Both methods are based on a finite–volume approach for solving the incompressible Navier–Stokes equations on general body–fitted, curvilinear grids. A non–staggered, cell–centered grid arrangement is used. In order to avoid the well known pressure field checkerboard problem, the momentum interpolation technique due to Rhie and Chow [13] is applied. The pressure–velocity coupling is achieved with the SIMPLE algorithm of [12]. In both codes the viscous fluxes are approximated by central differences of second order accuracy. The linear discretized system of equations is solved using the strongly implicit solution procedure of Stone [17] which can be accelerated by a FAS multigrid technique.

The main differences between both codes is given by the temporal discretization as well as the spatial discretization of the convective fluxes. Due to totally different goals, different time stepping schemes are used. An implicit decoupled solution method is prefered for the RANS code, because it guarantees a fast convergence to the desired steady state solution of the RANS equations. In a LES it is necessary to resolve turbulent fluctuations in time with at least an accuracy of second order. A LES further requires small time steps which can be treated much more efficiently by an explicit scheme. Therefore the temporal discretization of the LES code consists of a predictor–corrector scheme, where the predictor step is an explicit Adams–Bashforth scheme for the momentum equations (second order in time) and the corrector step involves the implicit solution of the Poisson equation for the pressure correction.

Another important requirement for LES is a higher order approximation of the convective fluxes where in general the numerical dissipation produced by a scheme is a much better measure for its accuracy than the order of the discretization itself. Central differences of second order accuracy have been found to be a reasonable discretization of the convective fluxes for LES. In the RANS model the HLPA (hybrid linear–parabolic approximation) second order low–diffusive and oscillation–free scheme of [21] is applied for the convective part.

6. BOUNDARY CONDITIONS

For the three RANS turbulence models, described in chapters 3.1 – 3.3, the boundary conditions at impermeable walls involve the well known wall functions [14]. In order to obtain a fully developed channel flow solution as initial condition, the inflow profiles are approximated using a logarithmic profile $U(y)/U_\tau = \kappa^{-1} * ln(y/y_0)$ for the velocity, and $k(y) = 1.5(I(y) * U(y))^2$ for the k profile respectively. y_0 is the roughness length and $I = \sqrt{u'^2}/U_B$ the turbulence intensity. The on-coming flow is assumed to have a dissipation rate according to fully developed channel flow, leading to $\varepsilon(y) = C_\mu^{3/4} k^{3/2}/\kappa L_u$. The turbulence length scale L_u is set equal to $0.1H$ [8]. At the lateral planes, symmetry conditions are applied.

In the case of LES, the inflow is fully developed turbulent channel flow, generated by LES of plane channel flow (same grid in the cross-sectional plane). For the lateral boundaries (x-y plane) at $\pm z/H = 3.5$ periodic boundary conditions are chosen. At solid wall the wall function approach of Werner/Wengle [18] is applied. A convective boundary condition is used at the outflow boundary.

7. GRIDS AND CALCULATION DOMAINS

For all RANS computations except the two–layer one, the same grid and computational domain is used. A mesh consisting of $110 \times 32 \times 66$ grid points (streamwise/normal to the channel walls/lateral) is applied to form a computational domain with an upstream length of $x_1/H = 3.5$, a downstream length of $x_2/H = 10$, and a width of $b/H = 9$. Here the grid covers the whole domain. The smallest cell volume in the vicinity of the obstacle walls has a size of $(0.01\,H)^3$. However, for the two-layer model simulation a finer grid with $142 \times 84 \times 64$ grid points is used, which covers only one half of the calculation domain taking into account the symmetry condition of the flow field (smallest cell $(0.001\,H)^3$).

For all LES a computational domain with an upstream length $x_1/H = 3$ and a downstream length of $x_2/H = 6$ is used. The width is set to $b/H = 7$. The restriction to a smaller integration domain in the LES case compared with the RANS case is necessary to achieve a sufficient resolution of the flow field in the vicinity of the obstacle. Of course LES cannot take advantages of the symmetry of the time-averaged flow field as RANS can do. All LES computations are performed on a stretched grid with $165 \times 65 \times 97$ grid points for the x, y and z directions. In the streamwise direction, 70 grid points are distributed in the region in front of the obstacle. On the surface of the obstacle, 31 grid points are used in all directions. The smallest cell volume in the vicinity of the solid walls has a size of $(0.0125\,H)^3$.

8. RESULTS

Fig. 1 shows a first qualitative comparison of the results ($Re = 40,000$). The streamlines in the plane of symmetry and at a horizontal plane close to the channel floor are plotted for the experimental and numerical results. In the LES the velocities are averaged over a long period of more than 100 dimensionless time units (H/U_B) to achieve good statistics. It appears clearly that the stagnation point is well simulated by the different numerical approaches, whereas the primary upstream separation location (labeled A in the experimental oil flow pattern) caused by the strong adverse pressure gradient imposed by the obstacle, is slightly shifted upstream vis a vis the experiment ($x/H = -0.9$). Moreover only LES and two-layer calculations produce a correct separation bubble on the roof, with a somewhat better agreement with the experiment for the LES (no reattachment of the time-averaged flow). The Kato-Launder modification of the standard $k - \varepsilon$ model seems to improve the results compared with the original version. However, the calculated separation bubble on the roof of the obstacle is still too flat compared with the experiment and reattachment takes place. Indeed, the standard $k - \varepsilon$ model as well as the RNG version simulation show a very poor description of the flow in this region. Furthermore the extension of the large separation region (x_r) behind the obstacle is highly overpredicted using the RANS models (standard $k - \varepsilon$: $x_r/H = 2.20$, RNG: $x_r/H = 2.08$)) compared with the experimental value $x_r/H = 1.62$. The use of both Kato-Launder model ($x_r/H = 2.73$) and two-layer approach ($x_r/H = 2.68$) shows an unexpected overprediction of the size of the recirculating zone, resulting from an underpredicted turbulent viscosity level. The agreement between the experiment and the time-averaged flow field calculated by LES is much better. The reattachment length behind the obstacle is only slightly overpredicted by the LES with the Smagorinsky model (LES-S) ($x_r/H = 1.69$) and underpredicted somewhat by the LES with the dynamic model (LES-D) ($x_r/H = 1.43$).

Fig. 1 displays also a comparison of the experimental versus the calculated time-averaged surface streamlines at the channel floor. The difference between the locations of the calculated primary separation line using the different calculation approaches is clear. Furthermore one can notice that amongst the different RANS models used, only the two-layer model allows to capture the secondary recirculation at the front base of the obstacle (C), as well as LES. In comparison with the other RANS models the two-layer approach reproduces more details of the flow structure near the walls. However, this is at least partly the result of a finer resolution in the vicinity of the solid walls for the two-layer model in contrast to the coarse grids used for the wall function approaches. The same figure shows that the horseshoe vortex generated between the primary and the secondary separation lines (B) is fairly well predicted by the different approaches. The flow patterns suggest also that the structure of the outer limit of the wake region formed by the lateral arms of the horseshoe vortex (line D), varies between the different calculation approaches. In the experiment, the width of this wake decreases up to approximately the reattachment point; then it increases again. This behavior is well described only by the two-layer and LES models. Both RANS and LES approaches seem to predict correctly the corner vortices (N_{12}) generated downstream of the vertical leading edges of the cube at the channel-body junction. The location of the simulated corner vortices behind the obstacle (N_{14}) shows clearly the differences between RANS and LES results. Except for the two-layer approach, the center of the vortices produced by the RANS methods is shifted downstream compared with the experimental observation. The LES results agree fairly well with the experiment.

Fig. 2 displays the calculated versus the measured streamwise velocity profiles U profiles at six different locations in the symmetry plane. One obstacle height H in front of the cube ($x/H = -1.0$) all streamwise velocity profiles agree fairly well with the measurements. However, large differences can be observed for the next profile in the middle of the roof ($x/H = 0.5$). Here the best result compared with experiment is provided by the LES with the dynamic model (LES–D). The size of the separation bubble and the magnitude of the reversed flow velocity is well reproduced. An attempt to classify the rest of the simulations results in: LES–S, two–layer model, Kato–Launder model, standard $k - \varepsilon$ model and finally RNG model. The last two do not show any separation at this position on the roof of the obstacle at all, while both the Kato–Launder modification and the two–layer model results show a better behavior. This latter observation allows us to believe that combining the Kato–Launder modification with the two–layer approach would provide a better description of the flow in this region. At the third location ($x/H = 1.$), the same tendency can be detected. Moving further downstream, the effect of the variations in the computed length of the recirculation region is clearly visible. In the wake region ($x/H = 1.5$), the computed velocity magnitudes are globally underestimated; here LES-S results are closest to experiment. Far from the reattachment point at $x/H = 4.$, again both LES give a better representation of the flow than all RANS models. The bad agreement between experiment and RANS computations concerning the recirculation length behind the obstacle confirms the unsatisfactory results in the velocity profiles at this position. This also demonstrates the low level of recovery of the flow field. Therefore the RANS computations would require a much longer distance to establish fully developed channel flow conditions again.

In Fig. 3 three profiles of the turbulent kinetic energy k are plotted in the symmetry plane. It should be noted that for LES only the resolved part of the turbulent kinetic energy is included. At $x/H = 0.5$ all simulations give similar peak values for k, however, the form of the profiles is different, e.g. the standard $k - \varepsilon$ model shows too large values above the separation bubble. In this figure the influence of the Kato–Launder modification can be clearly observed. The two–layer approach produces a k–profile at this location very similar to LES. At $x/H = 1.$ the differences in the peak values of k become larger. None of the simulations gives the experimentally observed peak at the right position; in all computations, it is located higher than in the measurements. Further downstream ($x/H = 2.$) the scatter in the computed profiles for k increases. Again, both LES are in closer agreement with the experimental values than all RANS results, even if the two LES provide a slightly different behavior. Overall the level of turbulent kinetic energy is much too small in the RANS computations which may cause the far too long recirculation region behind the obstacle.

9. CONCLUSION

A typical bluff–body flow, namely the three–dimensional turbulent flow around a surface–mounted cubical obstacle placed in developed channel flow, has been investigated by four different 2-equation RANS models as well as LES with two subgrid–scale models. This is a geometrically simple but physically very complex flow with multiple, unsteady separation regions and vortices. Concerning all quantities considered LES in general shows better results compared with the experiment than the RANS approaches. Some qualitative features of the flow field are not even captured by some of the RANS models, e.g. the large separation region on top of the roof without reattachment of the time–averaged flow. Here the standard $k-\varepsilon$ model and the RNG modification do a rather poor job. Only for the two–layer approach the agreement with the experiment is better in this region. The length of the recirculation region behind the obstacle is highly overpredicted by all RANS models. Both LES show better agreement with the measurements. Depending on the applied subgrid–scale model the recirculation length is slightly overpredicted (LES-S) or even underpredicted (LES-D). Taking the surface streamlines at the bottom wall as well as the profiles of mean velocity and turbulence kinetic energy as the basis of assessment, the tendency is always the same. However, the price for better agreement with experiments is rather high and has to be mentioned here. It is well known that LES is a very CPU–time consuming way of computing turbulent flows. This is on the one hand due to the requirements concerning the spatial resolution of the flow field. On the other hand the most expensive part especially for fully inhomogeneous flows is the necessity to simulate the instantaneous flow over a long period in time to achieve good statistical values. In our case the ratio between the CPU–time requirements can be approximated by 1 : 25 : 200(400) where the three RANS models (standard $k - \varepsilon$ model, RNG, Kato–Launder) with nearly similar values are taken as the basis of reference (approximately 15 CPU-min. on SNI S600/20). Switching to the two–layer approach already increases the costs by a factor of about 25 due to the necessary resolution in the vicinity of solid walls and lower rates of convergence. A factor of about 200 is present between standard RANS models and LES for the mean quantities, where the value in brackets is an estimation for reasonable higher order moments. However, if the instantaneous features of the flow field are more interesting than the time-averaged results, e.g. for fluid-structure aerodynamic coupling problems, LES may become a reasonable alternative to RANS models. Such simulations, however, still require powerful vector or parallel machines, whereas 3-D RANS simulations (except the two-layer approach) can be performed on workstations.

ACKNOWLEDGMENTS

The work reported here was sponsored by the Deutsche Forschungsgemeinschaft and the Human Capital and Mobility Programme of the European Union. The calculations were carried out on the SNI S600/20 vector computer of the University of Karlsruhe (Computer Center).

REFERENCES

[1] Breuer, M., Rodi, W.: *Large-Eddy Simulation of Turbulent Flow through a Straight Square Duct and a 180° Bend*, Fluid Mech. and its Appl., vol. 26, Direct & LES I, Sel. papers f. the First ERCOFTAC Workshop on Direct & LES, Guildford, Surrey, U.K., 27-30 March 1994, ed. Voke, Kleiser & Chollet, Kluwer Acad. pub., (1994).

[2] Breuer, M., Rodi, W.: *Large-Eddy Simulation of Turbulent Flow through Straight and Curved Ducts*, ERCOFTAC Bulletin, vol. 22, Sept. (1994).

[3] Breuer, M., Pourquie, M., Rodi, W.: *Large Eddy Simulation of Internal and External Flows*, 3rd International Congress on Industrial and Applied Mathematics, Hamburg, 3-7 July, (1995), to be published in ZAMM, (1996).

[4] Breuer, M., Pourquie, M.: *First Experiences with LES of Flows past Bluff Bodies,* accepted for the 3rd Intern. Symposium of Engineering Turbulence Modeling and Measurements, Crete, Greece, May 27-29, (1996).

[5] Breuer, M., Rodi, W.: *Large Eddy Simulation of Complex Turbulent Flows of Practical Interest*, in preparation for the final report of DFG Priority Programme: 'Flow Simulation with High-Performance Computers II', Notes on Numerical Fluid Mechanics, Vieweg Verlag, Braunschweig, (1996).

[6] Germano, M.; Piomelli, U.; Moin, P.; Cabot, W. H. : *A dynamic subgrid-scale eddy viscosity model*, Phys. Fluids A. 3 (7), pp. 1760-1765, (1991).

[7] Kato, M., Launder, B.E.: *The Modeling of Turbulent Flow around Stationary and Vibrating Square Cylinders*, Proc. 9th Symp. Turb. Shear Flows, Kyoto, 10-4-1, (1993).

[8] Lakehal, D.: *Simulation Numerique d'un Ecoulement Turbulent autour de Batiments de Formes Courbes*, Thesis at the University of Nantes, Ecole Centrale of Nantes, (1994).

[9] Lilly, D.K.: *A proposed modification of the Germano subgrid-scale closure method*, Phys. Fluids A 4 (3), pp. 633-635, (1992).

[10] Majumdar, S., Rodi, W., Zhu, J. : *Three-dimensional finite-volume method for incompressible flows with complex boundaries*, J. Fluid Eng., vol. 114, pp. 496-503, (1992).

[11] Martinuzzi, R. and Tropea, C.: *The Flow around surface-mounted, prismatic obstacle placed in a Fully Developed Channel Flow*, J. of Fluids Engineering, vol. 115, (1993).

[12] Patankar, S.V., Spalding, D.B.: *A calculation procedure for heat, mass and momentum transfer in three-dimensional parabolic flows*, Int. J. Heat & Mass Transfer, vol. 15, pp. 1778-1806, (1972).

[13] Rhie, C.M., Chow, W.L.: *A numerical study of the turbulent flow past an isolated airfoil with trailing edge separation*, AIAA-J., Vol. 21, pp. 1225-1532, (1983).

[14] Rodi, W.: *Turbulence Models and their Application in Hydraulics*, International Association for Hydraulic Research, Delft, The Netherlands, (1980).

[15] Rodi, W.: *Experience with two-layer models conbining the $k - \epsilon$ model with a one-equation model near the wall*, AIAA paper, AIAA-91-0216, (1991).

[16] Smagorinsky, J.: *General circulation experiments with the primitive equations, I, The basic experiment*, Mon. Weather Rev. 91, pp. 99-165, (1963).

[17] Stone, H.L.: *Iterative solution of implicit approximations of multidimensional partial differential equations*, SIAM J. on Num. Anal., vol. 5, pp. 530-558, (1968).

[18] Werner, H., Wengle, H.: *Large-Eddy Simulation of Turbulent Flow over and around a Cube in a plate Channel*, 8th Symp. on Turb. Shear Flows, (Schumann et al., eds.), Springer Verlag, (1993).

[19] Wilcox, D. C.: *Turbulence Modeling for CFD*, DCW Ind., ING, Las Canada, California, USA, (1993).

[20] Yakhot, V., Orszag, S. A., Tangham, S., Gatski, T. B., Speziale, C. G.: *Development of Turbulence Models for Shear Flows by a Double Expansion Technique*, Physics of Fluids A. 4, pp. 1510-1520, (1992).

[21] Zhu, J.: *A low-diffusive and oscillating-free convective scheme*, Communications in Applied Numerical Methods, vol. 7, pp. 225-232, (1991).

FIGURES

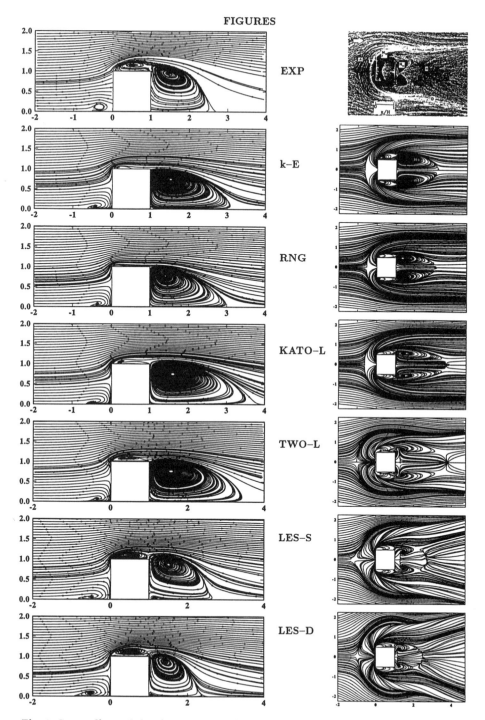

Fig. 1: Streamlines of the time–averaged flow in the symmetry plane of the 3–D obstacle and surface streamlines in the bottom wall of the channel, Re = 40,000

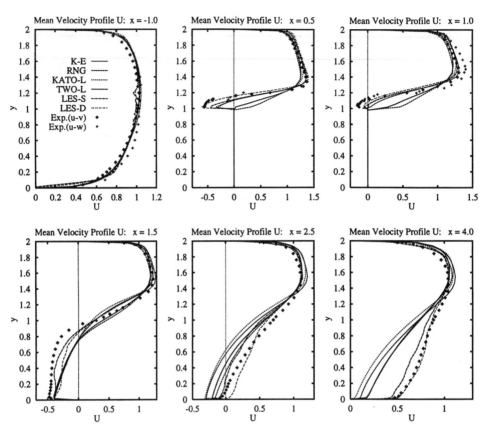

Fig. 2: Comparison of mean velocity profiles U of the time–averaged flow in the symmetry plane of the 3–D obstacle, Re = 40,000

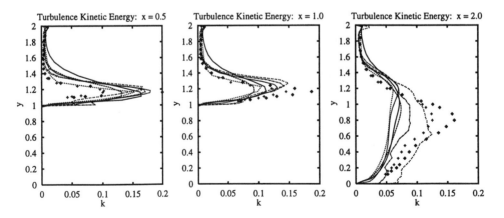

Fig. 3: Comparison of turbulent kinetic energy profiles k of the time–averaged flow in the symmetry plane of the 3–D obstacle, Re = 40,000, (same legend as in Fig. 2)

3D FLOW COMPUTATIONS UNDER A REACTOR VESSEL CLOSURE HEAD

by O. DAUBERT, O. BONNIN, F. HOFMANN, M. HECKER

Electricité de France, Direction des Etudes et Recherches, Laboratoire National d'Hydraulique, 6 quai Watier
78400 CHATOU, FRANCE

Summary

The flow under a vessel cover of a pressurised water reactor is investigated by using several computations and a physical model. The case presented here is turbulent, isothermal and incompressible. Computations are made with N3S code using a k-epsilon model. Comparisons between numerical and experimental results are on the whole satisfying. Some local improvements are expected either with more sophisticated turbulence models or with mesh refinements automatically computed by using the adaptive meshing technique which has been just implemented in N3S for 3D cases.

I. Introduction

The aim of the study is to determine the incompressible steady flow field under an hemispherical upper closure head of a Pressurised Water Reactor. A lot of vertical tubes, called " adapters " pass through this cover and are weld on it. The control drive rods can move inside the adapters, under the action of electro-magnetic mechanisms.
The fluid domain is partly occupied by vertical guide tubes and thermal sleeves protecting each control rod cluster (fig. 1).
It is necessary to determine the general flow under the cover as well as the local flow inside the adapters. But it is not possible to represent the overall phenomena in a single model. That is why several models are used: two of them are presented here.

II. Organisation of the study

In a first step we compute the general 3D velocity and pressure field, in a 45° of the cover. This geometry is called "COUVERCLE-45°".
The flow in this region is driven by peripheral jets which sweep the internal side of the cover. Flow outlets are located at the top of the 13 guide tubes represented in the model.
The second step is a local 3D computation around an adapter. Boundary conditions are given by the results of the first computation. The flow configuration is more complicated here (fig. 2): the incoming wall flow meets a set of two concentric cylindrical obstacles with an incidence of 48°. It is not a standard flow feature, so the numerical model needs a validation by a physical experiment on the same geometry (fig. 5, 6). This geometry is called "TRAVERSIN".
As the Reynolds number is lower in the experimental model than in the reactor, we use the following procedure:
    - in a first time, we try to reproduce the experimental data,
    - and in a second one, we impose the real boundary conditions to determine the velocity and pressure fields in and around the adapter.

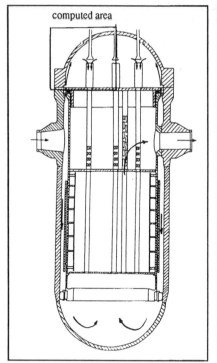 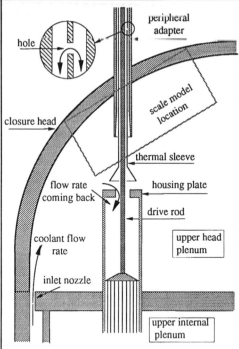

Figure 1 - General view of a pressurised water reactor

Figure 2 - Global view of the upper head geometry with one adapter

III. The N3S code

N3S is a finite element code developed by EDF for incompressible flow simulations in industrial studies [3]. For code assessment, a wide range of test cases is made under a Quality Assurance procedure for every main release of N3S. Code results are compared with analytical solutions when available or with literature experiments.
Since March 95, version 3.2 have been offering extended capabilities, mainly for compressible flows, coupled thermal computations in fluid and solid [1], and adaptive meshing [2].
In the present study classical Navier-Stokes equations are used with a turbulent viscosity coefficient:

$$\begin{cases} \rho \dfrac{Du}{Dt} = -\nabla p + \nabla \cdot \left[(\mu + \mu_t)(\nabla u + {}^t\nabla u)\right] \\ \nabla \cdot u = 0. \end{cases} \quad (1)$$

In these equations, D/Dt denotes the total derivative in time, u the velocity field, p the pressure, ρ the density, μ the dynamic viscosity, and μ_t the eddy dynamic viscosity coefficient (see below).

The turbulence model used is the $k - \varepsilon$ model for which the turbulent kinetic energy k and dissipation rate ε are the solutions of the following transport-diffusion equations:

$$\begin{cases} \rho \dfrac{Dk}{Dt} = \nabla \cdot \left[\left(\mu + \dfrac{\mu_t}{\sigma_k}\right)\nabla k\right] + P - \rho\varepsilon \\ \rho \dfrac{D\varepsilon}{Dt} = \nabla \cdot \left[\left(\mu + \dfrac{\mu_t}{\sigma_\varepsilon}\right)\nabla\varepsilon\right] + C_{\varepsilon_1}\dfrac{\varepsilon}{k}P + \rho C_{\varepsilon_2}\dfrac{\varepsilon^2}{k} \end{cases} \quad (2)$$

with: $\mu_t = \rho C_\mu \dfrac{k^2}{\varepsilon}$, $\qquad P = 2\mu_t \mathrm{tr}(\mathbf{d}{:}\mathbf{d})$, $\qquad$ and $\qquad d_{ij} = \dfrac{1}{2}\left(\dfrac{\partial u_i}{\partial x_j} + \dfrac{\partial u_j}{\partial x_i}\right)$.

Various values are given for the coefficients involved in this equations:
in the standard k - ε model, they are constant. In the RNG model [5, 6], derived according to the renormalization group theory, $C_{\varepsilon 1}$ is no longer a constant but a scalar function of the variable $\eta = (2d_{ij}d_{ij})^{1/2}k/\varepsilon$
Coefficients for both models are given below:

	C_μ	σ_k	σ_ε	$C_{\varepsilon 1}$	$C_{\varepsilon 2}$
standard k-ε	0.09	1	1.30	1.44	1.92
RNG	0.085	0.72	0.72	$1.42 - \dfrac{\eta(1-\eta/4.38)}{1+0.012\eta^3}$	1.68

Appropriate boundary conditions are also defined: for the solid boundaries, a Reichardt law is used (see [3]). Otherwise velocity vectors are prescribed at inlet nodes, and free slip conditions on symmetry planes. For the need of the present study, special treatments for multiple outlets with prescribed normal stress or a pressure-flow rate relation, have been tested in order to validate re-entering flows on some outlets. These tests are excellent for laminar flows, and are slightly penalised, for turbulent flows, by the lack of information on incoming k and ε. Moreover, they show that it is convenient to dispose outlet boundaries far from each others when using pressure-flow rate relations, to avoid instabilities.

These equations and boundary conditions are solved for transient or stationary flows, via a time evolution from initial conditions, generally the rest.
The time discretization is based on a fractional step method involving:
   - an advection step, for the non-linear convection terms of the Navier-Stokes, k-ε and eventually temperature equations, solved by a characteristics method,
   - a diffusion step for the remaining part of the k and ε equations : the finite element discretization leads to linear systems solved by a preconditioned conjugated gradient algorithm,
   - a generalised Stokes problem for the velocity and the pressure, solved by a Chorin algorithm.
The spatial discretization uses P2-P1 or isoP2-P1 tetrahedral elements : pressure is always linear (P1) by element and velocity is quadratic (P2) or piece wise linear (isoP2) ; other scalar variables may be linear, quadratic or piece wise linear.

Adaptive meshing

This method already tested in N3S for 2D flows has been recently available for 3D cases. It needs a local error indicator giving the places where the mesh is too coarse, and a geometrical procedure which refines or de-refines the mesh.
   - For turbulent flows, a *projection indicator* is used. It was first introduced in elasticity [4] and then transposed to fluid dynamics. It has been implemented with appropriate adjustments in N3S. It is based on the difference between discrete entities which are discontinuous at the boundary of the elements, and their projection on the functional space of discretization.

For instance, given that $\mathbf{d_h}$ is the discrete tensor calculated using u_h velocity, solution of the discrete problem, and $\mathbf{d_h}^*$ the L^2 projection in the velocity discretization space, the local velocity error indicator, defined for each element K is

$$I_{K,u} = \left(\int_K 2\mu (\mathbf{d_h} - \mathbf{d_h^*}) : (\mathbf{d_h} - \mathbf{d_h^*}) d\omega \right)^{\frac{1}{2}}. \tag{3}$$

As we cannot define a similar indicator by projecting the pressure which is continuous, a L^2 projection Gp_h^* of the discontinuous pressure gradient is calculated in the pressure discretization space. The pressure error indicator is written:

$$I_{K,p} = \left(h_K^2 \int_K \frac{1}{2\mu} (\nabla p_h - Gp_h^*)^2 d\omega \right)^{\frac{1}{2}}. \tag{4}$$

In the general turbulent case, we could also introduce indicators on the turbulent quantities k and ε.
The whole error indicator is written :.

$$I_K = \left(I_{K,u}^2 + I_{K,p}^2 + I_{K,k}^2 + I_{K,\varepsilon}^2 \right)^{\frac{1}{2}}. \tag{5}$$

- The 3D *refinement module* breaks down tetrahedra into eight sub-elements. Mesh conformity is obtained dividing tetrahedra into two or four parts.

IV. Global flow under the cover

- Numerical model COUVERCLE 45°
The upper region of the reactor vessel (fig.1) presents some geometrical symmetry planes, so it is possible to model only a 45° sector of the cover. Nevertheless, only major obstacles are represented in the finite element mesh (fig.3) : it has 140 000 elements and 207 000 velocity nodes (the same domain with all obstacles represented would have about 270 000 elements and 400 000 nodes). There are three peripheral narrow nozzles where the inlet flow is prescribed. Outlets are located inside the 13 guide tubes and their boundary conditions are normal stresses calculated with pressures measured in a previous experimental study (1978) and including the pressure drops of the housing plates. These boundary conditions allow to obtain the flow rate in each guide tube as a result.
The mesh construction is submitted to some criterion, more or less empirical :
- each cylindrical obstacle is discretized by at least 24 circumferential facets,
- along the rigid walls we impose one or more layers of flattened elements, and we take care to have more than one fluid element between two opposite walls,
- special attention is given to the 3 nozzles representation in order to get the right inlet flow rate as well as the real value of the velocities in the potential cone of the 3 jets ; this point leads to very narrow elements in the orthogonal direction of the jets, allowing sharp discontinuities of the incoming velocity profiles,
- but, in the axial direction of the jets, it is better to impose a longer mesh size because of the Courant criterion : $U.\Delta t/\Delta x \approx 1$, which gives an idea of the precision of convective terms. An isotropic mesh would result in a very small time step. In our case, Courant number calculated in the flow direction is less than 2, it is a reasonable value for N3S.

Starting from a null velocity field the simulation is led up to 3100 time steps. They are computed through successive sessions using a restart procedure. The total CPU time is of 32 hours on a CRAY C98. That means a value of 0.17s by time step per 1000 nodes. The steady flow is obtained after about 2000 time steps.

This model gives a good representation of the global flow in the upper head, as it can be seen on the figure 4.

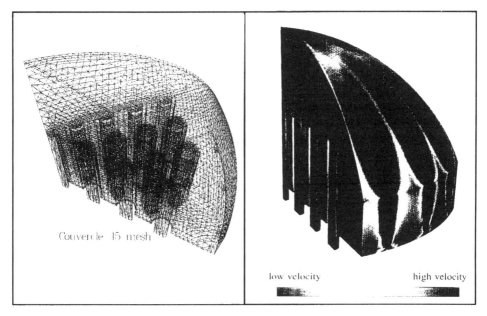

Figure 3 - Finite element mesh of a 45° sector of a reactor upper head

Figure 4 - Velocity field computed in this mesh

V. Local flow around an adapter

- *Physical and numerical models* TRAVERSIN
Using the results of the previous general flow computation, we have designed a local geometry around a peripheral adapter in order to get precise values of pressure and velocity fields in this area. We want particularly to determine the flow penetration in the annular space between the two concentric cylinders called adapter and thermal sleeve on figure 2.
So we have built a mesh and a scale model on this geometry (fig.5, 6). The scale of the physical model TRAVERSIN is 1 for the length and the velocity, and $1/8^{th}$ for the Reynolds number VD/ν, because of the water viscosity ν which is greater at room temperature than in the reactor. The inlet boundary condition is a velocity profile set by an adjustable guide flow followed by a honey comb which regularises the profile at the entrance of the test section. Pressure taps give the pressure elevation in the annular space with respect to the incident flow.

The mesh (fig.5) used for the numerical model TRAVERSIN has 78 000 tetrahedral elements and 116 000 velocity nodes. This model has the advantage of being free of limitations on the flow rate and the velocity profile at the inlet, but the turbulence model used in the computation has to be tested. From previous test cases, we know that turbulence modelling is not obvious in this type of flow (impinging jet on a cylinder). The comparison between experimental and numerical models is made in the following way:
velocity measurements by laser-Doppler anemometer are done at the entrance of the test section to get the exact 2D velocity profile, and in some planes around the adapter. A time analysis of the velocity measured in a point of the wake doesn't give any Strouhal frequency. So we hope to calculate the steady mean velocity and pressure fields by using a k-ε turbulence model.

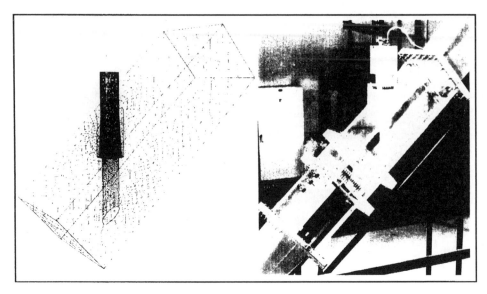

Figure 5 - Mesh of the TRAVERSIN model   Figure 6 - View of the scale model TRAVERSIN

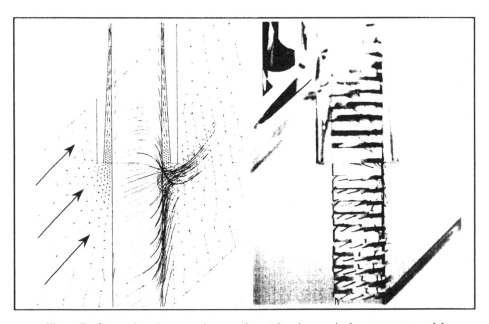

Figure 7 - Comparison between the experimental and numerical TRAVERSIN model

The measured inlet velocity profile is interpolated on the upwind boundary of the mesh with little adjustments to get the right flow rate.
Comparisons with measured velocities in the computed domain are quite good: a very stable horizontal vortex behind the adapter is clearly visible. It should perhaps become more precise with a finer mesh.

The flow penetration in the upper annular channel inside the adapter is visualised in the physical model with wool yarns representing the direction of the velocity vectors. In the numerical one, particle traces are calculated by the post-processor ENSIGHT from the computed velocity field (fig.7). They show the same behaviour, but might be closer on the upwind side. We note also that the computed pressure in the channel is a little lower than the measured one.

It is then interesting to see the both influences of turbulence modelization and mesh refinement. A test made on the same mesh using the modified (RNG) k-ε turbulence model doesn't improve significantly the results ; it will be probably more efficient on a finer mesh.

The adaptive meshing method applied on the previous steady state results gives a first mesh refinement according to the local error indicator The new mesh has 115 000 elements and 169 000 velocity nodes.

To go further in the process more data are needed for the transfer of the boundary conditions on the new nodes created. The only data present in the given input files for this computation are mainly defined by nodal values, and very few information are tight to boundary facets. It is necessary to reconsider the mesh definition in order to characterise more precisely each type of boundary facets and edges.

VI. Conclusion

Numerical 3D flow simulations made with N3S code, give consistent and very useful information about general and local flow under the closure heads of PWR. Thanks to the experimental study lead in parallel with the local flow calculation around an adapter of the cover, one can evaluate the precision of the results. They are really good everywhere, but may be probably enhanced locally by using a method of adaptive meshing with local refinement, and more sophisticated turbulence models as the "RNG k-ε turbulence model" or a Reynolds Stress model "Rij-ε", reducing the energy dissipation at the impact of incident flow on obstacles.

REFERENCES

[1] PENIGUEL C., RUPP I. :
A numerical method for thermally coupled fluid and solid problems.
Proc. of 8th Int. Conf. on Num. Meth. for Thermal Problems, Swansea, U. K., July 1993.

[2] BONNIN O., METIVET B., NICOLAS G., ARNOUX-GUISSE F., LEAL DE SOUSA L. :
Adaptive meshing for N3S fluid mechanics code.
Computational Fluid Dynamics 94, Wiley, p201-208. (1994).

[3] CHABARD J.-P., METIVET B., POT G., THOMAS B. :
An efficient finite element method for the computation of 3D turbulent incompressible flows.
Finite Elements in Fluids, Vol. 8, ed. Wiley (1992).

[4] ZHU J.Z., ZIENKIEWICZ O.C..
Adaptive techniques in the finite element method.
Comm. Appl. Num. Meth., Vol 24, 197-204, 1987.

[5] YAKHOT, ORSZAG :
Development of turbulence models for shear flows by a double expansion technique.
Physique of Fluids, July 1992.

[6] DURBIN, SPEZIALE :
Local anisotropy in strained turbulence at high Reynolds numbers
Fluids Engineering, n°113 - 1991.

FINITE ELEMENT ANALYSIS OF THREE-DIMENSIONAL TRANSIENT FREE CONVECTION PROCESSES IN POROUS MEDIA

H.-J. G. Diersch
WASY Institute for Water Resources Planning and Systems Research Ltd.,
Waltersdorfer Str. 105, D-12526 Berlin, Germany

SUMMARY

The paper discusses present effort in developing numerical schemes for a 3D finite-element groundwater transport code capable of simulating 3D buoyancy-driven flows in porous media. Different aspects are emphasized: (1) Basic formulations encompassing Boussinesq and Extended Boussinesq approaches for coupled mass and heat transport processes, (2) Proposed computational strategies for the nonlinear problem at high spatial and temporal resolution, (3) Pattern formation for a 3D multicellular 'Benard convection' in a porous layer and (4) Impact of the developed finite element model in practical tasks for the analysis of releasing contaminants from leach rock units arising in flooding an uranium pit.

INTRODUCTION

Density-driven convection processes in subsurface water resources are of growing concern in relation to contamination and geothermal problems. Hydrodynamic instabilities can play an important role in the spreading of heavy leachates with or without temperature influences. There is both a theoretical and practical need in modeling coupled flow, mass and heat (thermohaline) transport in three dimensions where free convection and transient influences dominate.

Theoretical and experimental studies on free convection phenomena in porous media started already in the middle of our century. They have concentrated exclusively on paradigmatic porous layers analogue of the Benard problem, also called the Lapwood problem [1]. Galerkin methods, finite differences and finite element techniques have been utilized to solve numerically the nonlinearly coupled balance equations, e.g. [2]. Most previous works refer to 2D problems. Classic solutions are here the Henry problem for the computation of the advance of a saltwater front in a confined aquifer and the Elder problem [3] for 2D pattern formation of the Rayleigh-Benard convection in Hele-Shaw cells. Henry's and Elder's works have become two of the standard tests to prove and benchmark variable density groundwater models. It is obvious the used numerical schemes and resolutions can dramatically influence the computational results. In this light previous numerical solutions have to be revised against most recent findings studying the 2D Elder problem at refined grids [4]. The extension to 3D free convection problems will increase the importance of both a physically equivalent process description in the discretized models and overcoming the numerical burden. In the following different aspects and solutions in modeling 3D transient free convection phenomena by using finite element techniques will be discussed.

BASIC EQUATIONS

The groundwater modeling system is based on the following equations written by using indicial notation and summation convention for repeated indices. Denoting time by t (T) and the Cartesian coordinates by x_i (L) the laws of fluid mass, linear fluid momentum, contaminant mass and energy conservations have the forms [5]:

$$S_o \frac{\partial h}{\partial t} + \frac{\partial q_i^f}{\partial x_i} = Q_p + Q_{EB}(C, T) \tag{1}$$

$$q_i^f = -K_{ij} f_\mu \left(\frac{\partial h}{\partial x_j} + \frac{\rho^f - \rho_o^f}{\rho_o^f} e_j \right) \tag{2}$$

$$\left. \begin{array}{l} \dfrac{\partial}{\partial t}(RC) + \dfrac{\partial}{\partial x_i}\left(q_i^f C - D_{ij}\dfrac{\partial C}{\partial x_j} \right) + R\vartheta C = 0 \quad \text{divergence form} \\[6pt] R_t \dfrac{\partial C}{\partial t} + q_i^f \dfrac{\partial C}{\partial x_i} - \dfrac{\partial}{\partial x_i}\left(D_{ij}\dfrac{\partial C}{\partial x_j} \right) + R\vartheta C = Q_C \quad \text{convective form} \end{array} \right\} \tag{3}$$

$$[\varepsilon \rho^f c^f + (1-\varepsilon)\rho^s c^s] \frac{\partial T}{\partial t} + \rho^f c^f q_i^f \frac{\partial T}{\partial x_i} - \frac{\partial}{\partial x_i}\left(\lambda_{ij} \frac{\partial T}{\partial x_j} \right) = Q_T \tag{4}$$

to be solved in the domain Ω for the hydraulic head $h = h(x_i, t)$ (L), the Darcy fluxes $q_i = q_i(x_i, t)$ (LT^{-1}), the contaminant concentration $C = C(x_i, t)$ (ML^{-3}) and the temperature $T = T(x_i, t)$ (Θ), with the constitutive equations:

$$\left. \begin{array}{l} \rho^f = \rho_o^f \left[1 + \dfrac{\alpha}{C_s}(C - C_o) - \beta(T - T_o) \right] \\[6pt] h = \dfrac{p^f}{\rho_o^f g} + x_k \qquad K_{ij} = \dfrac{k_{ij} \rho_o^f g}{\mu_o^f} \qquad \alpha = [\rho^f(C_s) - \rho_o^f]/\rho_o^f \\[6pt] f_\mu = \dfrac{\mu_o^f}{\mu^f(C, T)} = \dfrac{1 + 0.7063\varsigma - 0.04832\varsigma^3}{1 + 1.85\omega - 4.1\omega^2 + 44.5\omega^3} \qquad \varsigma = \dfrac{(T - 150)}{100}, \omega = \dfrac{C}{\rho_o^f} \\[6pt] D_{ij} = \left(\varepsilon D_d + \beta_T V_q^f \right) \delta_{ij} + (\beta_L - \beta_T) \dfrac{q_i^f q_j^f}{V_q^f} \\[6pt] R = \varepsilon + (1-\varepsilon) \kappa(C) \qquad R_t = \varepsilon + (1-\varepsilon) \dfrac{\partial \kappa(C)}{\partial t} \\[6pt] \lambda_{ij} = \lambda_{ij}^{cond} + \lambda_{ij}^{disp} \qquad Q_T = \varepsilon \rho^f Q_T^f + (1-\varepsilon) \rho^s Q_T^s \\[6pt] \lambda_{ij}^{cond} = [\varepsilon \lambda^f + (1-\varepsilon)\lambda^s]\delta_{ij} \qquad \lambda_{ij}^{disp} = \rho^f c^f \left[\alpha_T V_q^f \delta_{ij} + (\alpha_L - \alpha_T) \dfrac{q_i^f q_j^f}{V_q^f} \right]. \end{array} \right\} \tag{5}$$

The used symbols are summarized in the Appendix. The balance equations (1) - (4) are nonlinearly coupled through the fluid density and viscosity which depends on contaminant (solute) concentration and temperature.

BOUSSINESQ APPROXIMATION AND ITS EXTENDED FORMULATION

Commonly, the so-called Boussinesq approximation is employed to simplify the above model equations. It neglects density variations in the balance equations except in the buoyancy term of the momentum equation. However, it is known the Boussinesq approximation becomes insufficient for problems involving high salinity and temperature gradients. The main difference can by expressed by the additional term $Q_{EB}(C, T)$ in equation (1) as

$$Q_{EB}(C, T) = -\underbrace{\varepsilon\left(\frac{\alpha}{C_s}\frac{\partial C}{\partial t} - \beta\frac{\partial T}{\partial t}\right)}_{1} - \underbrace{q_i\left(\frac{\alpha}{C_s}\frac{\partial C}{\partial x_i} - \beta\frac{\partial T}{\partial x_i}\right)}_{2} \quad (6)$$

which is neglected in the Boussinesq approximation. The first term of (6) is only negligible if the temporal change in the concentration and temperature fields is small. However, the evolving features of a convection process can be thoroughly affected as soon higher density contrasts occur (e.g., bifurcation problems, causing hydrodynamic instabilities). The second term of (6) only vanishes if the density gradient is perpendicular to the flow which is not tolerable in all cases. Accordingly, in the present approach an '*Extended Boussinesq approximation*' is available where the term $Q_{EB}(C, T)$ is incorporated.

FINITE ELEMENT APPROACH

SPATIAL DISCRETIZATION

The above equations (1) to (4) are discretized by the FEM using prismatic pentahedral or hexahedral trilinear elements. It yields the following coupled matrix system:

$$\begin{aligned} O\dot{h} + S(h)h &= F(h, q, C, \dot{C}, T, \dot{T}) \\ Aq &= B(h, C, T) \\ P(C)\dot{C} + D(q, h, C, T)C &= R(C) \\ U\dot{T} + L(q, h, C, T)T &= W(T) \end{aligned} \quad (7)$$

where h, q, C and T represent the resulting vectors of nodal hydraulic head, Darcy fluxes, contaminant concentration and temperature, respectively. The superposed dot means differentiation with respect to time t. The matrices S, A, O, P and U are symmetric and sparse, while D and L are unsymmetric, however sparse too. The remaining vectors F, B, R and W encompass the right-hand sides of equations (1) to (4), respectively. The main functional dependence is shown in parenthesis. Alternatively to the standard Galerkin-based formulations (GFEM) streamline upwind techniques (SU) [2] are utilized to stabilize the solutions if desired.

TEMPORAL DISCRETIZATION AND NONLINEARITIES

Implicit two-step techniques are preferred for the present problem class. In general, for a more complex task it cannot be foreseen which time steps are allowable with respect to the accuracy requirements. Accordingly, a predefined time step marching strategy is often inappropriate and inefficient. Alternatively, an automatically controlled time stepping based on predictor-corrector techniques of first order as the *Forward Euler/Backward Euler* (FE/BE) and of second order as the *Adams-Bashforth/Trapezoid Rule* (AB/TR) becomes useful [2]. A full Newton method is embedded into the AB/TR and FE/BE predictor-corrector methods termed as *one-step Newton method*. It has proven to be a powerful and accurate technique, especially for strong nonlinearities and complex situations.

PRECONDITIONED ITERATIVE SOLVERS

To solve the resulting large sparse matrix systems (7) appropriate iterative solvers for symmetric and unsymmetric equations have to be applied. For the symmetric positive definite flow equations the conjugate gradient (CG) method is successful provided that a useful preconditioning is applied. Standard preconditioner such as the incomplete factorization (IF) technique and alternatively a modified incomplete factorization (MIF) technique based on the Gustafsson algorithm are used. Different alternatives are available for the CG-like solution of the unsymmetric transport equations: a restarted ORTHOMIN (orthogonalization-minimization) method, a restarted GMRES (generalized minimal residual) technique and Lanczos-type methods, such as CGS (conjugate gradient square), BiCGSTAB (bi-conjugate gradient stable) and BiCGSTABP (postconditioned bi-conjugate gradient stable). As preconditioner an incomplete Crout decomposition scheme is currently applied.

CRITERIA

The dimensional analysis provides characteristic numbers that control the coupled convection process: The Rayleigh number Ra written for the soluble mass and the thermal process, respectively:

$$Ra_s = \frac{\frac{\alpha}{C_s} \cdot \Delta C \cdot K \cdot d}{\varepsilon \cdot D}, \quad Ra_t = \frac{\beta \cdot \Delta T \cdot K \cdot d}{\Lambda} \tag{8}$$

where ΔC and ΔT represent concentration and temperature differences, respectively, through a porous layer of thickness d and hydraulic conductivity K, D is the mass diffusion and $\Lambda = (\varepsilon \lambda^f + (1-\varepsilon)\lambda^s)/(\rho^f c^f)$ corresponds to the thermal diffusivity of the layer. The relationship between mass and heat convection can be expressed by the Lewis number Le and the buoyancy ratio N, respectively:

$$Le = \frac{\Lambda}{\varepsilon \cdot D}, \quad N = \frac{(\alpha/C_s) \cdot \Delta C}{\beta \cdot \Delta T}. \tag{9}$$

The spatial and temporal discretization can introduce spurious dispersion effects where the physical amount D^{phys} is enlarged by the numerical dispersion D^{num} as

$$D^{effective} = D^{phys} + D^{num}. \tag{10}$$

To estimate the actual dispersion $D^{effective}$ effective in the numerical approach the truncation errors for the spatial and temporal discretization result

$$D^{num} = D^{num}_{spatial} + D^{num}_{temporal} \tag{11}$$

with

$$D^{num}_{spatial} \approx \alpha \frac{vl}{2} + O(l^2) \quad \text{and} \quad D^{num}_{temporal} \approx \left(\Theta - \frac{1}{2}\right)\Delta t_n v^2 + O\left(\Delta t_n^2\right) \tag{12}$$

where l is the characteristic length of a finite element, Δt_n is the time step increment at time level n, $v = q/\varepsilon$ is the pore velocity, α corresponds to the upwind parameter which is zero for the GFEM and unity at SU upwinding, and Θ represents a time weighting factor which is 0.5 for the 2nd order AB/TR scheme and unity for the fully implicit FE/BE scheme. Taking this into account the numerical solutions can now be characterized by an effective Rayleigh number in the form

$$Ra^{effective} = \frac{Ra}{1 + \frac{\alpha}{2}Pg + \left(\Theta - \frac{1}{2}\right)CrPg} \tag{13}$$

where $Pg = \frac{vl}{D}$ is the element (grid) Peclet number and $Cr = \frac{\Delta t_n v}{l}$ corresponds to the Courant number. It is obvious that especially upwind and fully implicit techniques can change the physical conditions actually simulated in the numerical model and sufficient spatial and temporal resolutions are needed ($\frac{\alpha}{2}Pg + \left(\Theta - \frac{1}{2}\right)CrPg \ll 1$).

RESULTS

3D BENARD CONVECTION

While previous studies [4] were devoted to stability criteria and the differences against the known 2D solutions a fully 3D analysis of the multicellular pattern formation is the subject of recent research. An appropriate test field concerns the 3D Benard problem for a porous layer where convective structures are caused by a heated or mass-intruded border area. This porous Benard problem can be considered as a 3D extension of the long-heater Elder problem [3] for different aspect ratios. As an example the 3D multicellular convection processes in a porous layer with an aspect ratio of 10 and a Rayleigh number of 200 (about five times of the critical Rayleigh number) are simulated. The domain is completely discretized by 250.000 hexahedral finite elements (257.761 nodal points). For a long-term simulation the Newton FE/BE scheme with upwind technique was performed to model successfully the multicellular 3D convective currents. Results at selected times are exhibited in Fig. 1.

CONTAMINANT RELEASE IN FLOODING AN URANIUM PIT

Complex 3D variable density contaminant transport simulations have been performed to study the contaminant release from leached rock units arising in flooding the Königstein uranium pit [6]. After flooding within the mine area very small flow gradients result and forced convection processes become extremely small. It has been found formations of either multicellular patterns or more global recirculation zones are responsible to a distinctly higher contaminant release from leach rock blocks compared with a diffusion-controlled effluent which is important for the spreading and fate of contaminants in the groundwater field for different flooding alternatives and proposed remediation strategies. Fig. 2 displays two finite element models which have been used to predict regional (large-scale) and near-field (experimental area) transport processes of the pit involving multiple free surfaces and density coupling.

CLOSURE

The coupled transport simulations have been performed by using the FEFLOW package [5] for workstations. The 3D free convection processes reveal sensitivity to parameter contrasts and numerical schemes. For 3D buoyancy-driven transient problems the numerical burden is high. Otherwise, practical applications such as exemplified for the uranium pit flooding have to govern the abundant multidimensional parameter fields and computational results over a large number of time steps. The applied simulation system necessarily incorporates Geographic Information System (GIS) data interfaces and specific visiometric tools for the 3D visual analysis.

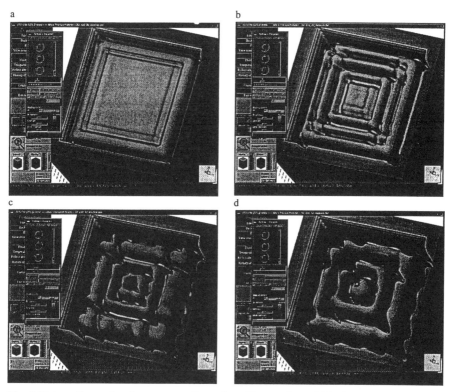

Fig. 1 Pattern formation for a porous Benard problem at *Ra = 200*: Concentration isosurfaces at dimensionless times (a) 0.013, (b) 0.026, (c) 0.039 and (d) 0.078.

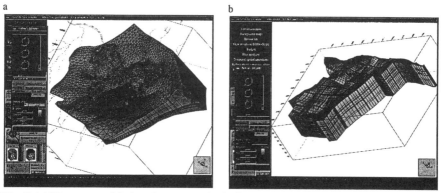

Fig. 2 (a) Unstructured finite element mesh (316.305 pentahedral elements) with high refinements at the mine area, (b) Model used for the 'experimental flooding area' simulation involving density effects and free surfaces.

REFERENCES

[1] NIELD, D. A., BEJAN, A.: "Convection in porous media", Springer Verlag, Berlin 1992.
[2] DIERSCH, H.-J.: "Finite element modelling of recirculating density-driven saltwater intrusion processes in groundwater", *Advances in Water Resources* 11 (1988) 1, pp. 25-43.
[3] ELDER, J. W.: "Transient convection in a porous medium", *J. Fluid Mech.*, 27, Part 3 (1967), pp. 609-623.
[4] KOLDITZ, O.: "Benchmarks for numerical groundwater simulations", In: *WASY Ltd: FEFLOW User's Manual*, WASY Ltd., Berlin, 1994.
[5] WASY Ltd.: "Interactive, graphics-based finite-element simulation system FEFLOW for modeling groundwater flow, contaminant mass and heat transport processes", *User's Manual Version 4.20*, December 1994, WASY Institute for Water Resources Planning and Systems Research Ltd., Berlin.
[6] DIERSCH, H.-J., ALBERT, H., SCHREYER, J., RICHTER, J.: "Three-dimensional modeling of flow and contaminant transport processes arising in flooding the Königstein uranium pit", In B. Merkel et al. (eds): *Proc. Intern. Conf. Uranium-Mining and Hydrogeology*, Verlag Sven von Loga, Köln, 1995, pp. 121-130.

APPENDIX: SYMBOLS

Symbol	Dimension	Description
C_s	ML^{-3}	maximum concentration;
C_o, T_o	ML^{-3}, Θ	reference values for concentration and temperature, respectively;
c^f, c^s	$L^2 T^{-2} \Theta^{-1}$	specific heat capacity of fluid and solid phase, respectively;
D_d	$L^2 T^{-1}$	molecular diffusion coefficient of fluid;
D_{ij}	$L^2 T^{-1}$	tensor of hydrodynamic dispersion;
e_j	1	components of the gravitational unit vector;
f_μ	1	viscosity relation function;
g	LT^{-2}	gravitational acceleration;
K_{ij}	LT^{-1}	tensor of hydraulic conductivity;
k_{ij}	L^2	tensor of permeability;
p	$ML^{-1}T^{-2}$	fluid pressure;
Q_C	$ML^{-3}T^{-1}$	sink/source of contaminant mass;
Q_{EB}	T^{-1}	extended Boussinesq approximation term;
Q_T	$ML^{-1}T^{-3}$	sink/source of heat;
Q_ρ	T^{-1}	sink/source of fluid;
R, R_t	1	specific retardation factor and its time derivative, respectively;
S_q	L^{-1}	specific storage coefficient (compressibility);
V_q	LT^{-1}	$\sqrt{q_i q_i}$ absolute specific Darcy fluid flux;
α	1	fluid density difference ratio;
α_L, α_T	L	longitudinal and transverse thermodispersivity, respectively.
β	Θ^{-1}	fluid expansion coefficient;
β_L, β_T	L	coefficients of longitudinal and transverse dispersivity, respectively;
ε	1	porosity;
ς	1	normalized temperature;
ϑ	T^{-1}	chemical decay rate;
$\kappa(C)$		linear (Henry) or nonlinear (Freundlich, Langmuir) sorptivity function;
λ_{ij}^{cond}	$MLT^{-3}\Theta^{-1}$	tensor of hydrodynamic thermodispersion;
λ_{ij}^{disp}	$MLT^{-3}\Theta^{-1}$	tensor of thermal conductivity;
λ_{ij}	$MLT^{-3}\Theta^{-1}$	tensor of thermodispersion;
λ^f, λ^s	$MLT^{-3}\Theta^{-1}$	thermal conductivity for fluid and solid, respectively;
μ^f, μ_o^f	$ML^{-1}T^{-2}$	dynamic fluid viscosity of fluid and reference one, respectively;
ρ^f, ρ_o^f	ML^{-3}	fluid density and reference one, respectively;
ρ^s	ML^{-3}	solid density;
ω	1	contaminant mass fraction;

NAVIER-STOKES 3D COMPUTATIONAL ANALYSIS OF INCOMPRESSIBLE TURBULENT FLOW IN A CURVED RECTANGULAR DUCT

P. Di Martino, A. Terlizzi, G. Cinque

Alfa Romeo Avio S.A.p.A.
Research & Development Department
Pomigliano d'Arco (NA)- ITALY

1. Summary

Developing turbulent flow simulations in 90° curved duct of rectangular cross section and an aspect ratio of 6 were performed by means of an in-house 3D computational tool. The analysis was aimed at establishing the capability of the code to predict complex flow features such as those that occur in similar devices. Two computations were carried out respectively, with and without a curvature effect correction to the turbulence model and results were compared with available experimental data.

Nomenclature

$C_\mu, C_{\epsilon 1}, C_{\epsilon 2}$ = constants of k-ϵ turbulence model, $C_\mu = 0.09$, $C_{\epsilon 1} = 1.44$, $C_{\epsilon 2} = 1.92$
C_l = 0.41 = Von Karman constant
k = turbulence kinetic energy
H = width of the duct, reference length, $H = 20.3$ cm
L = turbulent length scale
y = distance normal to wall
p, p_t = static and total pressure
S_{ij} = strain rate of mean field
U = streamwise velocity
U_i = mean field velocity component in tensor notation
u_i = velocity fluctuation   "   "
x_i = spatial coordinates   "   "
δ_{ij} = Kronecker delta, 1 for $i=j$, 0 for $i \neq j$
ϵ = rate of dissipation of turbulence kinetic energy
μ, μ_t = molecular and eddy viscosities
ρ = mass density
$\sigma_k, \sigma_\epsilon$ = constants of k-ϵ turbulence model, $\sigma_k = 1$, $\sigma_\epsilon = 1.3$
τ_w = wall shear stress
U_τ = friction velocity
V = velocity magnitude
U_0, p_0 = reference velocity and pressure at location $(0, 0.3H)$
C_p = $2(p-p_0)/\rho U_0^2$ = pressure coefficient
C_f = $2\tau_w/\rho U_0^2$ = friction coefficient

Subscripts

v = edge of sublayer
n = upper edge of turbulent boundary layer

2. Introduction

The fluid dynamics in curved ducts is such that the initially 2D boundary layers developing on the vertical lateral walls are subjected to strong streamwise curvature as long as to associated pressure gradients along the bend. Moreover, a longitudinal vortex on the convex wall forms due to the pressure driven secondary motion in the corner regions. The main interest in the inspected configuration[1] is that the duct aspect ratio yields these two features to develop nearly independently, without interaction. Furthermore, the availability of an extensive set of experimental data[2], ranging from mean flow to turbulent quantities measurements, allows to attain an exhaustive comparison between theoretical and experimental data about viscous effects, vorticous dynamics, and turbulence phenomena. The latter is limited to isotropic aspects, since the turbulence model here adopted has no Reynolds stress modeling incorporated. This work was primarily aimed to look at the turbulence features of such flows, particularly concerning the curvature effects on turbulence.

3. Mathematical Model

The mathematical model consists of the 3D incompressible, steady turbulent, Reynolds-averaged, Navier-Stokes (NS) equations, in cartesian tensor notation:

$$\frac{\partial(\rho U_k)}{\partial x_k} = 0$$

$$\frac{\partial(\rho U_k U_i)}{\partial x_k} = -\frac{\partial p}{\partial x_k} + \frac{\partial}{\partial x_k}\left(\mu S_{ij} - \rho \overline{u_i u_j}\right)$$

$$S_{ij} = \left(\frac{\partial U_i}{\partial x_j} + \frac{\partial U_j}{\partial x_i}\right)$$

$$-\rho \overline{u_i u_j} = \mu_t S_{ij} - \frac{2}{3}\rho k \delta_{ij}.$$

(1)

For turbulent closure the standard Jones-Launder k-ϵ turbulence model[3] is adopted:

$$\frac{\partial \rho U_i k}{\partial x_i} = \frac{\partial}{\partial x_j}\left[\left(\mu + \frac{\mu_t}{\sigma_k}\right)\frac{\partial k}{\partial x_j}\right] + \rho(P_k - \epsilon)$$

$$\frac{\partial \rho \epsilon}{\partial x_i} = \frac{\partial}{\partial x_j}\left[\left(\mu + \frac{\mu_t}{\sigma_\epsilon}\right)\frac{\partial \epsilon}{\partial x_j}\right] + \rho\frac{\epsilon}{k}N_\epsilon$$

$$\mu_t = C_\mu \rho \frac{k^2}{\epsilon}.$$

(2)

No-slip boundary conditions (BC) were applied at solid walls, whereas the wall function approach was used to accomplish the turbulence BC, that is the calculation of the k and ϵ production terms **near walls**. Namely, a local-equilibrium condition is applied ($P_k = \epsilon$), and the production of turbulent energy is obtained by integration over control volume[4]

$$\overline{P}_k = \frac{1}{y_n}\int_0^{y_n} -\overline{uv}\frac{\partial U}{\partial y}dy.$$

(3)

Assuming constant turbulent shear stress in turbulent region, zero in sublayer:

$$\bar{P}_k = U_\tau^2 \frac{(U_n - U_v)}{y_n} = \frac{U_\tau^4}{C_\mu^{1/4} k^{1/2} C_l y_n} \ln(y_n/Y_v)$$

$$U_\tau = \sqrt{\frac{\tau_w}{\rho}} \ .$$
(4)

The rate of dissipation production of k in near-wall cell must be handled analogously:

$$\bar{\epsilon} = \frac{1}{y_n} \int_0^{y_n} \epsilon \, dy = \frac{(kC_\mu^{1/2})^{3/2}}{C_l y_v} \left[1 + \ln\left(\frac{y_n}{y_v}\right) \right]$$
(5)

with k treated as a constant equal to the value at near-wall node. As near-wall value a constant level of ϵ within the viscous sublayer is implicitly assumed by the requirement that the length scale varies linearly with the distance from the wall:

$$\epsilon_p = \frac{(kC_\mu^{1/2})^{3/2}}{C_l y_p} \ .$$
(6)

Curvature effects were included as well, by means of the model presented by Childs and Caruso[5,6], which is based on interpreting the curvature effects as caused by centrifugal force in terms of a stability parameter β^*. Here p is an **effective** pressure (thermodynamic pressure + $2\rho\kappa/3$), $p_t = p + \frac{1}{2}\rho V^2$ is an **effective** total pressure, V is the velocity magnitude.

$$\beta^* = \frac{1}{\rho^2 V^2} \nabla p \cdot \nabla p_t \ .$$
(7)

The source term for the ϵ is modified to include a term that is a function of the stability parameter.

$$N_\epsilon = C_{\epsilon_1} P_k - C_{\epsilon_f} f_c \epsilon$$
$$f_c = \left(1 - C_c \frac{k^2}{\epsilon^2} \beta^* \right) \ .$$
(8)

The constant C_c is determined by empirical means to be

$$C_c = \begin{cases} 0.2 & \text{for } \beta^* > 0 \\ 0.1 & \text{for } \beta^* < 0 \end{cases}$$
(9)

where $\beta^* > 0$ is stabilizing and turbulence is suppressed, and $\beta^* < 0$ is destabilizing. $\beta^* = 0$ corresponds to no-curvature effects.

4. Numerical Model

The numerical model uses a finite-volume discretization on structured body-fitted meshes, with node-centered, no-staggered collocation of variables. A transformation from cartesian to generalized curvilinear coordinates was applied to the eqs. (1-2) in order to perform calculations on boundary-fitted meshes[7]. The steady solution is achieved via an iterative SIMPLEC algorithm[8]. The algebraic linear systems, coming from the discretization of the equations, are solved by the semi-implicit Line-S.O.R. method or by Stone's Strongly Implicit Method[9]. The numerical convective fluxes are spatially discretized by means of first-order and/or QUICK high-order schemes[10]; the latter was casted in a MUSCL-TVD fashion[11]. Central formulas are used for diffusive terms. In order to avoid the occurrence of checker-board instabilities the Rhie-Chow[12] interpolation is adopted to calculate velocities at interfaces of the elementary cells.

5. Discussion of Results and Comparison with Experiments

The tested geometry is sketched in fig. 1 (Test Case 5)[1]. The flow conditions were: Re_Y=224000, free-stream velocity U_0 = 16 m/s, kinematic viscosity ν=1.45 E-05 m²/s. The complete set of inflow conditions together with experimental results are to be found in references [1,2]. The computations were carried out on 121x61x21 and 161x61x31 grids, both with a Chebychev distribution of points to the wall, on half duct cross section, with symmetry BC on the axis. No noticeable difference was found between the respective results. The following figures refer to the finest grid (fig. 2). A converged computation (that is, a decay of five order of magnitude of the mass residual) was achieved in about 9½ h (CPU time on a single node of IBM SP2). The diagrams in figs. 3-6 refer to the cross-section D1, that is at half-duct-width distance after the bend, and show respectively, the U, ϵ, and k along the radial direction Y, at three different stations starting from the symmetry axis. The computed data (continuous lines) are there compared with experiments (open circles)[2]. The agreement is good for U and acceptable for ϵ; also quite accurate seem the computed k profiles. As well, a marked improvement can be noticed by the adoption of the curvature correction in the ϵ equation, both in k profiles (fig. 5-6) and C_f distribution (figs. 7). Fairly close to the experiments appear the distributions along the duct axes of C_p, on the convex (inner) as well as on the concave (outer) walls (fig. 8). Some vorticous structures are then shown, that illustrate secondary flows developing within (fig. 9) and after the bend (fig. 10).

6. Conclusions

Two computations by a NS solver were performed of an incompressible turbulent flow through a duct with a large aspect ratio of the cross section. It was assessed the capability of the code to represent the main features of the mean field and to grasp the principal isotropic turbulence effects in such flows. By comparing the computational results with the extensive set of experimental data it was also remarked a significative improvement in the description of the k and C_f distributions by inserting a curvature correction in the production term of the ϵ equation.

7. References

[1]   4th ERCOFTAC/IAHR Workshop on Data Bases and Testing of Calculation Methods for Turbulent Flows- Test Case Descriptions, University of Karlsruhe, April 3-7, 1995.

[2] W.J. KIM and V.C. PATEL- "Origin and Decay of Longitudinal Vortices in Developing Flow in a Curved Rectangular Duct"- Journal of Fluids Engineering, 116, 45, 1994.

[3] B.E. LAUNDER and D.B. SPALDING- "The Numerical Computation of Turbulent Flows", Computer Methods in Applied Mechanics and Engineering, Vol. 3, 1974, pp. 269-289.

[4] B.E. LAUNDER- "On the Computation of Convective Heat Transfer in Complex Turbulent Flows", Transactions of the ASME, Vol. 110, November 1988, pp. 1112-1127.

[5] R.E. CHILDS and S.C. CARUSO, "Assessment of Modeling and Discretization Accuracy for High Speed Afterbody Flows", AIAA paper N° 89-0531, Jan. 1989.

[6] R.E. CHILDS and S.C. CARUSO, "Turbulence Modeling for Complex Ground Effects Flows", SAE Technical Paper Series N° 901062, April 1990.

[7] S. MAJUMDAR, W. RODI, J. ZHU- "Three-Dimensional Finite Volume Method for Incompressible Flows With Complex Boundaries", Journal of Fluids Engineering, December 1992, Vol. 114, pp. 496-511.

[8] J.P. van DORMAAL, G.D. RAITHBY- "Enhancements of the SIMPLE Method for Predicting Incompressible Fluid Flows", Numerical Heat Transfer, Vol. 7, pp. 147-163, 1984.

[9] H.L. STONE- "Iterative Solution of Implicit Approximations of Multidimensional Partial Differential Equations", SIAM journal on Numerical Analysis, Vol. 5, pp. 530-558, 1968.

[10] B.P. LEONARD- "Locally Modified QUICK Scheme for Highly Convective 2-D and 3-D Flows", Proceedings, 5th International Conference on Numerical methods in Laminar and turbulent flow, Montreal, 1987, pp. 35-47.

[11] BRAM VAN LEER- "Upwind Difference Methods for Aerodynamic Problems Governed by the Euler Equations", Lectures in Applied Mathematics, Volume 22-2, pp. 327-335, American Mathematical Society, Providence, 1985.

[12] C.M. RHIE and W.L. CHOW, "Numerical Study of the Turbulent Flow past an Airfoil With Trailing Edge Separation", AIAA Journal, Vol. 21, pp. 1525-1532, 1983.

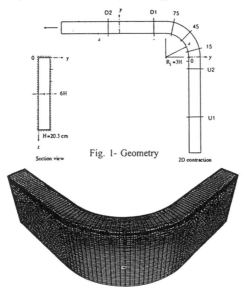

Fig. 1- Geometry

Fig. 2- Computational mesh

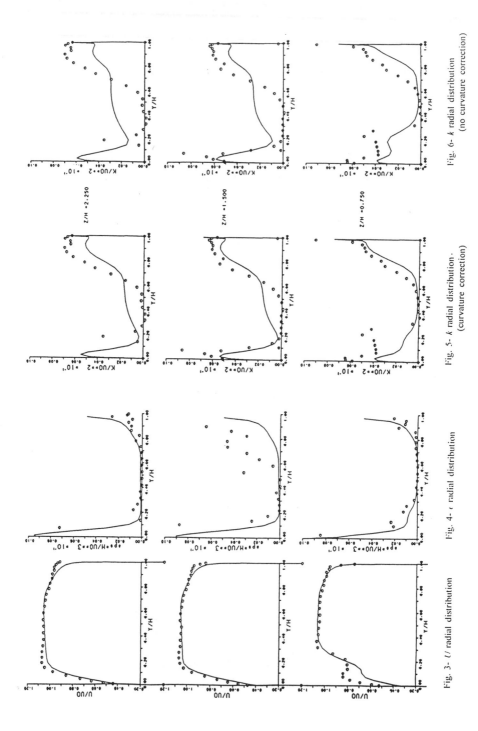

Fig. 3- U radial distribution

Fig. 4- ε radial distribution

Fig. 5- k radial distribution (curvature correction)

Fig. 6- k radial distribution (no curvature correction)

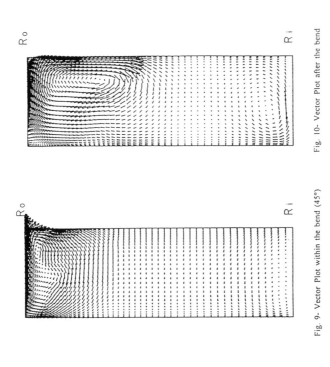

Fig. 9- Vector Plot within the bend (45°)

Fig. 10- Vector Plot after the bend

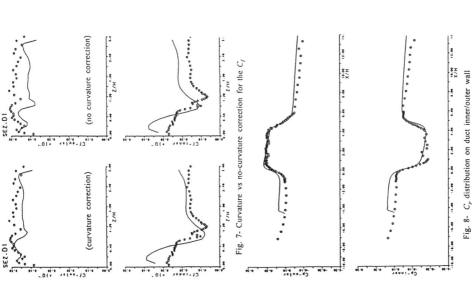

Fig. 7- Curvature vs no-curvature correction for the C_f

Fig. 8- C_p distribution on duct inner/outer wall

Study of Incompressible Flows in Rectangular Channels Using High Order Schemes and Parallel Computing

D. Drikakis [1]

UMIST
Department of Mechanical Engineering
PO Box 88, Manchester M60 1QD, United Kingdom

R. Zahner

University of Erlangen-Nuremberg
Department of Computer Science (IMMD 4)
Martenstr. 1, Germany

Results from the parallelization of a three-dimensional incompressible Navier-Stokes programme are presented. The programme is based on the artificial compressibility method for coupling the continuity with the momentum equations. The convection terms are discretized by a third order upwind characteristic based method. Second order central differences are used for the viscous terms. The time integration is obtained by an explicit Runge-Kutta scheme. The Navier-Stokes method has been parallelized using shared memory as well as message-passing models. Computations have been performed for a straight three-dimensional channel of rectangular cross-section. The efficiency of the parallel models is presented for different grid sizes and number of processors.

1 INTRODUCTION

Laminar and turbulent incompressible flows around and within different geometries have widely investigated in the past by several authors using different numerical methods. However, the last few years the requirements for solving more complex flow problems have provided a driving force for the extensive development of more accurate numerical methods. In addition the computer requirements are also increasing in terms of memory and computing time. In this respect parallel computing is becoming increasingly important.

The performance of parallel programmes depends on the hardware characteristics and the parallelization strategy. A user of parallel systems

---

[1] the initial part of the work was done when the author was still at the Department of Fluid Mechanics of the University of Erlangen-Nuremberg

cannot influence the hardware performance but he can improve the performance of the computations through the parallelization strategy. Many of the new parallel computers provide both shared memory and message-passing models. In addition the *parallel virtual machine* (PVM) model is nowadays a well established tool for parallelizing an algorithm.

The objective of the present paper is to compare the above models, in terms of the efficiency, when they are used for parallelizing a fluid flow solver. For this purpose a three-dimensional Navier-Stokes code [1] has been employed. The main features of the method are discussed in the next section. The code is currently applied in laminar and turbulent flows around and within different geometries. In the present paper some indicative results from the application of the method in rectangular channels will be shown.

2   NUMERICAL METHOD

The governing equations are the three-dimensional incompressible Navier-Stokes equations. The system of equations in matrix form using the artificial compressibility formulation and curvilinear co-ordinates can be written as:

$$(JU)_t + (E_I)_\xi + (F_I)_\eta + (G_I)_\zeta = (E_V)_\xi + (F_V)_\eta + (G_V)_\zeta. \quad (1)$$

The unknown solution vector is:

$$U = (p/\beta, u, v, w)^T$$

where p is the pressure and u, v and w are the velocity components. The parameter β is the artificial compressibility parameter, and it has to be chosen to ensure the fastest convergence to steady state. The terms E_I, F_I, G_I and E_V, F_V, G_V are the inviscid and viscous fluxes, respectively. J is the Jacobian of the transformation from Cartesian (x, y, z) to generalized (ξ, η, ζ) coordinates.

The convection terms of the Navier-Stokes equations are discretized by a characteristic based method [2]. The method has similar concept with the Riemann solvers used for solving the inviscid compressible equations. The characteristic based method constructs Riemann solutions on each flow direction and the primitive variables (p, u, v, w) are defined as functions of their values (p_l, u_l, v_l, w_l) on the characteristics $(l = 0, 1, 2)$. The velocity components and the pressure are computed at the cell face of the control volume using the following algebraic relations:

$$u = R\tilde{x} + u_0(\tilde{y}^2 + \tilde{z}^2) - v_0\tilde{x}\tilde{y} - w_0\tilde{x}\tilde{z} \qquad (2)$$

$$v = R\tilde{y} + v_0(\tilde{x}^2 + \tilde{z}^2) - w_0\tilde{z}\tilde{y} - u_0\tilde{x}\tilde{y} \qquad (3)$$

$$w = R\tilde{z} + w_0(\tilde{y}^2 + \tilde{x}^2) - v_0\tilde{z}\tilde{y} - u_0\tilde{x}\tilde{z} \qquad (4)$$

where

$$R = \frac{1}{2s}\left(p_1 - p_2 + \tilde{x}(\lambda_1 u_1 - \lambda_2 u_2) + \tilde{y}(\lambda_1 v_1 - \lambda_2 v_2) + \tilde{z}(\lambda_1 w_1 - \lambda_2 w_2)\right) \quad (5)$$

and

$$s = \sqrt{\lambda_0^2 + \beta}, \qquad \tilde{k} = \frac{\xi_k}{\sqrt{\xi_x^2 + \xi_y^2 + \xi_z^2}}, \qquad (k = x, y, z).$$

The pressure is computed using one from the following relations:

$$p = p_1 - \lambda_1\left(\tilde{x}(u - u_1) + \tilde{y}(v - v_1) + \tilde{z}(w - w_1)\right) \qquad (6)$$

or

$$p = p_2 - \lambda_2\left(\tilde{x}(u - u_2) + \tilde{y}(v - v_2) + \tilde{z}(w - w_2)\right). \qquad (7)$$

The fluxes are consequently defined by the above computed variables. In the formulae 2-7, $\lambda_0, \lambda_1, \lambda_2$ are the eigenvalues. The complete derivation of the above relations can be found in the references [1,2].
The variables u_l, v_l, w_l and p_l can be computed by third or fourth order upwind interpolation schemes [1,2]:
Third order

$$U^l_{i+\frac{1}{2}} = \frac{1}{6}(5U_i - U_{i-1} + 2U_{i+1}) \qquad (8)$$

$$U^r_{i+\frac{1}{2}} = \frac{1}{6}(5U_{i+1} - U_{i+2} + 2U_i) \qquad (9)$$

Fourth order

$$U^l_{i+\frac{1}{2}} = U^r_{i+\frac{1}{2}} = \frac{1}{12}(7U_i - U_{i-1} + 7U_{i+1} - U_{i+2}). \qquad (10)$$

The viscous terms are discretized by central differences. The time integration of the Navier-Stokes equations is obtained by an explicit Runge-Kutta scheme. The local-time-stepping technique is also used for steady state computations.

3 PARALLELIZATION

A grid-partitioning algorithm for the decomposition of the computational domain into several subdomains was initially developed. Using this algorithm each block can be decomposed in one (1D-partitioning), two (2D-partitioning) or three directions (3D-partitioning), respectively.

In the message-passing model the computational grid is subdivided into non-overlapping subdomains and each subdomain is assigned to one processor. Each processor stores its own data while computed values are exchanged between the processors during the numerical solution. For the parallelization of the Navier-Stokes programme subroutines for distributing the grid data to each processor, for exchanging computed values between the processors (i.e local and global communication), and for gathering data from all processors up to the host processor, were written. In the shared memory model the data are stored in a common memory. Each time a processor needs data from a neighbouring processor receives it from the shared memory. Synchronization is obtained by *synchronization primitive routines*. The shared-memory model is simpler than the message-passing one. The former does not need the development of complicated routines to achieve distribution and exchange of data between the processors. The *local* and *global* communication are obtained by receiving information from the shared memory. For the present method *local communication* between the processors is needed after each Runge-Kutta iteration for updating the values along the subdomain boundaries. *Global communication* is needed for checking the convergence at each time step.

4 RESULTS AND DISCUSSION

The three dimensional version of the method was initially validated for the entry flow in a straight 3D channel of square cross section. The Reynolds number is 100 and it is based on the average velocity through the channel and the channel width. The calculations were performed in a single quadrant of the channel due to the symmetry. The number of grid points in the streamwise direction (x) was 33 with 11 grid points in the other two directions. The grid points were slightly clustered in the x–direction close to the channel entrance and the grid was uniform in the other two directions. The computed pressure average over the cross sectional was compared (Fig. 1a) with the experimental data of Beavers et al. [3]. The results for the axial development of the streamwise velocity at the channel centerline were compared (Fig. 1b) with the corresponding laser Doppler velocimetry measurements of Goldstein and Kreid [4]. In addition the velocity profile at $X/(DRe) = 0.02$ was also compared (Fig. 1c) with the corresponding experimental results of reference [4]. These figures show satisfactory agreement between numerical and experimental results. The above computations were performed using the third order version of the method. Computations using the fourth order scheme showed no difference in the above results.

Investigations of the different parallelization models were performed on KSR-1 and Convex SPP systems with eight processors. On the KSR-1 system parallelization was obtained using the shared memory (SM) as well as the message-passing model. The latter is based on the "local instruction" procedure (MPLI) and it was programmed using the TCGMSG library from Argonne National Laboratories. The parallelization on Convex SPP was obtained by the PVM model. For investigating the efficiency of the parallel computations three different grids of 1000, 3993 and 8955 control volumes were used. In Tables 1 and 2 the computing times for computations on eight processors and the total efficiency of the computations on the finest grid are shown, respectively. The total efficiency E_n, when n processors are used, is defined as:

$$E_n = \frac{T_1}{nT_n} \qquad (11)$$

where T_1 and T_n are the computing times on one and n processors, respectively. An analysis of the total efficiency factor can be found in reference [5]. From Table 2 it is clearly seen that SM and PVM models provide better efficiency than the MPLI one. Convex SPP provides faster solution than KSR1 (see Table 1) due to the faster processor. Due also to the better efficiency of the shared memory model the computing time on KSR-1 is less when the SM model is used instead of the MPLI one. Results for the total efficiency ($E_n\%$) on different grids using the MPLI and SM models are shown in Tables 3 and 4, respectively. For fine grids and large number of processors the differences between shared memory and message-passing models are large. The low efficiency of the message-passing model on the KSR-1 is due to the high set-up time (t^{set}) needed to enable message passing. As in reference [5] has been shown, the most important factor affecting the performance of the computations, when message-passing is used, is the ratio of the (t^{set}) to the time needed for one floating point operation. The best results are always achieved for parallel systems with the smallest above ratio [3]. From the above Tables it is also seen that the efficiency is improved on finer grids.

ACKNOWLEDGEMENT

Parts of this work were financially supported by the CEC project Copernicus "Highly Efficient 3D CFD Codes for Industrial Applications" CP 94 1239. The CEC contribution is greatly appreciated.

REFERENCES

1. Drikakis, D., Durst, F., 1994, Numerical simulation of three-dimensional incompressible flows by using high order schemes, 1995, Proceedings of the Second Summer Conference on Numerical Modelling in Continuum Mechanics held in Prague, 22-25 August 1994, Edited by M. Feistauer et al., pp 79-88.

2. Drikakis, D., Govatsos, P., Papantonis, D., 1994, A characteristic based method for incompressible flows, Int. J. Num. Meth. Fluids, **19**, pp 667-685.

3. Beavers, G. S., Sparrow, E. M., Magnusson, R. A., 1970, Experiments on hydrodynamically developing flow in rectangular ducts, Int. J. of Heat Mass and Transfer, **13**, No. 4, pp. 689-693.

4. Goldstein, R. J., Kreid, D. K., 1967, Measurement of laminar flow development in a square duct using a laser-doppler flowmeter, J. of Applied Mechanics, Series E, **89**, No. 4, pp. 813-818.

5. Drikakis, D., Schreck, E., Durst, F., 1994, Performance analysis of viscous flow computations on various parallel architectures, ASME Journal of Fluids Engineering, 1994, **116**, 835-841.

Table 1: Computing time (in hrs) for different grids using eight processors (CV: computational volumes).

	KSR1 - SM	KSR1 - MPLI	Convex SPP - PVM
1000 CV	0.64	0.62	0.042
3993 CV	1.25	1.85	0.08
8995 CV	3.20	6.5	0.21

Table 2: Total efficiencies on KSR-1 and Convex SPP using shared memory (SM), message-passing based on local instructions (MPLI) and PVM.

	KSR1 - SM	KSR1 - MPLI	Convex SPP - PVM
1 proc.	100	100	100.
2 procs.	99.2	75.0	99.0
3 procs.	97.2	74.0	78.0
4 procs.	95.0	73.0	76.0
5 procs.	84.0	58.0	74.5
6 procs.	82.0	54.0	73.0
8 procs.	76.0	38.0	65.0

Table 3: Results on KSR-1 using shared memory.

	$E_n\%$, grid-1	$E_n\%$, grid-2	$E_n\%$, grid-3
1 proc.	100.	100.	100.0
4 procs.	49.0	75.0	95.0
5 procs.	41.1	61.0	84.0
6 procs.	33.0	57.0	82.0
8 procs.	23.0	43.0	75.0

Table 4: Results on KSR-1 using message-passing.

	$E_n\%$, grid-1	$E_n\%$, grid-2	$E_n\%$, grid-3
1 proc.	100.	100.	100.0
4 procs.	49.0	64.0	73.0
5 procs.	40.0	51.0	58.0
6 procs.	34.0	42.0	55.0
8 procs.	24.0	30.0	38.0

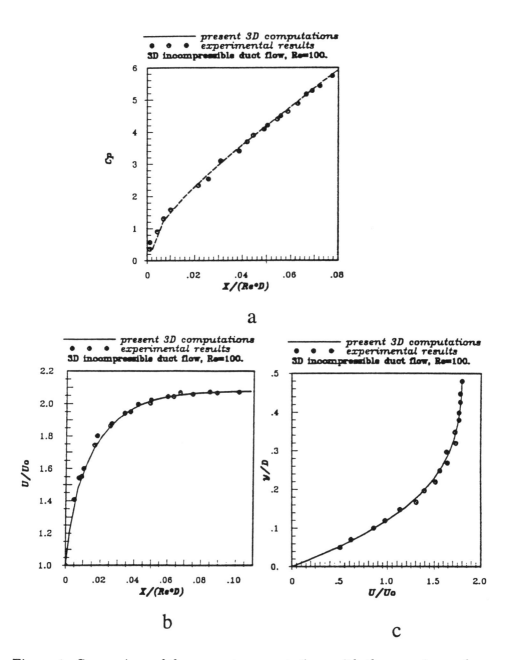

Figure 1: Comparison of the present computations with the experimental results for the three-dimensional channel flow ($Re = 100$); a. development of the pressure distribution at the centerline, b. development of the velocity at the centerline and c. streamwise velocity profile at $X/(DRe) = 0.02$.

Numerical Prediction of Abrasion for Francis Turbine Guide Vanes

Peter Drtina and Matthias Krause
Sulzer Innotec AG, Fluid Dynamics Laboratory,
CH-8401 Winterthur, Switzerland

Abstract

Three-dimensional numerical simulations of the hydro-abrasive wear of Francis turbine guide vanes are reported. The calculations discussed in this paper are based on recently performed investigations for a three-dimensional guide vane configuration including gap clearance effects. Three-dimensional simulations have been carried out following the promising results obtained in previously performed 2D calculations [1]. The present numerical simulations use the calculated turbulent 3D flow field to predict the trajectories of sand particles in the flow. Particle-wall interactions result in abrasive wear which is calculated by applying a simple wear model. Comparisons with the observed wear pattern obtained from a long duration field test show very good agreement.

1 Introduction and basic steps

One result of the increasing worldwide demand for electric energy is an intensified utilisation of available hydropower resources. Unfortunately more and more water resources with high particle content have to be used. The presence of abrasive particles in the flow often causes a degradation of certain parts of the hydraulic machinery particularly those which are exposed to the flow. In particular, guide vanes and turbine blades bear a high risk of being damaged by abrasion processes. This abrasion process directly leads to reduced lifetime and system performance. For the next generation of hydro power plants, abrasion needs to be taken into account in the early design and development phase.

Before the abrasion process on guide vane surfaces can be investigated in detail, the fluid flow has to be solved. In the present work a solution for the three-dimensional turbulent flow field is achieved by applying the TASCflow code which is based on a Finite-Volume method [2]. Once the flow field around the guide vane is available, particles are injected into the flow at the inlet section. In the present case, the inflow conditions have not been measured and constant inlet velocity and direction as well as local equilibrium of flow and particles had to be assumed. Particle motion is monitored until they leave the computational domain. In order to compute the particle trajectories the equation of motion derived by Basset, Boussinesq and Oseen is solved. For the present investigations both particle density and fluid density are of the same order and terms accounting for virtual particle mass, pressure gradients and the Basset integral cannot be neglected a priori. In order to take account of the effect of turbulent fluctuations in the flow on the particle motion, it is necessary to consider a large number of individual particles and to use statistical methods to derive the mean rates of impact of the particles with the surface.

Depending on its impact energy, the impingement of a particle on the guide vane leads to the removal of a certain amount of surface material and thus contributes to the wear. To describe this erosion process the Finnie model, which is discussed in more detail in section 4, is applied.

In a previously reported investigation [1] a small 2D strip representing the mid-span region of the guide vane was modelled. This reduced the computer run time and allowed parameter studies to be carried out. The present work focuses on the complete 3D flow problem including gap clearance effects and the complex support geometry.

The results of the particle dynamics simulations are presented in the form of erosion patterns which are compared to experimental results from field tests. Both, the overall erosion pattern as well as some local effects in the support and gap surfaces show very good agreement between numerical and field test data.

2 Field test

The field test was carried out in close collaboration with the plant services of a Swiss power plant [3]. The seasonal nature of the water source leads to a significant energy production only during the summer period. During winter the turbines are revised and prepared for the next operation period. For the duration of these tests the turbine was operated almost exclusively under 100% load, and the sand concentration was measured daily by filtering a sample of water taken from the underwater channel. Mineralogical analysis and measurements of the particle size distribution were carried out for several samples taken on different days.

The sand content measurements show that concentrations fluctuate considerably. The mean sand concentration was $0.106 g/l$ with a maximum of $2.6 g/l$. The hydro-abrasion was measured on the blade surface of three of the twenty inlet guide vanes. The wear pattern on the blades was obtained by comparing the blade surface profile measurements taken before and after the field test. The measured concentration profile and the results of the sediment analysis as well as the position of the guide vanes in the turbine are given in [1, 3].

3 Numerical simulation procedure

Significant attention has been devoted to the prediction of two-phase flows consisting of a carrier fluid and discrete particles (dispersed phase). These predictions involve the separate calculation of each phase with source terms accounting for the interaction between the phases. The flow of the continuous phase is predicted using a discretized form of the Navier-Stokes equations with a k, ϵ turbulence model. The most widely applied method available to determine the behaviour of the dispersed phase is to track several individual particles through the flow field. This method is called Lagrangian tracking. Each particle represents a sample of particles that follow an identical path. The behaviour of the tracked particles is used to describe the average behaviour of the dispersed phase.

While setting up the tracking model the following assumptions have been made:

- Particle-particle interactions are neglected. These interactions may become important in flows where the volumetric concentration of the discrete phase exceeds 1%.

- Any change of the flow turbulence caused by the particles is not accounted for.

- Only spherical and non-reacting particles are considered.

Consider a discrete particle travelling in a continuous fluid medium. The forces acting on the particle which affect the particle acceleration are due to the difference in velocity between the particle and the fluid and due to the displacement of fluid by the particle. The equation of motion for such a particle was derived by Basset, Boussinesq and Oseen for a stationary reference frame (see [4]):

$$m_p \frac{dv_p}{dt} = 3\pi\mu d C_{cor}(v_f - v_p) + \frac{\pi d^3 \rho_f}{6}\frac{dv_f}{dt} + \frac{\pi d^3 \rho_f}{12}\left(\frac{dv_f}{dt} - \frac{dv_p}{dt}\right) + F_e + \frac{3}{2}d^2\sqrt{\pi\rho_f\mu}\int_{t_0}^{t}\frac{\frac{dv_f}{dt'} - \frac{dv_p}{dt'}}{\sqrt{t-t'}}dt' - \frac{\pi d^3}{6}(\rho_p - \rho_f)\omega \times \left(\omega \times \vec{R}\right) - \frac{\pi d^3 \rho_p}{3}\omega \times v_p. \tag{3.1}$$

The last two terms which are present only in a rotating frame of reference are the centripetal and Coriolis force. In the present case they do not affect the calculations. A drag coefficient, C_{cor}, is introduced to account for experimental results on the viscous drag of a solid sphere. For moderate particle Reynolds numbers in the range $0.01 < Re_p < 260$, the drag correction in equation 3.1 is given by the relation

$$C_{cor} = 1 + 0.1315(Re_p)^{0.82-0.05\alpha} \quad Re_p < 20 \tag{3.2}$$
$$C_{cor} = 1 + 0.1935(Re_p)^{0.6305} \quad Re_p > 20 \tag{3.3}$$
$$\tag{3.4}$$

with $\alpha = \log(Re_p)$. The particle Reynolds number is defined by

$$Re_p = \frac{\rho_f |v_f - v_p| d}{\mu} \tag{3.5}$$

In this form the equation of motion has particle acceleration terms on both sides of the equation which would require an iterative solution method. To avoid this, equation (3.1) is rearranged to put all particle accelerations on the left hand side by applying an appropriate approximation [2].

The present investigation deals with a purely turbulent flow. In turbulent tracking the instantaneous fluid velocity is decomposed into mean and fluctuating components. The mean component of the fluid velocity affects the average trajectory of particles injected as in laminar flow. However, in turbulent flows particle trajectories are not deterministic and two identical particles injected from a single point, at different times may follow separate trajectories due to the random nature of the instantaneous fluid velocity. Therefore, the fluctuating component of the fluid velocity causes the dispersion of particles in a turbulent flow.

To account for the influence of turbulent fluid fluctuations on particle motion, the method originally developed by Dukowicz [5] and later by Gosman and Ioannides [6] and Faeth [7] has been applied in the flow code. This model has shown good agreement with fundamental dispersion data of Hinze [4], Snyder and Lumley [8] and Yuu et al. [9].

4 Erosion Model

The Lagrangian particle tracking model predicts the number of particles impinging on a surface in a given time, and their velocity and direction relative to the surface. In a second step the volume of surface material removed by each particle interacting with the wall has to be calculated. Finnie [11, 12] developed an erosion equation for ductile materials which is implemented in the present TASCflow version. Basically the erosion rate E depends on the kinetic energy of the impinging particle and its impact angle:

$$E = KV^n f(\Theta) \tag{4.6}$$

with

$$\begin{aligned} f(\Theta) &= \tfrac{1}{3}\sin^2(\Theta) & 0 &< \Theta < 0.4\pi \\ f(\Theta) &= \sin(2\Theta) - 3\cos^2(\Theta) & 0.4\pi &< \Theta < 0.5\pi. \end{aligned} \tag{4.7}$$

For the present simulations the parameter n has been set to 2. It has to be kept in mind that for a given mass of abrasive particles impinging on a surface the totally removed material calculated via equation 4.6 will be independent of the particle size. Finnie points out that this is true only for particles with diameters exceeding $100\ \mu m$. Below this value erosion becomes less effective with decreasing particle size. As we are concerned with particles in the range from $20\ \mu m$ to $400\ \mu m$, the model will tend to overestimate the influence of small particles in all cases where some particle size distribution function is applied. For this reason only distinct particle sizes have been used in the present calculations.

5 Computational grid and boundary conditions

The geometrical description of the vane contour has been taken from data available for the Francis turbine guide vanes installed in the Mörel hydraulic power plant (Switzerland). The topology was constructed in such a way to allow the generation of grids for different guide vane angles by simply changing the appropriate parameter in the pre-processing file. Only the half of the vane has been considered for the grid generation because all boundary conditions imposed at the inlet do not exhibit any variation over the span.

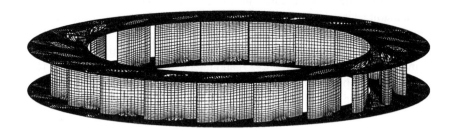

Figure 1 Computational grid for guide vane calculations - entire ring.

The entire computational domain was discretised by a single block mesh with distinct grid lines and grid planes associated to the given vane edges and surfaces including the support and the gap clearance (see fig. 1 and 2). Grid cells lying in the interior of the vane are blocked off during the boundary condition set-up procedure. This basic grid consists of 28,975 nodes and 25,920 cells, respectively. The grid has been refined successively by embedding subgrids in the vicinity of the vane surfaces and, in particular, at the leading edge. Including all nodes of the embedded subgrids, the total number of nodes increases to 200,025. The active number of nodes during the calculation is lower due to the block-off of cells lying inside the vane.

The computational domain of the present flow problem is bounded by a number of geometrically defined surfaces. The inflow and outflow regions are given by segments of a cylinder surface. On the inflow region a velocity vector is imposed with constant speed and circumferentially constant incidence angle. In addition, turbulence quantities - intensity and length scale - are defined in this region. The average pressure is set to be constant on the entire outlet region. The mid-span section of the blade is assumed to be a plane of symmetry. The boundaries which separate two adjacent vanes are linked to each other by periodic conditions. On all wall surfaces the logarithmic law of the wall is applied. A number of particle boundary conditions are available to simulate the behaviour of particles at the boundaries of the computational domain. At solid walls the particle may be reflected (as is assumed in the present case), collected or it may even

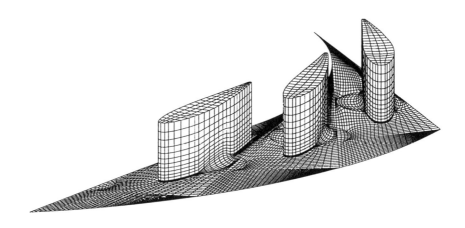

Figure 2 Computational grid for guide vane calculations - three vanes.

escape. At symmetry planes only reflection is possible. On inlet and outlet regions particles can escape. At periodic boundaries the particle is translated from the primary to the secondary surface, or vice versa, and the particle velocity is appropriately rotated.

Typically most of the particles are in the diameter range from $20\mu m$ to $100\mu m$. As representative classes the $30\mu m$, $60\mu m$ and $80\mu m$ particles have been chosen for our calculations. The vane orientation has been fixed to $\beta = 59°$ and the inlet flow angle to $\gamma = 29°$. A turbulence intensity of 5% is assumed in the inlet region.

All calculations were performed applying the TASCflow Navier-Stokes flow code implemented on an IBM RS/6000 workstation. This code is based on a Finite-Volume method described in detail by Raithby et al. [10]. It is applicable to fully three-dimensional, incompressible and compressible, laminar and turbulent flow problems. Additional code features, including the particle tracking and erosion modules, are described in reference [2].

6 Simulations and Results

In a first step the flowfield was calculated for geometrical and flow related boundary conditions given above. The most interesting flow related feature which appears due to the inclusion of the gap clearance, is the corner vortex generated on the pressure side. Figure 3 shows a slice half way between vane support and trailing edge. Following the pressure gradient the fluid penetrates the gap clearance on the left (suction side) and leaves it on the right (pressure side) forming the corner vortex. This flow situation leads directly to an increased abrasion along the vane edges where the particle laden flow is entering the gap.

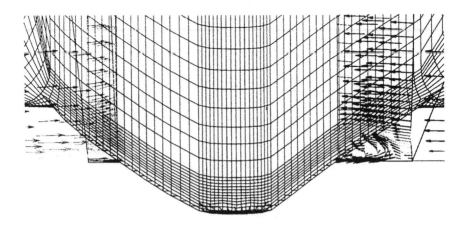

Figure 3 Corner vortex due to gap clearance flow.

Considering the entire vane all calculations show a similar typical abrasion pattern. On the pressure side there is a zone covering 25 – 30% of the surface, starting from the leading edge, which shows almost no deterioration (see figure 4 for the dark grey and black regions). The remaining part of the pressure surface experiences high abrasion (light grey). This is in agreement with our field tests (see figure 5). On the suction surface of the vane, the simulations indicate relatively low abrasion on the entire surface except some very small strips along the trailing edge as well as the above mentioned regions along the gap clearance edges. This is in close agreement with the field test where no significant surface deterioration was detected on the suction side with the exception of small strips along the gap and trailing edges. Regions exposed to high abrasion also occur on the vane surface which is faced towards the gap clearance. Unfortunately, no measurements exist for these locations. Photographs taken from the field test vanes show clearly that exactly at the locations indicated by the simulations increased abrasion occurs. Capturing these local flow and abrasion phenomena is a substantial improvement compared to the previously reported 2D simulations where the suction side results did not fit to the field test measurements and no 3D dependent features could be studied.

An important finding from these simulations is that if one wants to calculate not only the flow field but also the abrasion on surfaces exposed to particle impacts then the computational grid has to be much more refined than for pure flow calculations at least in the vicinity of the surfaces.

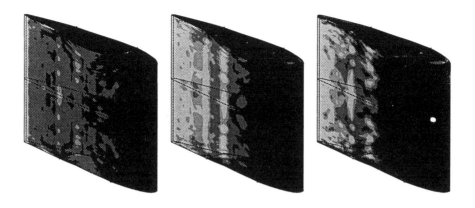

Figure 4 Numerically obtained abrasion pattern for $30\mu m$ (left), $60\mu m$ (center) and $80\mu m$ (right) particles - pressure side. Light grey regions indicate high abrasion, dark grey and black regions indicate vanishing abrasion.

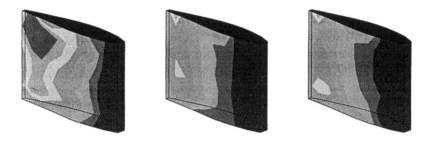

Figure 5 Abrasion pattern as measured on the field test sample vanes. Going from the leading edge to the trailing edge the grey scale steps indicate increasing abrasion.

7 Conclusions

Field tests have been performed to get complete and detailed information about the abrasion process under real operational conditions in a hydraulic turbine. All abrasion relevant parameters were acquired for the complete duration of the field test. The data from these field tests have been used to verify the simulated wear patterns.

A detailed fully three-dimensional numerical simulation of fluid flow and abrasion on a Francis turbine guide vane has been carried out. The complex geometry of the guide vane including tip clearance, support and fillets leads to a complex flow field (clearance flow, corner vortex) which, as a consquence, results in a complicated fluid-particle interaction strongly affecting the erosion pattern of the guide vane. The agreement of the numerically obtained erosion pattern and the field test measurements is very good even for local effects, i.e. on the gap surfaces and the support, although only the simple Finnie erosion model has been applied.

The encouraging good agreement with field test erosion patterns shows that for this application numerical simulation really can be used in a predictive manner. Such simulation results may serve as an input in an early stage of the design procedure to identify regions where special surface treatment is necessary in order to increase the guide vane life time.

References

[1] DRTINA, P., KRAUSE, M., Abrasion on a Francis Turbine Guide Vane - Numerical Simulation and Field Tests, 17th IAHR Symposium, Beijing (1994).
[2] TASCFLOW USER MANUAL, Theory Documentation - Version 2.4, Advanced Scientific Computing Ltd., Waterloo, Canada (1995).
[3] KRAUSE, M., Feldversuch Mörel 1992. Internal report STT.TB93.018, Sulzer Innotec AG (1993).
[4] J.O. HINZE, *Turbulence*, McGraw-Hill, New York (1975).
[5] DUKOWICZ, J.K., A Particle-Fluid Numerical Model for Liquid Sprays, J. Comp. Phys., Vol.35, pp.229-253 (1980).
[6] GOSMAN, A.D., IOANNIDES, E., Aspects of Computer Simulation of Liquid-fueled Combustors, J.Energy, Vol.7, pp.482-490 (1983).
[7] FAETH, G.M., Mixing, Transport and Combustion in Sprays, Prog. Energy Combust. Sci., Vol.13, pp.293-345 (1987).
[8] SNYDER, W.H., LUMLEY, J.L., Some measurements of particle velocity autocorrelation functions in a turbulent flow, J. Fluid Mech., Vol.48, pp.41-71 (1971).
[9] YUU, S., YASUKOUCHI, N., HIROSAWA, Y., JOKATI, T., Particle Turbulent Diffusion in a Dust Laden Round Jet, AIAA Journal, Vol.24, pp.509-519 (1978).
[10] RAITHBY, G.D., A 3D Code for Simulating Incompressible of Compressible Flows in Complex Geometries, ASC Report (1987).
[11] FINNIE, I., Some observations on the erosion of ductile materials, Wear, Vol.19, pp.81-90 (1972).
[12] FINNIE, I., Erosion of surfaces by solid particles, Wear, Vol.3, pp.87-103 (1960).

Simulation of turbulent flow in a cylindrical drum with multiple outlets

T. V. Dury and B. L. Smith
Thermal-Hydraulics Laboratory, PSI, CH-5232 Villigen PSI

1 SUMMARY

Two general-purpose CFD codes were employed to examine flow patterns in the upper header of a model condenser unit, and optimise the placement of a deflector plate to ensure an even flow distribution to the condenser tubes. Results are compared against experimental data. Aerosol tracking calculations have also been performed, to investigate deposition and re-entrainment behaviour.

2 INTRODUCTION

The motivation for this study is the need to assess the long-term cooling capabilities of the next generation of Advanced Light Water Reactors, featuring innovative, passive design concepts for long-term cooling of the reactor core following shut-down. One design employs drum-type condensers to transfer heat from the primary circuit and containment to large external water pools. Under severe accident conditions, fission products in the form of aerosols may escape from the pressure vessel into the various reactor containment compartments, possibly contaminating the condenser unit tubes. Formation of an aerosol layer inside these tubes may seriously reduce system heat removal capability, and plugging of some tubes may substantially degrade condenser efficiency. Consequently, a testing facility AIDA (Aerosol Impaction and Deposition Analysis), with an aerosol generation system, has been constructed at the Paul Scherrer Institute [1] to: (1) determine the degree of condensation degradation due to the presence of aerosols; (2) investigate aerosol behaviour under strong condensation conditions in tubes; and (3) provide the development basis of a physical model for aerosol transport for use in plant system analysis.

The AIDA test-section is a sectional model of a condenser unit, and consists of two thin drums, one above the other, joined by eight tubes (Fig. 1). Wet steam, containing aerosols, enters the top drum, is distributed across the connecting tubes, and passes to the lower drum, where the steam and water droplets are separated. With an upper drum containing no flow-diverting baffle, the incoming jet exits primarily through those tubes nearest the vertical centre-line of the drum. Within the drum, the spreading of the jet is similar to that of the developing region of a free jet in an infinite medium.

The goals of the present study were: (1) to determine the size and position of a flow baffle which intercepts the incoming steam jet and creates an even distribution of flow to the eight condenser tubes, without introducing excessive pressure drop; and (2) to obtain first estimates of aerosol deposition characteristics.

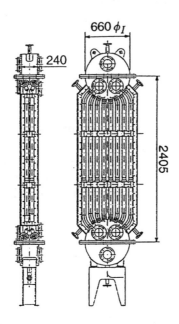 

Fig. 1 PCC Condenser Simulator   Fig. 2 Computer Model

3   NUMERICAL MODELS

Calculations are presented using two multi-purpose, transient fluid dynamics and heat transfer codes, ASTEC and CFDS-FLOW3D, both marketed by AEA Technology, based at Harwell [2, 3]. The codes have been developed over many years and are used extensively in Europe and North America for the simulation of practical flow problems and as bases for the development of advanced physical models and computational techniques. The custom-built pre- and post-processors are fully compatible for the two codes, ensuring that the same mesh structure is maintained for both the ASTEC and FLOW3D simulations, and that a meaningful comparison of results can be made.

A 3-D model of the test-section was created (Fig. 2), and parameter studies carried out, using ASTEC, for a range of sizes and positions of the baffle in the drum (Fig. 3), and for different mesh combinations. Pre-calculations showed that the computer model only needed to cover the upper drum and the condenser tubes, the lower ends of which were straightened, for simplicity, while preserving overall length (and pressure drop). Pressure loss over the test-section, with any baffle in the central region of the drum, was typically less than 5% of the inlet velocity head, and therefore insignificant. The final, finite-element-type mesh represents a compromise between accuracy and calculational efficiency, consists of about 34,000 elements, and is shown in cross-section in Fig. 4.

Although both codes ostensibly operate on the same mesh, the basic conservation equations are integrated over different control volumes (see Inset 1 to Fig. 4). For ASTEC, these are formed around the vertices of each hexahedral element, whereas FLOW3D uses the element itself as the control volume. This leads to somewhat different handling near fluid boundaries, but should have little effect in the bulk flow regions, except possibly in the condenser tubes, where the differences in representation could lead to erroneous conclusions regarding the velocity profiles. In the centre of an outlet tube, at any cross-section, ASTEC has only one non-zero *nodal* velocity (Inset 2 of Fig. 4), and any swirling motion down the tube can therefore not be represented. FLOW3D has four *elemental* velocities with the same mesh, so any swirl could be seen. (In fact, very little swirl was evident in any of the FLOW3D calculations, and even less in a fine-mesh ASTEC simulation.)

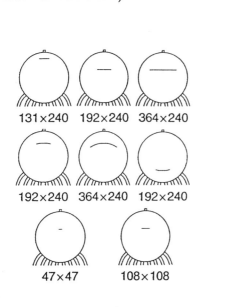

Fig. 3 Baffle Configurations

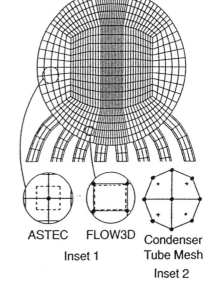

Fig. 4 Control Volumes and Final Mesh

All calculations were performed for a nominal wet steam flow of 1.4 kg/min — equivalent to a mean inlet velocity of around 13.2 m/s, and inlet Reynolds Number close to 6×10^4. The high Reynolds number k-ϵ turbulence model with standard coefficients was used throughout. Isothermal flow was assumed and, since the Mach number is low, the steam assumed to be incompressible, with density $1.7\,\text{kg/m}^3$ — corresponding to saturated steam at about 3 bar. FLOW3D was used to cross-check three cases, as discussed in the next Section.

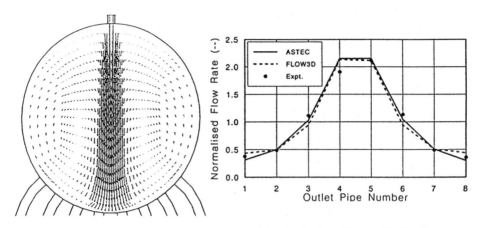

Fig. 5 ASTEC: No Baffle

Fig. 6 Outlet Flows: No Baffle

4 RESULTS

No Baffle

Figure 5 shows the velocity vectors in the centre-plane of the drum for this base-case ASTEC simulation. Both codes give essentially similar results and display perfect symmetry with respect to the centre-line of the drum, with a recirculation region on either side of the inlet jet. As expected, the flow distribution in the condenser tubes is strongly peaked towards the centre (Fig. 6) — the two central tubes (4,5) receiving 40% of the total flow, and the two outside tubes (1,8) about 12%. However, while the ASTEC solution is steady, no converged solution could be found using FLOW3D, and the flow distribution to the outlet tubes shown in Fig. 6 cannot be regarded as fully trustworthy. Nonetheless, there is very good agreement between the two codes.

To verify code predictions, an experiment was performed [4] in which a steady supply of air ($\rho = 1.19\,\text{kg/m}^3$) was fed to the AIDA inlet pipe (Fig. 1) and a hot-wire anemometer inserted in each condenser tube in turn to compare flow velocities. Data from the experiment are shown in Fig. 6. There appears to be an anomalous reading for Tube 4, thought to be due to a nearby obstruction in the drum (there is provision in the experiment for internally cleaning the viewports and the armature of the window wiper may be the cause), but otherwise there is almost perfect agreement between calculation and measurement.

47×47mm Baffle

The height of the AIDA rig, and the condenser tube diameter, pitch and spacing, are 1:1 with the full-size unit, though the number of tubes has been reduced from 496 to 8 (two rows of four tubes each). The same scaling factor (1:32) is applied to the inlet mass flow rate and, to preserve inlet velocity, the inlet pipe cross-sectional area is also scaled by the same ratio. Scaling the baffle plate similarly gives a plate diameter of 47mm, located 10cm above the centre of the drum. For convenience, however, a square plate of side 47mm was modelled instead.

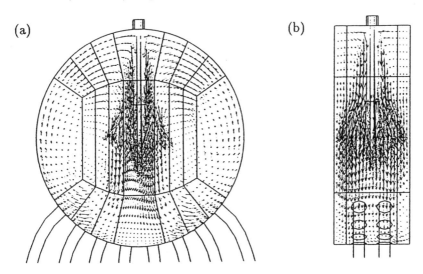

Fig. 7 FLOW3D: 47×47mm Baffle – (a) Front View (b) Side View

The ASTEC and FLOW3D predictions for the flow field in the drum display the same qualitative features: two major recirculation regions (vortices) either side of the inlet jet and by-pass flow around the baffle (Figs. 7a and 7b). Figures 8a and 8b show projected velocity vectors in a horizontal plane midway between the inlet and the baffle plate. There is much more evidence of three-dimensional motion with FLOW3D, seen as significant swirling around the inlet jet, which is broadened in relation to the drum confinement. The enhanced mixing that this implies becomes apparent if the distributions to the condenser tubes are compared (Fig. 9). There is a more even distribution according to FLOW3D, whereas with ASTEC there remains significant channelling of the flow towards the innermost tubes. There are no experimental data for this case.

108×108mm Baffle

In this variation, the baffle plate, in the same vertical location as before, was enlarged laterally to completely intercept the inlet jet, assuming this to have a spreading angle of 10°. This was the criterion employed to design the baffle for the actual full-size unit but, in the context of AIDA, means that different scaling criteria are applied for the inlet and baffle areas, so that relative geometric similarity is not preserved. Nonetheless, this arrangement provides an adequately even flow distribution to the condenser tubes, as will be shown below, with minimal increase in pressure drop (from $4.5\,\text{N/m}^2$ to $5\,\text{N/m}^2$).

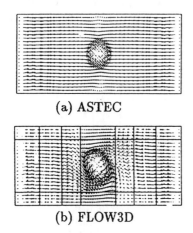

(a) ASTEC

(b) FLOW3D

Fig. 8 Velocities Between Inlet and 47×47mm Baffle

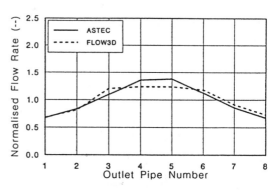

Fig. 9 Outlet Flows: 47×47mm Baffle

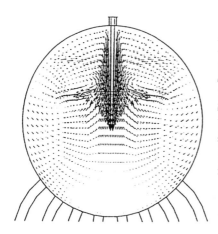

Fig. 10 ASTEC: 108×108mm Baffle

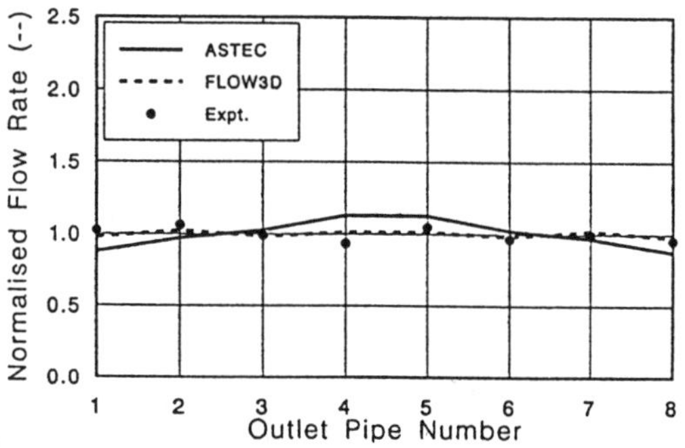

Fig. 11 Outlet Flows: 108×108mm Baffle

Figure 10 shows the flow distribution in this case. The baffle completely diverts the inlet jet, producing a double recirculation arrangement with separate vortices above and below the baffle level. Normalised mass flow rates to the condenser tubes are given in Fig. 11. As with the small baffle case, FLOW3D has predicted stronger 3-D motions and a more even outlet distribution. With ASTEC there is still significant channelling towards the two central tubes, even though the cross-sectional velocity fields in the vertical plane are visibly very similar.

To clarify this discrepancy a second validation experiment was performed [4], in this geometry and again with air, and the flows in the condenser tubes monitored as before. The data from the experiment are also included in Fig. 11. Again, there appears to be an anomaly with the Tube 4 reading, but otherwise the FLOW3D results are in excellent agreement with the measurements. It is very puzzling that a stable, converged ASTEC solution has produced an incorrect flow distribution for this case, but a seemingly perfect solution when there is no baffle (Fig 6) — a situation for which the ASTEC code developers have no answer. A possible mechanism is a more realistic prediction of swirl by FLOW3D, with better distribution due to its presence.

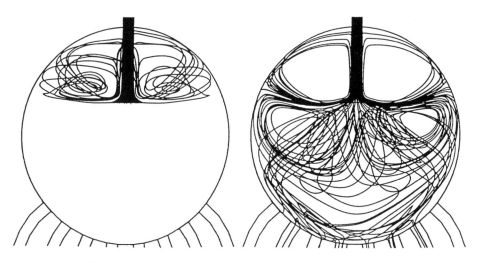

Fig. 12 FLOW3D: 108×108mm Baffle "Primary" particles

Fig. 13 ASTEC: 108×108mm Baffle "Primary" particles

5 AEROSOL DEPOSITION

Knowledge of the behaviour of aerosols in the AIDA facility may be obtained from the CFD calculations using Lagrangian, particle-tracking techniques, and the information later integrated into aerosol models regarding their deposition, agglomeration and re-entrainment characteristics. As a first step in this direction, aerosol tracking calculations were performed by interfacing the ASTEC and FLOW3D codes with a particle tracking model specially written for this purpose. (The standard particle-tracking model in FLOW3D proved to be unreliable for this application and there was no equivalent model in ASTEC.)

All calculations assumed that every particle reaching a surface would be deposited, and remain, on that surface. Thus, aerosols entering the drum with the inlet stream would preferentially deposit and stay on the baffle plate ("primary") particles). However, the two codes exhibit very different deposition behaviour. According to FLOW3D, 98% of the aerosols which enter the drum this way will deposit on the plate, with the remainder circulating in, and finally depositing on, the upper part of the drum (Fig. 12). There is no aerosol contamination below the level of the baffle. In contrast, ASTEC predicts that only 58% of the aerosols impact the baffle plate. The remainder are sufficiently diverted by the flow stream to by-pass the plate and circulate in the drum. Of these, 8% actually enter the condenser tubes, though not symmetrically (Fig. 13).

As a step further in the simulation, it was assumed that some of the aerosol build-up on the plate would eventually "spill-over" from the edges and re-entrain in the flow stream ("secondary" particles). The subsequent motion was then computed and the aerosol tracks are displayed in Figs. 14 and 15. This time, the code predictions are more consistent. The majority of the re-entrained aerosols deposit on the inner surface of the drum (FLOW3D 71%, ASTEC 81%). There is also deposition on the underside of the baffle itself (FLOW3D 14%, ASTEC 8%), while the remaining particles are swept into the condenser tubes, more or less symmetrically.

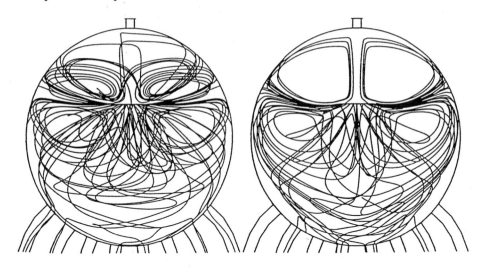

Fig. 14 FLOW3D: 108×108mm Baffle "Secondary" particles

Fig. 15 ASTEC: 108×108mm Baffle "Secondary" particles

6 CONCLUSIONS

Numerical simulations have been carried out to support the AIDA steam/aerosol test programme, using the codes ASTEC and FLOW3D. Three upper-drum configurations have been simulated and results compared, in two cases also against experimental data. With the upper drum empty (no deflector plate), no converged, steady-state solution could be found with FLOW3D, though the averages of several non-converged runs were in good agreement with the (converged) ASTEC results, themselves in direct accord with experimental data.

The remaining calculations featured flat baffle arrangements, of two sizes, at the same vertical location in the drum. In both cases, the codes produced steady-state solutions which were qualitatively similar, and quantitatively close enough for design purposes. Nonetheless, differences in code predictions were apparent. These appeared to be a consequence of the three-dimensional nature of the flow field, as calculated by FLOW3D, compared with the predominant two-dimensionality of the ASTEC results. The enhanced mixing in the third dimension – including some swirling around the inlet jet – resulted in FLOW3D predicting a more even mass flow distribution to the condenser tubes than ASTEC, which still exhibited significant bias to the central tubes. For the larger baffle, the FLOW3D results were in excellent agreement with the experimental data.

With the small baffle (47×47 mm), the outlet distribution was still too peaked at the drum centre-line. However, the larger baffle (108×108 mm) produced an adequately even flow distribution to the condenser tubes, with a pressure loss of only about $5 \, \text{N/m}^2$, and it was concluded that this design should be adopted for the subsequent steam/aerosol tests.

A Lagrangian, particle-tracking model was interfaced to the two codes and used to give first information regarding aerosol deposition in the test section. Very different behaviour was predicted by the two codes for those aerosols entering the drum co-flowing with the inlet stream. With FLOW3D, almost all aerosols impact the baffle plate while, according to ASTEC, 40% will by-pass the baffle and deposit elsewhere in the system, including in the condenser tubes. Results emphasise that very accurate fluid flow data are a necessary prerequisite for reliable aerosol tracking computations.

REFERENCES

[1] CODDINGTON, P. et al.: "ALPHA – The Long-Term Passive Decay Heat Removal and Aerosol Retention Program", 5th Int. Top. Mtg. on Nuc. Reac. Therm. Hyd. (NURETH-5), Salt Lake City, Utah, 21–24 Sept. 1992.

[2] ASTEC Release 4.2 User Manual, AEA Technology, Harwell, UK (Aug. 1992).

[3] CFDS-FLOW3D Release 2.4 User Manual, AEA Technology, Harwell, UK (Aug. 1991).

[4] LEUTE H., GÜNTAY. S.: Private communication (March 1995).

Secondary instabilities and transition to turbulence in wakes and jets

J. Dušek†, Ph. Fraunié‡, C. Dauchy* & I. Danaila*

†Institut de Mécanique des Fluides
2, rue Boussingault
F-67000 Strasbourg

‡Laboratoire de Sondages Electromagnétiques de l'Environnement
Université de Toulon et du Var,
B.P. 132,
F-83957 La Garde

* Institut de Recherche des Phénomènes Hors d'Equilibre,
12, avenue du Général Leclerc
F-13003 Marseille

Abstract

Secondary instabilities in the wakes of finite cylinders (circular and NACA 0012 with $34°$ of incidence) are simulated by directly solving the Navier-Stokes equations. It is shown that the secondary instability is again of Hopf type and leads to the onset of a new incommensurate frequency. This frequency is shown to be, in the same way as the primary one, uniform throughout the flow field at the instability saturation: The flow field describes a limit T^2 torus. Similar T^2 torus is evidenced in an unconfined round homogeneous jet. An alternative to the direct solution of the Navier-Stokes equations by low order time stepping methods is suggested for the analysis of subsequent stages of the transition.

1 Introduction

Since Landau [1] it has been believed that turbulence sets on via Hopf bifurcations. Although the original Landau theory could not withstand the progress in the understanding of dissipative dynamical systems, one of the most likely scenarios of transition, that of Ruelle, Takens and Newhouse [2], is still based on the onset of a limited number (3 at most) of Hopf bifurcations. It has been shown in the meantime that several other scenarios of transition are possible. E.g. extensive research of the Rayleigh - Bénard convection [3] allowed to evidence practically all of the proposed scenarios (transition via a T^2 and T^3 torus,

Feigenbaum period doubling cascade and type III intermittency scenario) in various configurations of this special type of flow. The mentioned results concerning the Rayleigh - Bénard convection indicate that no single universal scenario will probably explain the transition in all types of flow configurations.

Nevertheless, recent results of our numerical simulations of secondary instabilities of wakes and jets seem to show that if the transition via a T^2 torus is not universal in open flows it is likely the most typical scenario. In this paper we focus mostly to direct simulations of the secondary instabilities in wakes of infinite and finite cylinders showing that these two configurations behave quite differently. Latest results of simulations of the round jet are mentioned as well. Finally, we suggest a promising method of numerical investigation of the, probably, ultimate stage of the transition to turbulence: the transition from the T^2 torus to chaos.

2 Primary instability

It is a well known experimental fact that the primary instability in the wakes of finite 3D objects resembles strongly to that of infinite cylinders. The fact that the vortex shedding in the wake of a long cylinder remains largely two-dimensional has made it possible to achieve excellent agreement between experimental data and 2D simulations in this case ([4] and references therein). However, even for a finite cylinder a qualitative agreement with the 2D case exists. Quantitative laws relating some important wake characteristics such as the critical Reynolds number and the vortex shedding frequency to the aspect ratio of a circular cylinder were given by Mathis [5]. Appropriate tools for the theoretical investigation of the Bénard von Kármán instabilities have been sought in accordance with available computer power. It appears that the local instability theory is inadapted for wakes because if the strongly non parallel character of such flows [6]. A fully two dimensional linear theory was presented by Jackson [7], however, due to the specificity of the used computational method this approach never spread wider. As a result, the most frequent way of numerical investigation of wakes is a direct numerical solution of the unsteady Navier - Stokes equations. This approach has the advantage of taking into account the complete non linear dynamics of the flow. Its theoretical and numerical limitations may, however, present serious problems discussed in a subsequent section.

Due to their similarity with real experiments the direct numerical simulations have been used mostly to study the time behavior of the wakes. Little attention has, by contrast, been paid to the spatial structure of the instability. The concept of Hopf bifurcation in low dimensional dynamical systems applied to flows explains that the instability has to be expected to reach a limit cycle understood as a one dimensional manifold in the functional space in which the flow is described. Otherwise stated, all the flow characteristics, such as all velocity components and pressure, shall have strictly the same period throughout the flow-field. This property of the instability is not trivial and, to our knowledge, no rigorous mathematical proof based on the Navier-Stokes equations exists. In the absence of such a proof we submitted, in a recent paper [4], this widely accepted fact to a numerical test and confirmed its accuracy. The next logical step is then to characterize the limit cycle. The time parametrization yielded by direct numerical simulations gives huge amounts of data and requires some data reduction technique to be of practical use. Often the flow field evolution is approximated simply by storing a limited number of instantaneous flow field data. A more natural and efficient way consists, however, in computing the most relevant Fourier amplitudes of the periodic flow field. These amplitudes (complex space functions) are directly related to fundamental theoretical concepts such as the linear unstable mode. The Fourier decomposition allowed to explain, on the basis of Navier-Stokes equations, in a simple way the non linear effects leading to the rise of higher harmonics and the instability saturation [4]. The structure of the individual Fourier modes is of interest by itself. It allows to understand that the symmetry breaking at the bifurcation is in fact a transition from a simpler to a more complicated representation of the symmetry group.

3 Secondary instabilities in wakes of infinite cylinders

Substantial effort has been spent to study the secondary instability in the wake of infinite cylinders. This instability is accompanied by the onset of three dimensionality. An important theoretical issue consists in establishing the nature of the corresponding bifurcation because it may give a hint on the scenario of transition to turbulence in such a configuration. Since the controversial experimental work of Sreenivasan [9] a series of experimental [10] and numerical [8, 11, 12] results proved that the RTN scenario reported by Sreenivasan was due to aeroelastic effects and that the transition to three dimensionality in such wakes is accompanied by a period doubling.

Independently of the fact that a first period doubling does not prove the existence of the whole subharmonic cascade and that it would be technically extremely difficult to follow the cascade beyond a very limited number of bifurcations there arises the question whether the problem of the transition in infinite wakes is pertinent in most 3D wakes. Indeed, the difficulty of realization of a parallel vortex shedding shows that 2D wakes represent a negligible exception among the general 3D wakes. If the period doubling proves to be related only to the onset of three-dimensionality in 2D wakes then its relevance for the understanding of the transition in wakes, in general, is limited. This, indeed, seems to be the case.

4 Numerical simulation of wakes of finite cylinders

Recent simulations carried out at IRPHE in Marseilles [13] had the objective to investigate the case of wakes of finite cylinders. The idea to investigate this configuration arose as a consequence of our observation that numerical simulations of wakes of finite objects yielded a complicated, probably chaotic, behavior at Reynolds numbers lying substantially below the threshold of the secondary instability for a similar but infinite object.

Simulations were carried out for the following configurations: a half NACA 0012 wing of chord 1 and length 6 with 34 degrees of incidence representing a complete wing of aspect ratio 12, a circular half cylinder of diameter 0.56 and of the same length (see Fig. 1) and a full circular cylinder of the same diameter and of length 12. Both last configurations represent a cylinder of aspect ratio 21.4. The half wing and the half cylinder configurations were simulated with a symmetry condition in the symmetry plane normal to the cylinder axis assuming a conservation of the symmetry with respect to this plane. This assumption was tested by the last mentioned configuration above the threshold of the secondary instability. It appeared that the symmetry with respect to the symmetry plane brakes, however, for objects of the chosen large aspect ratio the deviation from symmetry was found of the order of about 0.001 of the primary instability amplitudes. This result and obtained reduction of computing costs justified fully this simplification. The used Navier-Stokes solver was Nekton, a code based on the spectral element discretization. Some optimization was carried out as far as the spectral element distribution was concerned. The number elements and their collocation points was kept rather low to allow simulation of sufficiently long time intervals. These tradeoffs explain why the primary instability threshold for the cylinder was found to lie at $Re_1 = 43$, i.e. about 10% below the expected critical Reynolds number, which should be slightly above 46, the value found experimentally and numerically for the infinite cylinder. In this communication, we primarily focus to the results obtained for the half cylinder.

As expected, the primary instability does not differ very much from the classical Bénard - von Kármán instability observed in the wake of infinite cylinders. The typical periodic signal obtained at $Re = 49$ is presented in Fig. 2. The spatial structure of the fundamental of the oscillations in the symmetry plane is the same as that obtained in usual 2D simulations (see Fig. 3). In what follows we shall place the axes in the following way: the origin lies at the intersection of the symmetry plane with the cylinder axis, the x-axis is directed along the direction of the inflow velocity (perpendicular to the cylinder), the z-axis coincides with that of the cylinder. (The y-axis completes a direct orthogonal system of axes.) In accordance with the expectations the maximum of the instability lies in the $z = 0$ plane (see Fig. 4). The most interesting three-dimensional feature of the flow is the spatial distribution of the velocity deficit of the mean flow (which does not differ very much of the basic flow of the instability as low above the instability threshold as in this simulation). In Fig. 5 we can see that the recirculation around the cylinder end amplifies the recirculation zone so that the maximum velocity deficit lies near the the cylinder end (at about $z=4$).

Rather low above the primary instability threshold, beatings of the original periodic signal set on (Figs. 7, 8). The analysis of the evolution of the amplitude and frequency of the beatings as a function of the Reynolds number allow to estimate the threshold of the onset of the beatings and the nature of this instability (Figs. 9 and 10). The threshold of the secondary instability lies as low as at $Re = 53$ which is in sharp contrast with the infinite configuration where the secondary instability (period doubling with the transition to three-dimensionality) sets on at $Re \approx 180$. The instability is clearly of Hopf type. Detailed investigation of the uniformity of its frequency at saturation at various points of the flow showed that local frequencies of beatings were indistinguishable.

The new frequency is incommensurate with that of the primary instability. This can be seen in Fig. 7 where we clearly see that the computed signal of the velocity is not periodic and that the ratio of both frequencies is about 13.5. More simulations have been carried out in the case of the NACA wing and could establish that the ratio was decreasing continuously between Re=90 and R=120 from 14.1 to 9.3. (The secondary instability threshold for the wing was at Re=87.) We conclude that the flow describes a two-dimensional torus in the functional space of flow characteristics. This conclusion is confirmed by the spectrum of the velocity fluctuations (Fig. 8).

The fact that the ratio of the period of the beatings to that of the primary oscillations approaches very roughly the aspect ratio of the cylinder suggests that the cylinder tips are responsible for the onset of the secondary instability. The analysis of its space structure shows (Fig. 6) that the maximum amplitude of the beatings lies in the zone of the maximum velocity deficit of the flow near the cylinder tip. Similar behavior is observed in the case of the NACA wing with incidence. (These results indicate that the wake of finite and elongated objects transites via a T^2 torus.)

The computing costs of simulations make an analysis of subsequent stages of the transition difficult. At this moment we can state that, in the case of the NACA wing with incidence, a more complicated behavior sets on at about twice the primary instability threshold (found to lie at $Re = 73.5$) - Fig. 11. In the case of the completely non symmetric configuration corresponding to the wing with incidence the onset of this complicated behavior appears earlier than for the circular cylinder.

5 Simulation of a homogeneous round jet

The simulations of unconfined homogeneous round jets focused so far mostly to the investigation of turbulent structures at turbulent Reynolds numbers (e.g. [14]) or to secondary instabilities (e.g. [15]). In both cases the turbulence or secondary instabilities are investigated in periodic domains and triggered by strong perturbations. As far as the spontaneous (triggered by weak perturbations) transition in a homogeneous round jet is concerned we were unable to find any numerical or experimental data. The reason might be that the local instability theory yields a local absolute axisymmetric instability region only for light jets with a ratio of the density of the inner core to the outer flow of less than 0.72 [16]. Flows that do not contain a pocket of absolute instability are not expected to display intrinsic dynamics and to be suited to a nonlinear dynamical system approach to the onset of turbulence [17].

In non-parallel flows there is, however, no direct relation between the local and global instability properties [6]. The results of local analysis may just mean that no bifurcation, conserving axisymmetry, can be observed in the zone of convectively unstable Reynolds numbers [18]. We undertook 3D simulations of an unconfined homogeneous round jet for an increasing Reynolds number by the same numerical method as that used for the 3D wake above. We found that no instability arose until rather high Reynolds numbers. Only at about $Re = 240$ we could observe an onset of instability accompanied by a breaking of the axisymmetry.

Fig. 12 characterizes rather well what happens. The figure represents the azimuthal velocity characteristic of the axisymmetry breaking at a point lying off the jet axis. The left hand side of the graph on the top figures represents a clear exponential development of a Hopf instability accompanied by the breaking of axisymmetry. The slow growth shows that the Reynolds number lies closely above the instability threshold. The instability, however, never reaches a limit cycle characteristic of an isolated Hopf bifurcation. Instead, a beating sets on in the right part of the graph. The Strouhal number of the primary oscillations is roughly 0.18, the ratio primary frequency/frequency of beatings is about 13. We conclude that the threshold of the secondary instability lies extremely close above the threshold of the primary one. It will be possible to refine the step of increase of the Reynolds number to determine the exact position of this threshold in simulations but it is not surprising that a pure Hopf bifurcation is difficult to be observed experimentally in this type of flow. Nevertheless, the T^2 torus attractor analogous to that observed in the finite cylinder wake seems to be an intermediate stage of transition also in this type of flow. Preliminary hot wire velocity measurements of a low Reynolds number ($Re \approx 700$) homogeneous weakly confined air jet carried out at IRPHE in Marseilles [19] allowed to evidence a practically biperiodic velocity signal. It appears that there is probably a very small interval of Reynolds numbers between the onset of the first instability and the onset of turbulence.

6 Limitations of classical direct simulations

When carrying out the described simulations we realized a certain number of limitations making the investigation of the nature of the transition scenario very difficult. These limitations can be summed up as:

1) impossibility to obtain linear characteristics of the instabilities with sufficient accuracy,

2) high, if not prohibitive, computing costs needed to reach attractors and to determine their characteristics (e.g. necessity of long simulations to reach the T^2 torus in the finite cylinder wake and to determine the frequency ratio primary frequency/beatings with sufficient accuracy)

3) difficulties of analysis of the transition from the T^2 torus to turbulence (extremely long simulations needed for classical methods of analysis such as power spectra and Poincaré sections.)

In a recent paper we proposed [20] a method of simulation of flows with a dominant periodicity based on the Fourier decomposition of the Navier-Stokes equations with respect to the dominant period. It has been shown that this method allows to simulate Hopf bifurcation transients by computing only the slowly varying envelopes of rapid oscillations. Usually a quite limited number of harmonics suffices to reach a much higher time accuracy than that achieved by usual low order time marching techniques. This allows to obtain not only significant CPU time reductions but also to increase the overall time discretization accuracy. The method allows the computation of the basic flow and the linear analysis of the instability as special cases. Extension of the system to a higher number of harmonics allows to reach an arbitrarily high accuracy. At saturation the s-derivatives vanish and the c_n modes become steady characterizing the reached, strictly periodic limit cycle. Formally, we recover the steady formulation of the steady laminar flow before the onset of instability with the only difference that the space on which the solution is sought is more complicated: instead of one real vector and scalar field we have several complex vector and scalar fields.

To investigate the destabilization of the limit cycle the envelopes of the primary oscillations can first be calculated directly. Later on, if biperiodic behavior is suspected, the procedure explained above can be generalized to a biperiodic Fourier development. In this way, the attractor - T^2-torus - can be reduced to a point in a generalized Fourier space. Its description may be reduced just to a very small number of relevant Fourier components. As a result, the relevant dynamics of the decay of the T^2 torus might be isolated and simulated at acceptable costs with sufficient accuracy.

7 Conclusion

The presented results concerning the secondary instabilities in 3D wakes and in the round homogeneous jet suggest that a biperiodic T^2 torus is a very frequent stage of the transition to turbulence. Such behavior could indicate that the

Ruelle-Takens-Newhouse or Curry-Yorke ([3]) scenario is a very frequent way of transition to turbulence in open flows. To understand the details of the decay of the biperiodicity and eventually to investigate the nature of the arising attractor classical direct simulation methods might prove inadequate. The adequacy and feasibility of implementation of the spectral time discretization method [20] will be investigated.

Acknowledgments :

This project was supported by the "Réseau Formation-Recherche: Simulation numérique des écoulements turbulents" of the french MESR - DREIFRT. Jan Dušek's participation was supported by the CEC grant number ERB 3510 PL92 1099 and an invited professor position in Université de Toulon et du Var (march 1995). Computational facilities have been partly provided by the "Centre Régional de Calcul et Télécommunications scientifiques " PACA and CRT-MAI, Université de Toulon et du Var. The work of I. Danaila is sponsored by Institut National de l'Environnement Industriel et des Risques.

Bibliography

[1] L. D. Landau & F. M. Lifchitz, *Fluid Mechanics, Course of Theoretical Physics, Vol 6.*, Pergamon Press, 1959.

[2] S. Newhouse, D. Ruelle & F. Takens, "Occurence of strange axiom-A attractors near quasi-periodic flows on T^m, $m \leq 3$", *Communications in Mathematical Physics*, **64** (1978) 35.

[3] P. Bergé, Y. Pomeau & Ch. Vidal, *L'ordre dans le chaos*, Hermann, Paris, 1988.

[4] J. Dušek, P. Le Gal & Ph. Fraunie, A numerical and theoretical study of the first Hopf bifurcation in a cylinder wake, *J. Fluid Mech.*, **264** (1994) 59.

[5] C. Mathis 1983, Propriétés de vitesse transverses dans l'écoulement de Bénard von Kármán aux faibles nombres de Reynolds, Thèse, Université Aix-Marseille.

[6] J. Dušek, Ph. Fraunié & P. Le Gal, Local analysis on the onset of instability in shear flows. *Phys. Fluids* **6** (1994) 172.

[7] C.P. Jackson, A finite-element study of the onset of vortex shedding in flow past variously shaped bodies, *J. Fluid Mech.* **182** (1987) 23.

[8] D. Barkley & R.D. Henderson, Three-dimensional Floquet stability analysis of the wake of a circular cylinder. *submitted to J. Fluid Mech.*

[9] K.R. Sreenivasan, Transition and turbulence in fluid flows and low-dimensional chaos, In *Frontiers in Fluid Mechanics.* S.H. Davis and J.L. Lumley, p 41, Springer, 1995.

[10] C. Van Atta & M. Gharib, Ordered and chaotic vortex.

[11] G.E. Karniadakis & G.S. Triantafyllou, The three-dimensional dynamics and transition to turbulence in the wake of bluff objects, *J. Fluid Mech.* **238** (1992) 1.

[12] J. Dušek, Ph. Fraunié & S. Seror Mise en évidence du doublement de période dans le sillage d'une aile NACA à la deuxième bifurcation de Hopf. *C. R. Acad. Sci. Paris* **319**, Série II (1994) 1271.

[13] C. Dauchy, Etude numérique d'une instabilité secondaire dans le sillage d'un obstacle tridimensionnel, Ph.D. thesis in preparation, IRPHE, Université de la Méditerranée, Marseille.

[14] R. Verzicco & P. Orlandi, Direct simulation of the transitional regime of a circular jet *Phys. Fluids* **6** (1994) 751.

[15] P. Brancher, J.M. Chomaz & P. Huerre, Direct numerical simulation of round jets: Vortex induction and side jets, *Phys. Fluids* **6** (1994) 1768.

[16] P.A. Monkewitz & K.D. Sohn, Absolute instability in hot jets. *AIAA J.* **26** (1988) 911.

[17] P. Huerre & P.A. Monkewitz, Local and global instability in spatially developing flows. *Ann. Rev. Fluid Mech.* **22** (1990) 473.

[18] X. Yang and A. Zebib, Absolute and convective instability of a cylinder wake, *Phys. Fluids A* **1** (1989) 689.

[19] M. Amielh, IRPHE, Marseilles, private communication.

[20] G. Carte, J. Dušek and Ph. Fraunié, A Spectral Time Discretization for Flows with Dominant Periodicity, *J. Comp. Phys.* **120** (1995) 171.

FIGURES

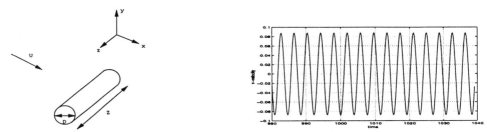

Figure 1 (Left) Simulated half-cylinder of diameter 0.56 and length 6 and the orientation of axes

Figure 2 (Right) Time signal of the transverse (v) velocity at the point $x = 3.515$, $y = 0$ $z = 3.213$ at $Re = 49$

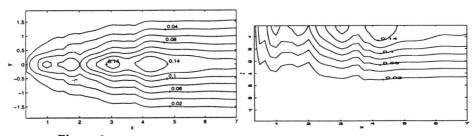

Figure 3 (Left) Isovelocity lines of the amplitude of the fundamental at $Re = 49$ in the symmetry plane $z = 0$ (The cylinder of diameter 0.56 is situated at $x = 0$ and $y = 0$.)

Figure 4 (Right) Isovelocity lines of the amplitude of the fundamental at $Re = 49$ in the $y = 0$ plane containing the cylinder axis (The cylinder extends from the symmetry plane $z = 0$ to $z = 6$ along the z-axis.)

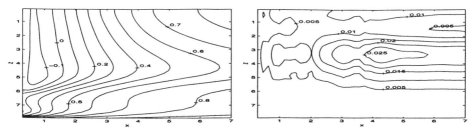

Figure 5 *(Left) Isovelocity lines of the mean flow at $Re = 49$ in the $y = 0$ plane containing the cylinder axis*

Figure 6 *(Right) Spanwise $(y = 0)$ cross section of the amplitude of the secondary instability at $Re = 54$.*

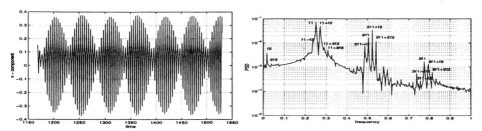

Figure 7 *(Left) Beatings in the v-velocity signal at $x = 3.515$, $y = -0.2859$, $z = 3.213$ at $Re = 62$*

Figure 8 *(Right) Spectrum of the v-velocity at the same point and for the same Re.*

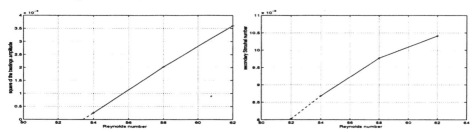

Figure 9 *(Left) Amplitude of beatings vs. Reynolds number at a point of the symmetry plane $(x = 3.515, y = -0.2859, z = 0)$*

Figure 10 *(Right) Strouhal number of the beatings vs. Reynolds number at the same point*

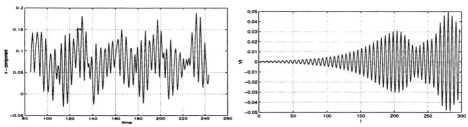

Figure 11 *(Left) Transverse velocity signal in the wake $(x=1.627, y=-0.3877, z=5.883)$ of the NACA wing at $Re = 150$ (The primary instability threshold was found to lie at about $Re = 73.5$.)*

Figure 12 *(Right) Onset of azimuthal oscillations in a round homogeneous jet at $R = 240$*

HEAT TRANSFER AND DRAG AUGMENTATION OF MULTIPLE ROWS OF WINGLET VORTEX GENERATORS IN TRANSITIONAL CHANNEL FLOW: A COMPARISON OF NUMERICAL AND EXPERIMENTAL METHODS

M.Fiebig, W.Hahne, D. Weber
Institut für Thermo- und Fluiddynamik
Ruhr-Universität Bochum, D-44780 Bochum, Germany
Tel. (0234) 700 3083, Fax (0234) 7094 162

SUMMARY

The 3-D self-oscillating flow and temperature fields in channels with periodically arranged winglet vortex generators (WVG) in transitional flow, Re = 500, are presented and discussed. The necessary conditions for streamwise and spanwise periodicity of flow and temperature field are investigated experimentally. Since experimental heat transfer data for thermally periodically developed transitional flow in channels with WVG were not available before, numerical and experimental values for time averaged local heat transfer coefficients and flow losses are directly compared for the first time. Good agreement of heat transfer coefficients and flow losses will be shown.

INTRODUCTION

An important heat-exchanger design problem arises when the flow inside ducts and channels falls in the transition region. This problem is of significant practical importance, because various enhanced heat transfer surfaces reduce the critical Reynolds number Re_{crit} by one order of magnitude compared to reported experimental results for Re_{crit} of plane channel flow, $1300 \leq Re_{crit} \leq 1700$, based on channel height H and average velocity U [1]. Hence, the flow inside these channels is transitional under most operating conditions. The mechanism of flow destabilization and transition in compact heat exchanger flow was studied extensively for <u>surfaces with multiple rows of wing-type vortex generators</u> (WVGs). This work is reviewed in references [2,3].

Wing-type vortex generators are slender rectangular winglets with their chords attached to one channel wall. Strong longitudinal vortices are generated by the flow separation at the side edges of the winglets. For the geometry shown in Fig. 1 the critical Reynolds-number is $Re_{crit} \approx 120$. Above Re_{crit}, self sustained flow oscillations occur. The 3-D unsteady flow <u>and</u> temperature fields for Re > Re_{crit} were predicted numerically. Direct comparison with experimental results for the local heat transfer coefficients and flow losses was not performed, because experimental heat transfer data for periodically fully developed transitional flow at Re < 1000 were not available, before.

Hence, the aim of this work is to test the numerical quantitative prediction of flow losses and local heat transfer coefficients for periodically developed flow in channels with wing-type vortex generators by means of direct comparison with experimental results.

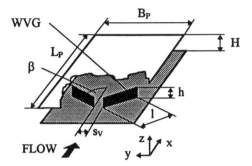

Fig. 1 Periodic element with WVG and computational domain: angle of attack $\beta=45°$, $L_P/H=5$, $B_P/H=4$, $l/H=2$, $h/H=0.5$, $s_v/H=0.3$; thickness of WVG: $\delta/H=0.04$;

METHODS OF INVESTIGATION

Numerical Procedure

In the transitional regime the numerical investigations were performed with a finite volume code for direct numerical simulation (DNS) of the three-dimensional unsteady Navier Stokes equations and continuity equation for a constant property Newtonian fluid [5]. The numerical simulation uses a grid of 102 x 82 x 22 and is limited to one periodic element, Fig. 1. Periodicity conditions for flow and temperature field are assumed at the entrance and the exit of the periodic element and at the lateral sides of an element. The channel walls and the WVGs are at constant temperature T_W.

Experimental Procedure

The extended analogy between mass and heat transfer is used to determine the local heat transfer coefficients for constant wall temperature. The Ammonia-Absorption-Measuring-Technique (AAM) is based on a color reaction of absorbed ammonia on the surface of the test model, coated with MnO_2 wetted filter paper. The experimental setup, the calibration procedure, and the high resolution image processing technique are described in detail in reference [6].

The test model consists of a channel with 11 rows of winglet vortex generator pairs in the streamwise direction and 7 pairs in the spanwise direction mounted on one channel wall. Its total dimensions are L/H=60, B/H=28, and H=10mm. Channel walls and WVGs are made out of aluminum. The WVG are not active mass transfer surfaces. The spatial resolution of the mass transfer image is 100dpi in the streamwise and spanwise direction (39 data points per channel height H). Based on 95% confidence, the calculation of the local Nusselt number from the distribution of the absorbed ammonia leads to an uncertainty of 8.1%. This uncertainty does not include variation of Nusselt number due to positioning errors of the WVGs. These errors must be minimized by averaging the Nusselt distribution over several periodic elements. The surface is exposed to the ammonia for more than 100 seconds. This time interval is at least 3 orders of magnitude larger than the time scale of the lowest frequency velocity fluctuations that were observed in self-sustained oscillatory ribbed channel flow [4].

RESULTS

Periodicity of flow and temperature fields
The static pressure distribution at the smooth channel wall was measured for $100 \leq Re \leq 2000$. For $Re \leq 500$, streamwise periodicity of the flow field was indicated by a linear static pressure gradient downstream of the 4th row at $20 \leq x/H \leq 45$. For $Re > 500$ the area of linear pressure gradient shifted downstream of the 5th row. Spanwise periodicity was shown by zero spanwise pressure gradient for the 3 center rows.
In Fig. 2, we present the spanwise averaged Nusselt number distribution on the top and bottom wall between the 4th and 8th streamwise WVG-row, $20 \leq x/H \leq 40$. The Nusselt number is based on the channel height H. The data clearly indicate thermally fully developed flow within this area. We also compared Nu(x) distributions, spanwise averaged over one, three, five, and seven periodic elements. For all Re investigated, the Nu distributions, spanwise averaged over one, three, and five periodic elements were almost identical. This indicates spanwise periodicity for the 5 center rows of the $\beta=45°$ geometry.

Local and global heat transfer and drag for hydrodynamically and thermally periodically developed flow
For $25 \leq x/H \leq 40$ the experimental data, plotted in Fig.2, give an average Nu number increase for one periodic element of $Nu/Nu_0 = 2.88$ for the ribbed wall and $Nu/Nu_0 = 2.63$ for the opposite wall without WVGs. Here, $Nu_0 = 3.77$ is the value of plane Poiseuille flow with constant wall temperature. The time and space averaged numerical data show excellent agreement. The numerical data give an increase of $Nu/Nu_0 = 2.84$ for the ribbed wall and $Nu/Nu_0 = 2.62$ for the smooth wall. Global Nusselt numbers for ribbed and smooth wall are plotted vs. Reynolds number in Fig. 4 a).

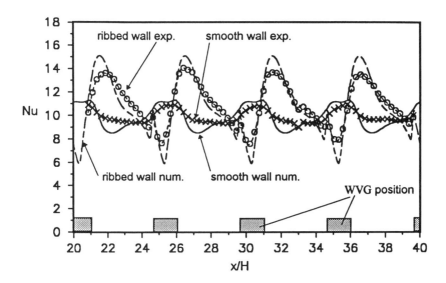

Fig. 2. Comparison of experimentally and numerically determined spanwise and time averaged Nusselt number distributions for periodically fully developed flow, $\beta=45°$, Re =500. Experimental data are spanwise averaged over 5 periodic elements.

The deviations of local numerical heat transfer data from the experimental data are significantly higher than the deviations of the global values, and rise up 30%. This cannot be explained by experimental uncertainties due to noise and positioning errors of the WVG, only. The numerical Nusselt number distributions on the ribbed and smooth wall show much sharper gradients in both the spanwise and the streamwise direction than the experimentally determined distributions. This is indicated clearly by the spanwise averaged numerical data, included in Fig. 2. Maxima and minima of the spanwise averaged Nusselt numbers are over- and underestimated, respectively, up to 20% by the numerical data. The variations of the experimental data for the four different WVG-rows, plotted in Fig. 2, are less than 5%.

From comparison of the numerical results of the local time dependent velocity vectors with hot-wire measurements [4], it is a known phenomenon that the DNS reduces the amplitude and frequency of the unsteady flow motions and Reynolds shear stress at Re = 500. The reason for this is the given grid resolution, limited by computing space and speed of currently available workstations. This leads to an underestimation of the flow losses, see Fig. 4 b), but this cannot explain the deviations of the numerical and experimental Nusselt number data on the ribbed wall, plotted in Fig. 2.

At the ribbed wall, at $x/H \approx 25, 30$, etc., the numerical data (Fig. 2) show a distinct Nusselt number minimum that is due to a recirculation zone behind each WVG, see Fig.3. Downstream, in the region of $x/H \approx 26.5, 31.5$, etc, the numerical data indicate fluid flow onto the wall ('downwash zones') due the longitudinal vortices, generated by WVG. These are regions of absolute maxima of the Nusselt number. The experimental data in Fig. 2 indicate both smaller recirculation zones and smaller absolute Nusselt number maxima. This deviation may be explained by the fact that both the recirculation zones and the longitudinal vortices are superimposed by additional fluid motions and dynamic vortex shedding at the lateral sides of the WVG which are suppressed by the DNS at the given grid resolution.

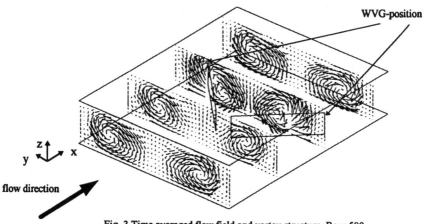

Fig. 3 Time averaged flow field and vortex structure, Re = 500

The deviations between numerical and experimental data on the <u>top wall</u> without WVGs cannot be clearly explained. We observed that the Nusselt number distribution for one periodic element on the smooth wall was strongly influenced by the distribution of its spanwise adjacent elements. For example, we found the Nusselt maxima on the smooth wall of all seven spanwise elements of one streamwise WVG-row shifted sidewards, whereas the distribution on the ribbed wall was strongly symmetrical. Generally speaking, the lateral boundary conditions do influence the Nusselt number distributions on the smooth wall much more than on the ribbed wall.

Fig. 4 b) shows drag augmentation of the WVG geometry. Numerical and experimental data are directly compared. For Re=500, the numerical data underestimate the experimental values by 20%. For Re < 250 the numerical data are within the uncertainty interval of the experimental data.

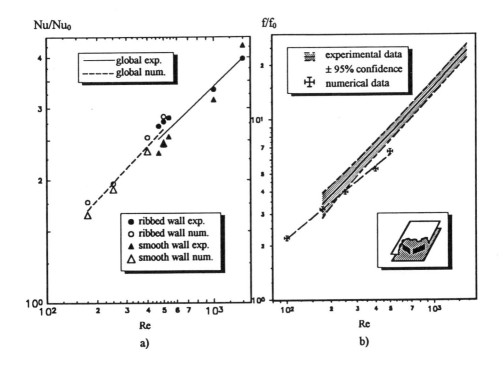

Fig. 4 Comparison of a) Nusselt number for ribbed and smooth wall, and b) drag augmentation, for periodically fully developed flow vs. Re. The results are related to plane Poiseuille flow, $Nu_0=3.77$, $f_0 = 12/Re$.

SUMMARY AND CONCLUSIONS

From the results we can draw the following conclusions:

1. Thermally periodically fully developed flow in channels with winglet vortex generators $h/H = 0.5$, and $L_P/H=5$ is not achieved at shorter entrance lengths than fully developed flat duct flow based on correlations [1].
2. Excellent agreement of global Nu for $Re = 500$ indicates that the numerical investigation of a single periodic element with any arbitrary chosen starting field matches the experimentally investigated transitional heat transfer for an array of WVG-pairs very well.
3. The lateral boundary conditions do influence local Nu on the smooth wall much more than on the ribbed wall. From this we draw the general conclusion that transitional effects, which can be expressed as self-sustained oscillations or turbulence level, do influence heat transfer on the smooth side of the duct much more than on the ribbed side.
4. Local Nu-distribution for $Re > 500$ still requires experimental reference data.

Hence, detailed numerical investigations of complex 3-D non-stationary flow have reached a quality that is not only sufficient for optimization of geometrical parameters but also for prediction of Re_{crit}, and local and global Nu.

ACKNOWLEDGMENTS

This research was supported by the Deutsche Forschungsgemeinschaft (DFG), research group *Wirbel und Wärmeübertragung*. The authors gratefully acknowledge this support.

REFERENCES

[1] Kakac, S., Shah, R.K., Aung, W.:"Handbook of Single-Phase Convective Heat Transfer", Wiley & Sons (1987).

[2] Fiebig, M.:"Vortex Generators for Compact Heat Exchangers", Journal of Enhanced Heat Transfer, Vol. 2., (1995), pp. 43-61.

[3] Fiebig, M., Müller, U., Weber, D.: "Flow Destabilization and Transition through Coherent Vortices", Proc. CCFD3 and LASFM3, Caracas, Venezuela (1995).

[4] Grosse.-Gorgemann, A., Weber, D., Fiebig, M..: "Experimental and Numerical Investigation of Self-Sustained Oscillations in Channel with Periodic Structures", to be published in Experimental Thermal and Fluid Science, (1995).

[5] Kost, A., Mitra, N.K., Fiebig, M.:"Calculation Procedure for Unsteady Incompressible 3-D Flows in Arbitrarily Shaped Domains", Notes on Numerical Fluid Mechanics, Vieweg Verlag, Vol. 35, pp.269-278 (1992).

[6] Weber, D., Schulz, K., Fiebig, M.:"Comparison of a high spatial resolution mass transfer technique with a transient Infrared-Thermography technique for determenation of local heat transfer coefficients on a surface with winglet vortex generators" Eurotherm 32 Seminar Proceedings, pp. 76-81, Oxford (1993).

COMPUTATIONAL DISPERSION PROPERTIES OF 3-D STAGGERED GRIDS FOR HYDROSTATIC AND NON-HYDROSTATIC ATMOSPHERIC MODELS

Michael S. Fox-Rabinovitz
University of Maryland/NASA/JCESS
College Park, MD, USA

SUMMARY

Computational dispersion properties in terms of frequency and group velocity components for centered difference schemes using all possible, practically meaningful 3-D staggered grids, have been examined for hydrostatic and non-hydrostatic atmospheric systems. The most advantageous 3-D staggered grids have been obtained by combining the best horizontal and vertical grids. The time-staggered grids have been also included into consideration. The best 3-D staggered grids provide twice the effective spatial resolution of the regular (unstaggered) 3-D grid. The obtained results, along with other computational considerations, provide the practical guidance for the optimal grid choice for various atmospheric models.

INTRODUCTION

The large-scale global climate and numerical prediction models use hydrostatic primitive equation systems. The mesoscale regional and local models employ non-hydrostatic anelastic (with the Boussinesq approximation) or fully compressible systems. Model dynamics include large-scale nonlinear advection (convection), and adjustment or wave processes whereas model physics provides diabatic forcing. The linear adjustment or gravity and gravity-inertia (with Earth rotation or the Coriolis force) wave subsystems considered below are most sensitive to model spatial resolution. The adjustment subsystems are used for investigating computational dispersion properties and identifying the best staggered grids.

COMPUTATIONAL DISPERSION PROPERTIES OF STAGGERED GRIDS FOR VARIOUS ATMOSPHERIC MODELS

The computational dispersion properties are investigated in terms of frequency and group velocity components for centered difference approximations of adjustment systems. The complete set of the 3-D staggered grids, combining all possible horizontal and vertical grids is used. The hierarchy of linear atmospheric adjustment systems is ranging from the 2-D and 3-D hydrostatic systems to the 3-D non-hydrostatic, namely anelastic and fully compressible systems [5 - 7].

The following adjustment or wave subsystems are used for studying the computational dispersion properties of horizontal, vertical and 3-D grids. For horizontal grids for hydrostatic systems we consider the linear shallow-water system:

$$\partial u/\partial t + \partial h/\partial x - fv = 0,\ \partial v/\partial t + \partial h/\partial y + fu = 0,\ \partial h/\partial t + H(\partial u/\partial x + \partial v/\partial y) = 0, \quad (1)$$

where u, v are velocity components, h = gz geopotential, z height, g the gravity constant, f the Coriolis parameter, and H=constant.

For vertical grids for hydrostatic systems we used the linear baroclinic adjustment, or wave system in $\zeta = \ln p$ coordinates:

$$\partial u/\partial t + \partial h/\partial x - fv = 0,\ \partial v/\partial t + \partial h/\partial y + fu = 0,\ \partial^2 h/\partial t\partial\zeta + c^2\Omega = 0,$$
$$\partial u/\partial x + \partial v/\partial y + \partial\Omega/\partial\zeta = 0, \quad (2)$$

where $\Omega = d\zeta/dt$, and c^2 = constant.

For the 3-D grids for non-hydrostatic anelastic systems, the following linear baroclinic system (in the Boussinesq approximation) in z-coordinates, is considered:

$$\partial u/\partial t + \partial P/\partial x - fv = 0,\ \partial v/\partial t + \partial P/\partial y + fu = 0,\ \alpha\cdot\partial w/\partial t + \partial P/\partial z - g\theta = 0,$$
$$\partial\theta/\partial t + Sw = 0,\ \partial u/\partial x + \partial v/\partial y + \partial w/\partial z = 0, \quad (3)$$

where $P = p'/\rho_0$, $p(x, y, z, t) = \bar{p}(z) + p'(x, y, z, t)$, $p(z)$ is the hydrostatic base-state pressure, ρ_0 the constant representative value ρ for the Boussinesq approximation, p pressure, ρ density, Θ potential temperature, $\theta = \Theta/\Theta_0$ a fractional perturbation from the base state $\Theta(z)$, $\Theta' = \Theta - \bar{\Theta}$, Θ_0 a constant representative value of $\bar{\Theta}$, w = dz/dt the vertical velocity component, $S = d\ln\bar{\theta}/dz = S(z)$, $\bar{\theta} = \bar{\theta}(z)$ the mean θ profile, the parameter $\alpha = 1$ for a non-hydrostatic anelastic system, and $\alpha = 0$ for a hydrostatic system.

For the 3-D grids for non-hydrostatic fully compressible systems we used the following equations:

$$\partial\bar{u}/\partial t + \partial p'/\partial x = 0,\ \partial\bar{v}/\partial t + \partial p'/\partial y = 0,\ \partial\bar{w}/\partial t + \partial p'/\partial z + g\bar{\rho}' = 0,$$
$$\partial p'/\partial t + \partial\bar{u}/\partial x + \partial\bar{v}/\partial y + \partial\bar{w}/\partial z = 0,\ \partial p'/\partial t - c^2\partial\rho'/\partial t + c^2\bar{S}\bar{w} = 0, \quad (4)$$

where $\bar{u} = \bar{\rho}u'$, $\bar{v} = \bar{\rho}v'$, $\bar{w} = \bar{\rho}w'$, the symbols " - " and " ' " stand respectively for the components of the basic flow and small deviations from them, or disturbances.

All possible grids are considered and the best 3-D spatial and/or temporal-spatial staggered grids with the advantageous computational dispersion properties have been determined. They provide twice the effective horizontal and vertical resolution, and the appropriate signs of group velocity components for all resolvable scales for discrete systems.

The considered 3-D grids include combinations of the horizontal Arakawa C [1, 9] and Eliassen [4] grids (Fig. 1), and the vertical Lorenz [8] and Charney-Phillips [2] grids and their time-staggered versions are (Fig. 2) introduced by the author [6]. The 3-D grid combining the Eliassen and Charney-Phillips grids is an example of a best grid for the hydrostatic system, and for the non-hydrostatic anelastic system with and without the Coriolis terms (Table 1). The example is important for the case of coupling the large-scale and meso-scale models using the same 3-D grid. We would like to emphasize that the Charney-Phillips grid should be at least as widely used as the currently most popular Lorenz grid (Fig. 2).

Other examples of the best 3-D grids (see Figs. 3 and 4) are the combinations of: 1) the Arakawa C and Charney-Phillips grids (only for the anelastic system without the Coriolis terms); 2) the Eliassen and time-staggered version of the Lorenz grids for the same system; and 3) the Arakawa C and the time-staggered versions of the Lorenz or Charney-Phillips grids for the same system and for the fully compressible system (Table 1). The summary of other grid properties is also presented in Table 1.

The group velocity components for the anelastic system (3) discretized with the 3-D regular (unstaggered) and the best staggered grids (Fig. 4 and Table 1), and for the exact differential form of the system (3), are presented in Figs. 5 and 6.

For the regular grid, the group velocity components have a wrong sign for small scales (negative for the horizontal component and positive for the vertical component), and exhibit other erroneous features (Figs. 5b and 6b). Using the best grids allows us to obtain the group velocity components (Figs. 5c and 6c) that are very close to those of the differential case (Figs. 5a and 6a).

For the CDSs with temporal-spatial staggered grids, all spatial averaging operators may be replaced by temporal or temporal-spatial averaging (convection) operators that result in introducing two-time-level semi-implicit schemes [5 - 7].

The best time-staggered grids also provide twice the effective spatial resolution for approximation of first derivatives in vertical advection (convection) terms as well as second derivatives in horizontal or vertical diffusion terms. Hence, the use of time-staggered grids provides a more uniform approximation with twice the effective resolution for various terms of atmospheric models [5 - 7].

The introduction of higher-(fourth) order vertical approximation results in a relatively moderate overall improvement mostly for the small vertical and large horizontal scale range [5 - 7].

The study provides the guidance for the optimal choice of grids for various hydrostatic and non-hydrostatic atmospheric systems. Other computational considerations such as conservation properties and boundary condition problems should also be taken into account.

REFERENCES

[1] Arakawa, A. and V. R. Lamb: "Computational design of the basic dynamical processes of the UCLA general circulation model." *Methods in Computational Physics*, J. Chang, Ed., Academic Press, (1977), pp. 173-265.

[2] Charney, J. C. and N. A. Phillips: "Numerical integration of the quasi-geostrophic equations for barotropic and simple baroclinic flows." *J. Meteor.*, 10, (1953), pp. 17-29.

[3] Du Fort, E. C. and S. P. Frankel: "Stability conditions in the numerical treatment of parabolic differential equations." *Mathematical Tables and Other Aids to Computation*, V. 7, (1953), pp. 135-152.

[4] Eliassen, A.: "A procedure for numerical integration of the primitive equations of the two-parameter model of the atmosphere." *Sci. Rep. No. 4*, Dept. of Meteorology, UCLA, (1956), pp. 56

[5] Fox-Rabinovitz, M. S.: "Computational dispersion properties of horizontal staggered grids for atmospheric and ocean models." *Mon. Wea. Rev.*, 119, (1991), pp. 1624-1639.

[6] Fox-Rabinovitz, M. S.: "Computational dispersion properties of vertically staggered grids for atmospheric models." *Mon. Wea. Rev.*, Vol. 122, No. 2, (1994), pp. 377-392.

[7] Fox-Rabinovitz, M. S.: "Computational dispersion properties of 3-D staggered grids for the non-hydrostatic anelastic system." *Mon. Wea. Rev.*, (1995) accepted.

[8] Lorenz, E. N.: "Energy and numerical weather prediction." *Tellus*, 12, (1960), pp. 364-373.

[9] Mesinger, F. and A. Arakawa: "Numerical methods used in atmospheric models." *WMO/ICSU Joint Organizing Committee*, GARP Publ. Series, (1976), pp. 64.

Table 1. The summary of the first and second choice 3-D grids for approximating the hydrostatic and non-hydrostatic systems (see Figs. 3 and 4).

System	The best, first choice grids providing twice the effective spatial resolution	The second choice grids providing twice the effective spatial resolution except for small horizontal or vertical scales
Hydrostatic [5, 6]	EL/CP, EL/L	C/CP, C/L, C/LTS, C/CPTS EL/LTS, EL/CPTS
Anelastic [6] ($f \neq 0$)	EL/CP	C/CP, EL/L, C/LTS, C/CPTS, EL/LTS
Anelastic ($f = 0$) [6]	EL/CP, C/CP, C/LTS, C/CPTS, EL/LTS	C/L, EL/L, EL/CPTS
Fully compressible	C/LTS, C/CPTS	C/L, C/CP, EL/L, EL/CP EL/LTS, EL/CPTS

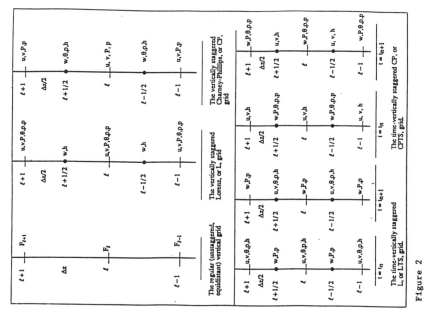

Figure 1

Figure 2

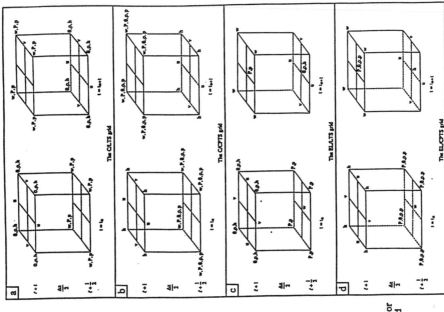

Fig. 3 The 3-D grids for an anelastic system as the combinations of the horizontally staggered Arakawa C grid, and the vertically staggered Lorenz (L) grid, or the C/L grid (a), and the Charney-Phillips (CP) grid, or the C/CP grid (b); and the time-horizontally staggered Eliassen (EL) grid with the same vertically staggered grids, or the EL/L grid (c), and the EL/CP grid (d). l and $\Delta z/2$ are the vertical index and half vertical interval respectively, n is the time index.

Fig. 4 Same as in Fig. 3 but with the use of the time-staggered versions of the Lorenz (LTS) and Charney-Phillips (CPTS) grids, namely, the C/LTS grid (a) the C/CPTS grid (b), the EL/LTS grid (c), and the EL/CPTS grid (d).

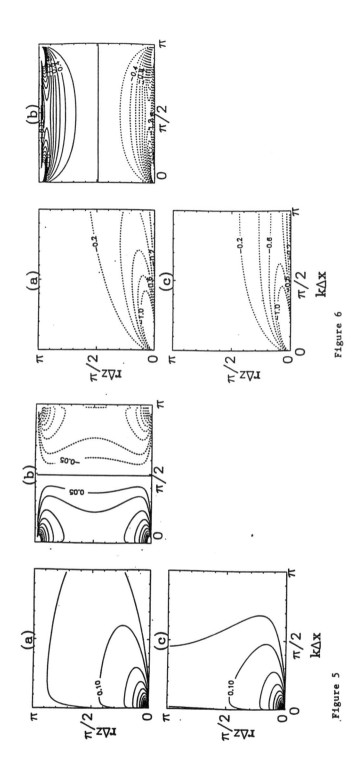

Fig. 5 (a) The horizontal group velocity component for the anelastic system (3) (the exact differential case); and for the system (3) discretized with: (b) the 3-D regular (non-staggered, equidistant) grid, and (c) the best 3-D grids (Fig. 4 and Table 1). k and r are horizontal and vertical wave numbers correspondingly. Δx and Δz are the spatial increments. Negative values are shown by the dashed lines.

Fig. 6 Same as in Fig. 5 but for the vertical group velocity components.

Numerical Simulation of Oscillatory Regimes in Vertical Cylinders Heated from Below

A. Ivančić, A. Oliva, C.D. Pérez Segarra, H. Schweiger

Laboratori de Termotècnia i Energètica
Dept. Màquines i Motors Tèrmics. Universitat Politècnica de Catalunya
Colom 9, 08222 Terrassa, Barcelona (Spain)
email: alex@termo1.mmt.upc.es

Summary

Three-dimensional and transient calculations have been performed in order to investigate unsteady fluid flow and heat transfer in vertical cavities of cylindrical geometry. The fluid flow is assumed laminar, incompressible and the Boussinesq approximation is considered. The governing equations are solved in primitive variables using the SIMPLEC numerical algorithm. The problem parameters are fixed to A=1.5, Pr=$2 \cdot 10^{-2}$, adiabatic side-wall and Ra is changed acting as a control parameter. The bifurcations from steady state to periodic flow, from periodic to quasiperiodic flow and from quasiperiodic to turbulent flow have been detected. The transition may be identified as the quasiperiodic or Ruelle-Takens-Newhouse route to turbulence. The frequency locking f1/f2=3 is present in the quasiperiodic regime. The influence of the flow history has been observed.

1. Introduction

The present paper talks about buoyancy convection in fluids framed in vertical cylindrical enclosures with different end-wall temperatures. Special attention is paid to time dependent flow range. Time dependent instability occurs beyond the critical Rayleigh number (Ra_{cr}), between the steady laminar regime and the turbulent one. The oscillatory flow is characterized by its frequencies.

The classical Rayleigh-Bénard problem, i.e. the buoyancy driven flow in a fluid layer heated from below, is one of the fluid dynamic topics widely analysed. Today, almost a hundred years after the first Bénard's publication on the theme, the basic mechanisms of the fluid motion are well known. The convection threshold is established within the stretch limit for different geometries, theoretically by means of linear analysis, by experimental work and also by numerical simulation of full Navier-Stokes equations.

The problem of oscillatory convection in cylindrical cavities arises in different technical applications and industrial processes; however it is widely recognized that the knowledge of the transition, oscillatory phenomena, helps very much in understanding the creation of turbulent flows. The time dependent instability occurs beyond the convective onset, between the steady laminar regime and the turbulent one. When the Rayleigh number is increased over a certain limit the steady laminar three-dimensional convective flow becomes unstable with respect to the time dependent disturbances.

The unsteady laminar flow in moderate and tall cylinders has been previously studied by various authors; experimental observations and measurements of time-dependent convection motion have been performed by Olson and Rosenberger [1], Müller et al. [2,3], Fauve and Libchaber [4] and recently by Kamotani et al.[5]. Olsaen and Rosenberger [1] worked with a cylinder of A=3 filled with different gases (Pr≈0.7) and found oscillatory onset for Ra_{osc}=5.86Ra_{cr} and aperiodic fluctuations at about 7.39 Ra_{cr}; route to the turbulence has been described as period doubling. Müller and Neumann [3] examine the flow in a tall cylinder (A=5) finding transition Ra values and describing the corresponding flow structure. In [2] Müller et al. detect the onset of oscillatory and turbulent flow for a range of aspect ratios (A=0.5-5) using water (Pr=7) and liquid gallium (Pr≈2·10^{-2}). Fauve and Libchaber[4] study the transition from steady laminar to turbulent flow in moderate aspect ratio cylinder (A=0.5) filled with liquid mercury (Pr≈3.1·10^{-2}). They have described two different routes to turbulence conditioned by the flow history.

In some papers the numerical methods have been applied in order to investigate the phenomenon. But the numerical simulation of oscillatory flows in cylinders appears sporadically, together with other aspects of the Rayleigh-Bénard phenomena (Crespo et al. [6], Neumann [7], Ivancic et al. [8]) . To our best knowledge, there is no systematic study of transition to turbulence in the cylindrical cavities by numerical means as there is the case for rectangular cavities. In our opinion the problem deserves more attention in order to deepen knowledge about the phenomenon.

The buoyancy convection phenomenon in vertical cylinders is determined by boundary conditions and the following dimensionless parameters: Rayleigh number, Prandtl number and aspect ratio. The simulation has been performed for the low Prandtl number fluid, Pr=2·10^{-2}, using the Rayleigh number as a control parameter; the Rayleigh number has been increased gradually looking for the onset of oscillatory flow. The oscillatory flow onset has been found. The dynamic behavior of time-dependent flow has been examined in the usual way by the power spectra and the phase trajectories. Numerical results of flow patterns and temperature distribution in a cylindrical cavity are presented.

2. Mathematical formulation and numerical method

We consider a fluid framed in a vertical cylindrical enclosure characterized by height to diameter aspect ratio A=1.5. The upper and lower base are kept at the uniform constant temperature, T_C and T_h respectively, while the lateral wall is assumed adiabatic. After the usual introduction of the Boussinesq approximation we get the next set of governing equations:

$$\nabla \cdot \vec{v} = 0 \qquad (1)$$

$$\rho_0 \frac{D\vec{v}}{Dt} = -\nabla p + \mu \nabla^2 \vec{v} + \vec{g} \rho_0 [1 - \beta (T - T_0)] \qquad (2)$$

$$\rho_0 c_p \frac{DT}{Dt} = k \nabla^2 T. \qquad (3)$$

The set of equations (1-3) is subject to the following boundary conditions:
-no slip boundary condition at the walls for Navier-Stokes equation (V=0)
-isothermal base walls
-adiabatic lateral wall

The differential equations are discretized using the finite volume method and solved by SIMPLEC [9] algorithm. Different discretization schemes, both second and third order accurate, have been used to approximate the convective term: QUICK[10] and occasionally SMART[11] and central difference, while the diffusion term is calculated by central differencing. The first order numerical schemes such as upwind and blending schemes based on upwinding have been rejected as insufficiently precise. Further details about used numerical code can be found in [8]

3. Code validation

To demonstrate the validity of the numerical code, the results are compared with data available from the literature. The code was validated for steady regime with the experimental data from [1,2,3,5] and predictions of linear stability theory published in [12,13,14]. A summary of these comparisons is given in [8] and here, because of space limitation, only the most significant results are shown. The numerical predictions agree closely with the experimental and theoretical data found in the literature (figure 1). The best coincidence is found with the results of Buell and Catton [14] (difference less than 4%). The code was previously used to simulate mixed convection in cylinders [15]

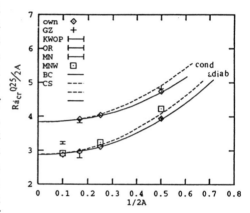

Figure 1. Ra_{cr} number vs A, stability limit curves for adiabatic and conducting wall; GZ[12], CS[13], BC[14]-linear stability theory, OR[1], MN[3], MNW[2], KWOP[5] experimental

4. Oscillatory flow

When the Rayleigh number is increased to reach a certain value (Ra_{osc}) the phenomenon becomes time-dependent. Over this limit the flow can be laminar (periodic or quasiperiodic) or turbulent. There are several possible scenarios of transition from steady state to turbulence; a good survey of this topic is given by Yang [16].

The transition to time-dependent flow in vertical cylinders is affected by the aspect ratio, the Prandtl number and the lateral wall thermal boundary condition. The aspect ratio and the lateral wall boundary condition and the Prandtl number are fixed (A=1.5, adiabatic boundary condition, Pr=$2 \cdot 10^{-2}$) and the Rayleigh number is used as a control parameter. The calculation has been made on grid 10x20x30 (in r, θ, z direction). The non-dimensional time step for calculus of unsteady range has been $4 \cdot 10^{-4}$; the influence of the time step has been examined by doubling and halving it. Since the difference in frequencies obtained is less then 2% it was concluded that the time step is of quite minor importance in that range. In order to find the oscillatory limit, the Rayleigh number has been increased gradually, in steps of $5 \cdot 10^3$ and later once the oscillatory flow has been detected, in steps of $1 \cdot 10^3$. Since the numerical calculation produces unavoidable round-off error it has been considered that this error is enough to provoke oscillatory flow; no other disturbance has been introduced.

It is very expensive to calculate directly the critical Rayleigh number (Ra_{osc}) because the nearer Ra to Ra_{osc} the larger the transition period. The critical Rayleigh number lies between 32 000 and 33 000. Since the amplitude of oscillation grows with the Rayleigh number and for Ra_{osc} amplitude tends to zero, the critical Rayleigh number may be estimated extrapolating the amplitude to zero; this estimation gives $Ra_{osc} \approx 32\ 400$.

The dynamic behaviour of the flow can be observed by means of power spectra and phase trajectories. The power spectra was calculated by the Fourier transform of the Nusselt number or temperature time series. The phase trajectories plot the temperature in two fixed points: t (6,10,1) and t (6,10,30).

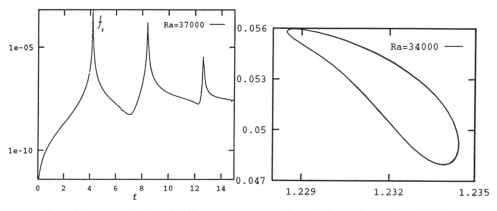

Figure 2. Power spectra, Ra = 37 000   Figure 3. Phase trajectory, Ra = 34 000

Up to the Rayleigh number of 39 000 the flow is periodic (P). The power spectra consist of one fundamental frequency f_1 and its harmonics (figure 2). The phase trajectory in this range (Ra=33000-39000) is a limit cycle which again confirms periodic behaviour of the flow.

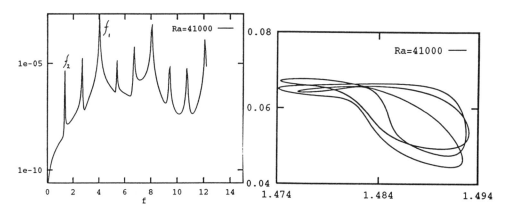

Figure 4. Power spectra, Ra = 41 000   Figure 5. Phase trajectory, Ra = 41 000

Between the Rayleigh number of 39 000 and 40 000 the periodic regime loses its stability and motion becomes quasiperiodic (QP2). The second fundamental frequency f_2 appears with intensity of $f_2 \approx f_1/3$. The limit cycle from figure 3 transforms into a torus. For Ra= 40 000 the frequencies are incomensurate, $f_1/f_2=2.911$; while for Ra=41 000 the ratio f_1/f_2 becomes 3. The phase rajectory, shown in figure 5, is closed which indicates frequency locking $f_1/f_2=3$.

When the Rayleigh number is increased to Ra=42 000 a third frequency f_3 is present (QP3). The new frequency is of an order of magnitude lower than the two previously detected (figure 6); phase trajectory shows even more complex flow dynamics. Soon after, around Ra=44 000 the flow becomes turbulent. This kind of transition from steady state laminar flow to turbulent flow is identified as the Ruelle-Takens-Newhouse route to chaos.

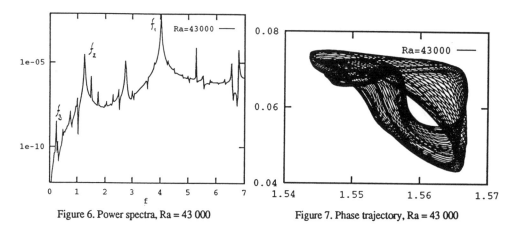

Figure 6. Power spectra, Ra = 43 000   Figure 7. Phase trajectory, Ra = 43 000

Figure 8 shows the transition map and non-dimensional frequencies. The first fundamental frequency f_1 is of the same order of magnitude as those measured by Kamotani et al.[5] in cylinders of aspect ratio A=2 and 3 filled with liquid gallium ($Pr=2.07 \cdot 10^{-2}$).

In figure 9 the mean Nusselt number versus Rayleigh number is depicted. The stagnation of Nu near the quasiperiodic bifurcation can be observed. As can be expected, amplitude growth is more intense for the quasiperiodc regime. It was found that the flow depends on flow history, i.e. the solution to be established depends on the initial conditions. It is such a common phenomenon for nonlinear problems that more than one solution exists for the given parameters. For example, there is a steady solution for Ra=35 000 (figure 9).

4.1 Mechanism of periodic flow

In order to get knowledge about the oscillatory nature the result for Ra=34 000 has been analysed. Three characteristic moments have been chosen. The amplitudes are much more pronounced in the symmetry plane than in the orthogonal one, which remains almost motionless. On the various heights of the symmetry plane the axial velocity oscillations are in phase and there is no difference in amplitude. The mean Nusselt numbers on the top and on the bottom of the cavity are in phase. All these facts indicate that the motion in question is back and forth rolling of the convective structure as shown in figure 10.

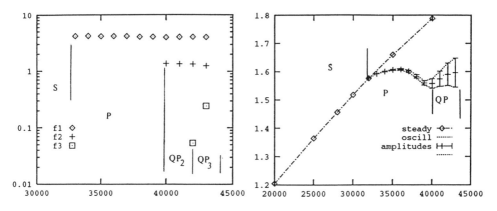

Figure 8. Frequency vs Rayleigh number;   Figure 9.Nusselt number vs Rayleigh number;
S- steady, P -periodic, QP2 quasiperiodic with two frequencies, QP3 quasiperiodic with three frequencies

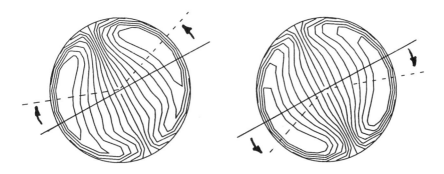

Figure 10. Axial velocity isocontours at two characteristic moments; symmetry plane rolling indicated

Conclusions

The buoyancy driven flow in cylinder of aspect ratio 1.5 filled with low Prandtl number fluid ($Pr=\cdot 10^{-2}$) has been simulated by numerical means. The attention has been focused on unsteady regime. The transition follows the Ruelle-Takens-Newhouse scenario i.e firstly the periodic flow (P) appears, then the quasiperiodic one with two frequencies (QP2) followed by QP3 which bifurcates to nonperiodic flow. In the periodic range the oscillations correspond to the rolling motion around the axis. There is a need for comparison of numerical results with experimental data.

Nomenclature

$A = H/D$	aspect ratio	$R = g\beta H^3(T_h-T_c)/\nu\alpha$		Rayleigh number
$Pr = \nu/\alpha$	Prandtl number	$Nu = \int_a \left(V_z T - \frac{\partial T}{\partial z}\right) ds$		average Nusselt number
$t = t^*/(H^2/\alpha)$	dimensionless time	$f = f^*/(\alpha/H^2)$		dimensionless frequency

cp	specific heat at constant pressure	p	pressure
g	gravitational acceleration	T	temperature
k	thermal conductivity	V	velocity

Subscripts: Greek:

c	cold		α	thermal diffusivity
cr	critical value		β	thermal expansion coefficient
h	hot		μ	dynamic viscosity
osc	oscillatory		ρ	density

References

[1] J.M. Olson, F. Rosenberger, 'Convective Instabilities in a Closed Vertical Cylinder Heated from Below. Part 1. Monocomponent Gases', J. Fluid Mech., 92, 609-629 (1979).
[2] G. Müller, G. Neumann, W. Weber, 'Natural Convection in Vertical Bridgman Configurations', J. Crystal Growth, 70, 78-93 (1984).
[3] G. Müller and G. Neumann, 'Investigation of Convective Flows in Model Systems of Directional Solidification Configurations', ESA SP-191, pp 285-293., Madrid (1983).
[4] S Fauve and A. Libchaber, 'Rayleigh-Benard Experiment in a Low Prandtl Number Fluid, Mercury', Chaos and Order in Nature, Springer-Verlag, Berlin (1981).
[5] Y. Kamotani, F.B. Weng, S. Ostrach, J. Platt, 'Oscillatory Natural Convection of a Liquid Metal in Circular Cylinders', J. of Heat Transfer, 116, 627-632 (1994).
[6] E. Crespo Del Arco, P. Bontoux, R.L. Sani, G. Hardin, G.P. Extremet, A. Chikhaoui, 'Finite Diference Solution for Three-Dimensional Steady and Oscillatory Convection in Vertical Cylinders- Effects of Aspect Ratio', Natural Convection in Enclosure ASME HTD, 99, 647-676 (1988).
[7] G. Neumann,' Three-Dimensional Numerical Simulation of Buoyancy-Driven Convection in Vertical Cylinders Heated From Below', J. Fluid Mech. 214, 559-578 (1990).
[8] A. Ivančic, A. Oliva, 'Heat Transfer Simulation in Vertical Cylindrical Enclosure for Supercritical Rayleigh Number and Arbitrary Sidewall Conductivity', to appear.
[9] J.P. Van Doormaal, G.D Raithby, 'Enhancements of the Simple Method for Predicting Incompressible fluid flow' Numerical Heat Transfer, 7, 147-163 (1984).
[10] B.P. Leonard, 'A Stable and Accurate Convective Modelling Procedure Based on Quadratic Upstream Interpolation', Comp. Meth. in Appl. Mech. and Eng. ,19, 59-95 (1979).
[11] P.H.Gaskell and A.K.C.Lau, 'Curvature-Compensated Convective Transport: Smar, a New Boundedness-Preserving Transport Algorithm', Int. J. Numerical Methods Fluids, 8, 617-641 (1988).
[12] G.Z. Gershuni and E.M. Zhukovitskii, 'Convective Stability of Incompressible Fluids', Keterpress Enterprisies, Jerusalem (1976).
[13] G.S. Charlson and R.L. Sani, On Thermoconvective Insability in a Bounded Cylindrical Fluid Layer', Int. J. Heat Mass Transfer, 14, 2157-2160 (1971).
[14] J.C Buell and I. Catton, 'The Effect of Wall Conduction on the Stability of a Fluid in a Right Circular Cylinder Heated From Below', J. of Heat Transfer, 105, 255-260 (1983).
[15] A. Ivančic, A. Oliva, C.D. Pérez Segarra, H. Schweiger, 'Three-dimensional Numerical Study of Mixed Convection in Cylindrical Cavities' Numerical Methods in Thermal Problems, Vol. 8 part 1, 561-572, Pineridge Press, Swansea (1993).
[16] K.T. Yang, 'Transitions and Bifurcations in Laminar Buoyant Flows in Confined Enclosure', J. of Heat Transfer, 110, 1191-1204 (1988).

SLIDING MESH SIMULATION OF TRANSITIONAL, NON-NEWTONIAN FLOW IN A BAFFLED, STIRRED TANK

Z. Jaworski*, M.L. Wyszynski,
School of Manufacturing and Mechanical Engineering, University of Birmingham, Birmingham, UK

R.S. Badham, K.N. Dyster, I.P.T. Moore, N.G. Ozcan-Taskin, A.W. Nienow,
School of Chemical Engineering, University of Birmingham, Birmingham B15 2TT, UK

J. McKemmie
APV Fluid Handling Industrial Pump and Mixers, East Kilbride Glasgow G73 5BJ, UK

SUMMARY

A fully predictive model of single phase flow in a stirred bioreactor configuration is presented. To do so, a sliding mesh numerical technique has been applied to simulate the interaction between the rotating impeller and the stationary tank on shear-thinning fluid mimicking the rheological properties of a fermentation broth. The basic physical models and their implementation are presented here together with details of the numerical solution procedure. In addition, the results of the modelling are compared with LDA experimental data obtained in equipment of the same geometry using a transparent carboxymethylcellulose solution of the same rheological properties as the working fluid used in the numerical simulation. In general, the model and the experimental data are in good agreement.

INTRODUCTION

Stirred tanks are widely used in the chemical. food and bioprocess industries. In such tanks, a significant increase in the rate of transport processes, both in single- and multiphase systems is usually achieved and this increase is associated with the very complex flow fields produced in the tanks. The complexity of flow patterns in such systems results from the complicated geometry of the tank, usually with vertical baffles and with a rotating impeller (see figure 1 for example), and it is often further magnified as the fluid being processed may exhibit non-Newtonian behaviour. The resulting flow is 3 dimensional, swirling, periodic (due to the passage of the blades on the impeller) and with viscous, highly shear dependent fluids, the flow regimes range from highly turbulent in the vicinity of the impeller to laminar at the vessel wall. Simulation of such a class of unsteady flows has not yet been satisfactorily explored, nor have the predictions been tested. At present, therefore, numerical modelling of such stirred tanks requires extensive validation, even for single phase flows.

Modelling of flows in such systems involves a rotating fluid boundary at the impeller. In order to reproduce that moving boundary, a very effective technique is the use of a rotating frame of reference for the fluid associated with the impeller sliding over a stationary one for the fluid in the remainder of the tank. The latter thus satisfies the no slip condition at the cylindrical wall and its associated baffles. Such a sliding mesh technique was used in this study.

* on leave from the Technical University of Szczecin, PL

In a previous study, the present authors [1] used that approach and concluded that the CFD predictions of velocities and torque were accurate enough for laminar, Newtonian flow in stirred tanks. For turbulent or non-Newtonian flows, however, the predictions were too coarse. No other validated modelling of transitional, shear dependent flow in stirred tanks has been published in the literature.

EXPERIMENTAL SETUP

Typical tank and impeller configurations were used in this study. The stirred tank system consisted of an open top cylindrical vessel with 4 vertical baffles and a radial flow impeller, called a Rushton turbine, mounted on a vertical shaft. The tank was filled with a fluid up to the level, H, equal to the internal tank diameter, $T = 0.15$ m. The impeller, which rotated at 600 rpm, was located centrally in the tank and had 6 vertical blades fixed to a horizontal disc. The impeller diameter, $D = T/3$, is defined here as the diameter of a circle enscribed by the rotating blade tip. The elevation of the Rushton turbine disc above the tank bottom was equal to $H/2$. In order to mimic the rheology of a typical non-Newtonian fermentation broth, a medium concentration sodium carboxymethylcellulose (CMC) solution was chosen as the stirred fluid. It exhibited shear thinning behaviour with the flow index, n, of 0.58 and consistency index, K, of 0.4 [Pa s^n].

The LDA data for the CMC solution were acquired using the ensemble averaging mode. All 3 components of the mean and rms velocities were measured in one plane only, it was the 45° mid-plane between two baffles. These data were used to validate the numerical modelling. Detailed description of the experiments can be found in [2]. The radial and tangential mean velocities were highest in the discharge flow from the impeller, the axial component had maxima in the inflow stream to the impeller region.

NUMERICAL MODELLING

Recent progress in the sliding mesh techniques rendered it possible to simulate the flow in stirred, baffled tanks without having to make serious simplifying assumptions in the impeller region. This way the zero velocity condition at a surface could be met at both the impeller and the wall. The commercially available FLUENT™ structured code with the sliding mesh option was chosen for this study. The code uses a control (finite) volume discretization method of the differential transport equations. In that version two mesh regions; one rotating with the impeller and one attached to the stationary tank, slide relative one to another along a cylindrical surface and are not distorted. The flow variables and fluxes across the slip surface are obtained by conservative interpolation. Control cells on either side of the slip surface can be split and joined into fictitious control volumes to match their counterpart on the other side. Next, a solver is used sequentially to the rotating and stationary grids until the variables of the fictitious cells converge. Then the procedure is repeated for the next time step, and if periodicity is present, the time-marching method should be chosen.

The computations were carried out for a stirred tank that had the same diameter and proportions as in the LDA measurements. Non-compressible flow of the shear-thinning CMC solution was assumed. Cylindrical formulation of the set of the transport equation was chosen, as it gives reduced numerical errors for the rotating flow considered. In order to save time and memory, only one half of the symmetric tank was considered. The lateral domain boundaries in the angular direction were defined as rotationally cyclic. Free surface of the stirred liquid and tank axis were treated as symmetry boundaries. No-slip conditions were imposed at all solid boundaries. Location of the slip planes of the cylindrical, rotating mesh was chosen roughly mid-way between the impeller blades and the nearest solid, stationary boundaries on the free surface. The angular velocity of the rotating mesh was equal to that of the impeller.

The standard code version did not allow the use of a turbulence model and a non-Newtonian model at the same time. However, the code version with user defined subroutines enabled us to model the non-Newtonian viscosity along with turbulence. The power law viscosity (equation (2)) for each cell was calculated in the user defined subroutine and returned to the main program. Because of the transitional flow, a turbulence model had to be used. The general indication for high swirl flow is to use the Reynolds stress model [3], but in the code version used that option was not available. The RNG k-ε model [4] was selected. It usually yields better results than the other two-equation models, when used for modelling of time dependent, high swirl, transitional flows. The effective viscosity, v_{eff}, was computed from equation (1),

$$v_{eff} = \left(v_m^{0.5} + v_t^{0.5}\right)^2 , \qquad (1)$$

in which the apparent molecular viscosity, v_m, and the turbulent viscosity, v_t, were calculated from equations (2) and (3), respectively.

$$v_m = \frac{K\dot{\gamma}^{(1-n)}}{\rho} \qquad (2)$$

$$v_t = \frac{C_\mu k^2}{\varepsilon} . \qquad (3)$$

In those equations, $\dot{\gamma}$ and ρ denote shear (strain) rate and liquid density, respectively, and $C_\mu = 0.0845$. The shear rate, $\dot{\gamma}$, is defined in equation (4).

$$\dot{\gamma} = \left(\frac{\partial u_i}{\partial x_j} + \frac{\partial u_j}{\partial x_i}\right) . \qquad (4)$$

Typical range of Reynolds number, Re, for transitional flow in stirred tanks extends from 10 to 10000 [5]. The value of Reynolds number in this study was 430. The definition of Reynolds number used in this study follows that given in [5] for non-Newtonian flows in stirred tanks.

RESULTS AND DISCUSSION

All computations were carried out in cylindrical coordinates using the IBM RS6000 model 375 workstation with 64 MB RAM and performance of 118 SPECfp92. The collocative grid had 48x19x25 cells in the angular, radial and axial directions, respectively. The simulation was started from a stationary liquid and impeller. The number of cells on the interface of the rotating and stationary domains passed in each time step was varied from 4 at the start to 0.5 at the end of iterations. About 1500 time marching steps were necessary to reach a practically converged periodic flow in the tank. Convergence was regarded as achieved after the total normalised residuals fell below 10^{-3} and also the history graphs of all velocity components at 2 selected points showed no visible change over last 100 time steps.

SIMPLE or SIMPLEC algorithms were used for coupling between mean velocity components and pressure. SIMPLEC gave faster convergence as it could be used with higher underrelaxation parameters for pressure. However, very high axial velocities were computed in a narrow area close to the vessel axis. One of the possible reasons for it is that for such intensive fluid rotation a high degree of coupling between the momentum equations exists. This requires low values of the underrelaxation parameters on the velocities and pressure. Therefore, it was concluded that the more conservative SIMPLE algorithm should be used.

The power law discretization scheme along with the line-by-line Gauss-Seidel solver were applied. For the sliding mesh computations, higher order interpolation schemes and multigrid solvers were currently not available. The average number of internal repetitions (sweeps) was 20 for the continuity equation, and 2 for k, ε and also for u, v, w components of momentum. Good results were achieved in the final iterations with 40 sweeps and 3 sweeps, respectively. The number of the main loop iterations per one time step was 10 at the beginning and 20 for the final time steps. Underrelaxation factors were chosen to be rather low and amounted to 0.2; 0.4; 0.5; 0.7 for velocities, pressure, turbulence and viscosity, respectively.

An example of the CFD solution obtained is presented in figure 1 which shows the distribution of the radial and axial components of the mean velocity and also contours of the effective viscosity. The typical radial discharge flow from the impeller was obtained. The mean velocity components in the discharge flow rate were lower than those for water reported in [1]. Also the extent of the radial-axial circulation loop diminished. Highest values of turbulence quantities were computed in the discharge stream (figure 2), where the apparent non-Newtonian viscosity had its lowest value because of the high velocity gradients existing there. However, the apparent viscosity was predominant throughout the tank and was practically equal to the effective viscosity. The average apparent viscosity in the tank was of the order of 0.1 [Pa s]. Maxima of the effective viscosity were computed at the lower and upper corners of the tank, where shear rates were very small.

As can be seen from figure 3, good prediction of the general features of the flow was obtained. In order to quantitatively validate the numerical model against the ensemble averaged LDA data it was necessary also to angle average the model velocities for several impeller positions. Generally, the agreement between the modelled and experimental mean velocities was satisfactory, with smallest deviations in the tangential component, and slightly higher for the axial and radial components. Large discrepancies are discussed below. When analysing figures 3a and 3b, it should be noted that the velocity scale for the profile of the radial and tangential components at the impeller level was decreased by 10 or 5 times, respectively. The simulation results were very close to those obtained without the use of turbulence models. Because of the very small size of the equipment, there was a low contribution of turbulence.

There was met, however, a problem which should be addressed before this kind of modelling can be advised for design purposes. It was associated with the appearance of a peculiar area at the tank axis, between the lower slip surface and the impeller. The axial velocity component modelled in that area was much higher than that measured (cf. figure 3c). The kinetic energy of turbulence showed also unexpected higher values of the kinetic energy of turbulence (cf. figure 2). No justification for the turbulence generation can be found in that region. It may be due to a combined effect of the high swirl, non-linearity introduced by the shear thinning behaviour of the stirred liquid and the interpolation algorithm for the slip surfaces which was positioned just below this anomalous region. High axial velocity in the wall jet near the impingement point was obtained, as expected (figure 3c). The significant discrepancy with the experiments for the axial component in that region was caused probably by erroneous LDA data at the tank wall.

It is anticipated that the single-phase flow modelling presented will be followed by more complicated flow simulations of aerated stirred tanks.

CONCLUSIONS

Results of the fully predictive simulations of 3 dimensional, transitional, non-Newtonian flow in a stirred tank were reported for the first time. The results showed a good agreement with experiments and were best for the tangential component of mean velocity. In order to use that modelling in process equipment design, the modelling of both mean velocities and turbulence quantities needs to be improved especially, as noted above, because of the discrepancies in a small volume at the tank axis below the impeller.

REFERENCES

[1] JAWORSKI Z., WYSZYNSKI M.L., BADHAM R.S., DYSTER K.N., MOORE I.P.T., NIENOW A.W., OZCAN-TASKIN N.G., MCKEMMIE J.: "Sliding mesh CFD flow simulation for stirred tanks", Mixing XV, North American Mixing Conference, June 1995, Banff, Canada.

[2] COSSOR G.: "Mixing of viscous and non-Newtonian fluids in stirred tanks: A study by laser-Doppler velocimetry", PhD Thesis, University of Birmingham, UK, 1995.

[3] MARKATOS N.C.: "The mathematical modelling of turbulent flows", Appl. Math. Modelling, 10 (June 1986) pp. 190-220.

[4] YAKHOT V., SMITH L.M.: "The renormalization group, the ε-expansion and derivation of turbulence models", J. Sci. Comp. 7 (No.1, 1992) pp. 35-61.

[5] NAGATA S.: "Mixing, Principles and applications", Halsted Press, New York 1975.

ACKNOWLEDGMENTS

Gratefully acknowledgments are due to BBSRC/DTI Link in Biochemical Engineering for funding the research project, to Dr. F. Boysan of Fluent Europe for providing us with the User Defined Subroutines and to Dr. M. Makhlouf for her valuable help in their implementation.

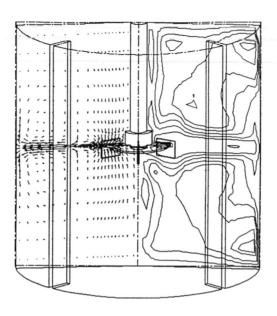

Figure 1. A half section through a Rushton turbine agitated bioreactor showing vectors of mean velocity (left) and contours of effective viscosity (right)

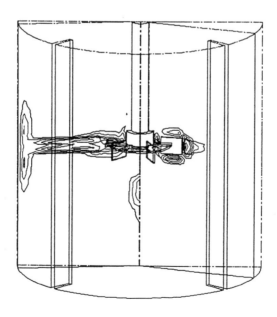

Figure 2. Contours of turbulence quantities: kinetic energy of turbulence (left), energy dissipation rate (right)

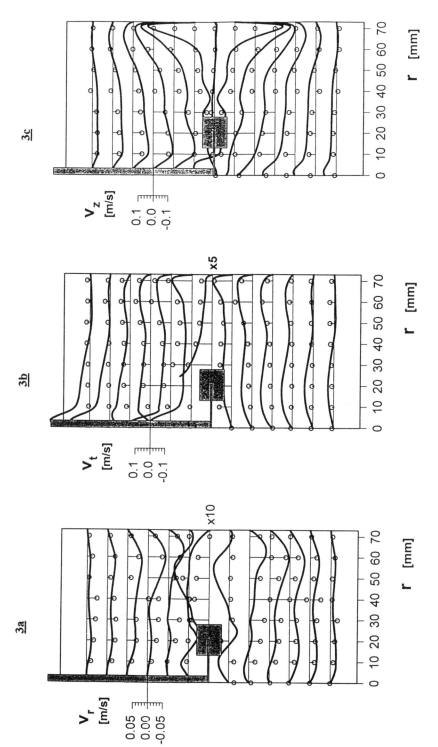

Figure 3. Comparison of mean velocity modelling results (———) with LDA measurements [2] (○);
3a - radial component, 3b - tangential component, 3c - axial component.

Computation of Three-Dimensional Laminar and Turbulent Shock Wave/Boundary Layer Interactions

Simon KEGHIAN [†], Michel GAZAIX [‡]

[†] AEROSPATIALE-MISSILES
91370 Verrières-le-Buisson (France)

[‡] Office National d'Etudes et de Recherches Aérospatiales,
BP 72 92322 Châtillon Cedex (France)

Abstract

In this paper we study laminar and turbulent shock wave / boudary layer interaction using FLU3M code. This is the first use of Osher / Van Leer Hybrid Upwind Scheme (HUS) [2] for 3D Navier-Stokes computations. The laminar computation is in good agreement with the experiment, and the turbulent case allows us to validate the 3D implementation of Baldwin-Lomax model [4] .

1 Introduction

Scramjet inlet flows involve complex phenomena such as large separated regions and shock wave/boundary layer interactions. Navier-Stokes computations can greatly improve our understanding of such flows, provided that detailed validation over basic three-dimensional geometries are made prior to more complex configurations. In this work, two 3D test cases have been selected, for which good experimental results are available: a (laminar) swept compression ramp and a (turbulent) double fin on a flat plate. These test cases have been selected by the AGARD Fluid Dynamics Panel (Working Group 18) to provide a critical assessment of the state-of-the-art in numerical simulation of shock wave/boundary layer interaction. We describe briefly the numerical method, and the computed results obtained with FLU3M code are discussed and compared with experimental data.

2 Numerical Method and Turbulence Modeling

The flow is described by the compressible time-averaged Navier-Stokes equations with a turbulence model. The flow governing equations can be written in the following form :

$$\frac{\partial \mathbf{f}}{\partial t} + \text{div}(F - F_v) = 0. \tag{2.1}$$

in which $\mathbf{f} = (\rho, \rho \mathbf{V}, \rho E)$ and where $\rho, \mathbf{V}, E$ denote the density, the mean velocity, and the total energy per unit of mass. The fluxes F and F_v represent the convective and the dissipative terms. The eddy viscosity μ_t is evaluated with Baldwin-Lomax algebraic turbulence model [4] . The 3D implementation of the model consists in computing μ_t in relation to the closest wall.

The numerical method uses a MUSCL upwind finite volume scheme on a cell vertex structured mesh. We choose the HUS Osher/Van Leer scheme [2] to evaluate the convective fluxes and a 3D-ADI algorithm is implemented in FLU3M to increase convergence speed.

3 Laminar Interaction

3.1 Experimental configuration

The laminar shock wave/boundary layer interaction test body is a swept compression ramp on a blunt flat plate ($R = 2.5mm$) with a sweep angle $\Psi = 60°$, and a ramp angle $\beta = 30°$. Detailed experiments took place at ONERA Chalais-Meudon center in R3CH wind-tunnel [1]. The free stream conditions are summarized in Table 1.

Table 1 Laminar interaction : free stream conditions

Mach number	10.
Stagnation pressure	120. 10^5Pa
Stagnation temperature	1050.K
Wall temperature	290.K
Reynolds number	8.68 10^6 m^{-1}

Experimental data consisted of wall pressure, heat fluxes, and surface oil flow visualization.

3.2 Numerical results

Three cell-vertex, NS-laminar calculations are carried out with FLU3M. The computational domain is divided in two blocks : one containing the blunt leading edge and one containing the obstacle. The separation between the two blocks is located so that the first block, which is computed with a 2D calculation, is ahead of the interaction. Three different meshes are used to compute the second block (Table 2 and Figure 1).

Table 2 Laminar interaction : mesh points

Mesh 1	26x21x25
Mesh 2	51x41x25
Mesh 3	101x81x25

The results on the first mesh (resp. the second mesh) is used to initialize the flow on the second one (resp. the third one). For mesh 3, the computation requires 60 Mwords and 20000 seconds on a CRAY-YMP at CFL number 5.

Figure 1b shows the skin friction lines computed on mesh 3. We note two separation lines (S1) and (S2) and three attachment lines (A1) (A2) and (A3). For topological reasons, there must be a separation line between (A2) and (A3) which cannot be seen on the visualization.

For $Z = 0.05$ line, (Figure 2b) we note the decrease of the heat flux due to (S1) and (S2) at $X = 0.13$ and $X = 0.175$. The decrease at $X = 0.185$ is probably due to the interpolation at the corner of the configuration. The maximum corresponding to (A2) is also clearly seen, and the peak heat flux discrepancy goes from 50% to 20% with mesh refinement.

Figure 3a shows the evolution of the wall pressure for $Z = 0.05$. As for the heat flux, the first separation zone is clearly seen ((S1) - (A1)). The maximum pressure discrepancy with the experimental values is 6% for mesh 3.

The influence of the mesh refinement is quite significant, especially for the prediction of the maximum value of the wall heat flux. The good agreement for the viscous coefficient before and after the interaction suggests that no turbulent transition occured during the experiment, and confirms that it is a good laminar test case for code validation. This work is the first 3D Navier-Stokes computation using HUS method and demonstrates the ability of this scheme to describe accurately such flows.

4 Turbulent interaction

4.1 Experimental configuration

The turbulent interaction studied is that of Kussoy and Horstman [3]. It consists of a 15° double fin on a flat plate. The free stream conditions are summarized in Table 3 :

Table 3 Turbulent interaction : free stream conditions

Mach number	8.28
Static pressure	$430 N/m^2$
Reynolds number	$5.3\ 10^6$
Wall Temperature	300K

Experimental data were obtained for surface pressure, surface heat flux, pitot pressure flowfields, and yaw angles for three transversal sections.

4.2 Numerical results

Computations are done using HUS fluxes. The search of the maximum of the Baldwin Lomax function is limited to 1/4 of the width of the channel to avoid the fin shock. The configuration is divided into four blocks. The first block, corresponding to a 162 cm flat plate, is computed with a 162x81 mesh and is used as entry condition for the 3D computation. A laminar computation is done to initialize the flow field. The mesh point repartition is summarized in Table 4:

Table 4 Turbulent interaction : Mesh point repartition

Block 2	33x41x17
Block 3	33x41x80
Block 4	33x41x51

Figure 4 shows the computed skin friction lines. The separation line (S1) due to the strong interaction between the fin-shock and the initial flat plate boundary layer is clearly seen and the separation node is situated at $X = 0.097$. The flow reattachment line is close to the fin.

The pressure at the center line is well predicted ahead of the interaction (Figure 4b). The maximum value is overpredicted and a second maximum is seen. This second maximum does not appear in the experiment and is due to a secondary separation. This phenomenon is probably due to the underestimation of the turbulence level. To check the 3D adaptation influence of the model we have performed calculations using an inverse blending method [5]. This adaptation consists in evaluating the eddy viscosity for each wall independently. The two values obtained at each point are then blended with a function following the formula :

$$\mu_t = \frac{\frac{\mu_{t1}}{y_1^+} + \frac{\mu_{t2}}{y_2^+}}{\sqrt{(y_1^+)^{-2} + (y_2^+)^{-2}}} \ . \tag{4.2}$$

With this model, the spurious maximum of the pressure is still detected.

5 Conclusion

Two numerical computations, using the HUS upwind scheme and a fully implicit 3D ADI time integration, have been compared with the experimental data, and a limited study of the grid influence has been presented. To compute the turbulent test case, modifications of the Baldwin-Lomax turbulence model have been introduced, to take into account several walls, and to avoid spurious effects associated with the presence of strong shocks inside the flow. This study demonstrates the capability of our code to compute hypersonic laminar/turbulent 3D flows.

6 Acknowledgements

This work was supported by the "Direction de la Recherche et de la Technologie" (French Ministry of Defense).

References

[1] M. C. Coët and B. Chanetz, "Sur les Effets Thermiques dans les Interactions Onde de Choc/Couche Limite en Écoulement Hypersonique", *La Recherche Aérospatiale, n° 4, 251-268, 1994.*

[2] F. Coquel, M. S. Liou, *"Hybrid Upwind Splitting (HUS) by a Field by Field Decomposition", NASA TM 106843 , 1995.*

[3] M. I. Kussoy, K. C. Horstman, *"Intersecting Shock-Wave/ Turbulent Boundary-Layer Interactions at Mach 8.3",NASA TM 103909 , 1992.*

[4] B. S. Baldwin, H. Lomax, *" Thin Layer Approximation and Algebraic Model for Separated Turbulent Flows", AIAA Paper 78-257 , 1978.*

[5] J. J. Gorsky, D. K. Ota, S. R. Chakravarthy, *" Calculation of Three Dimensional Cavity Flow Fields ", AIAA Paper 87-117 , 1987.*

[6] D. Gaitonde, J. S. Shang, M. Visbal, *" Structure of a Double-Fin Turbulent Interaction at High Speed ", AIAA Journal Vol. 33, n° 2, 1995.*

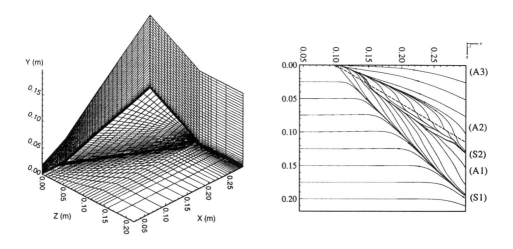

Figure 1 Swept ramp : (a) mesh 2, (b) skin friction lines on mesh 3.

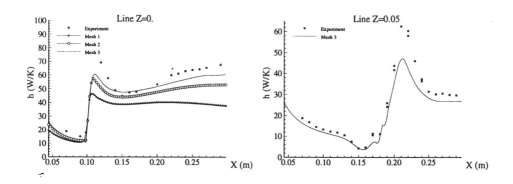

Figure 2 Swept ramp : heat flux coefficient (a) $Z = 0.$, (b) $Z = 0.05$.

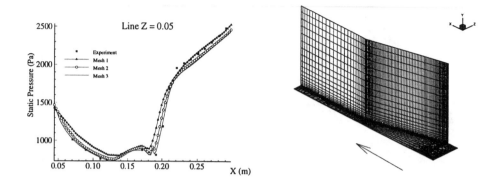

Figure 3 (a) Swept ramp : pressure $Z = 0.05$, (b) Corner interaction : mesh.

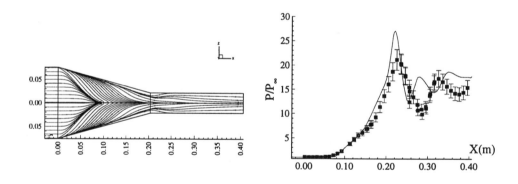

Figure 4 Corner interaction : (a) skin friction lines, (b) pressure at $Z = 0.0$.

COMPLEX THREE-DIMENSIONAL JET FLOWS: COMPUTATION AND EXPERIMENTAL VALIDATION

K. Knowles and M. Myszko

Aeromechanical Systems Group
Royal Military College of Science, Cranfield University
Shrivenham, Swindon, Wiltshire. SN6 8LA. U.K.

SUMMARY

Computational and experimental studies have been carried out on turbulent, round, normally impinging jets covering a range of nozzle heights and pressure ratios. Wall jet growth is seen to depend on both these parameters. Turbulence production in the wall jet is dominated by work done against shear stresses. Numerical modelling has been based on the k-ε turbulence model and modifications to this. These all underpredict the growth of the wall jet to an extent and the predicted velocity profiles are seen to peak too close to the wall. A truly 3-D test case is proposed based on a jet impinging on a moving surface.

NOTATION

b	Wall jet thickness (see Fig 1)
D_n	Nozzle exit diameter (see Fig 1)
H_n	Height of nozzle above ground (see Fig 1)
H_p	Height of measuring probe above ground
k	Turbulent kinetic energy
NPR	Nozzle pressure ratio = p_0/p_a
p_0	Nozzle stagnation pressure
p_a	Ambient static pressure
r	Radial co-ordinate from nozzle centre-line (see Fig 1)
$r_{1/2}$	Free jet thickness to half peak velocity
T_{int}	Turbulence intensity at nozzle exit
U	Free jet mean axial velocity
u	Fluctuating velocity in the U-direction
V	Mean velocity in the r-direction
y	Co-ordinate perpendicular to ground
$Y_{1/2}$	Wall jet thickness to half peak velocity (see Fig 1)
ε	Rate of dissipation of turbulent kinetic energy
μ	Laminar viscosity
π	Production of turbulent kinetic energy
π_1	Component of π due to normal stresses
π_2	Component of π due to shear stresses

INTRODUCTION

Impinging jets occur in a number of areas of engineering, including their use for heat transfer enhancement and, at a quite different scale, under jet-lift V/STOL aircraft operating near the ground. Many previous studies have been concerned with the growth of the radial wall jet formed by the impingement of a round jet [1-6], whilst others have concentrated on turbulence characteristics [7].

Our own studies started with impinging jets in cross-flow, where we were interested in parametric effects on the ground vortex formed by wall jet separation [8]. Our initial computational work was limited to axisymmetric calculations, representing some of the characteristics of the flow in the plane of symmetry of this highly three-dimensional problem [9]. This showed that many parametric trends could be correctly predicted but not the details of the flowfield. It also revealed areas where improved turbulence modelling could offer benefits. Our subsequent work has gone in two directions: parametric CFD studies of more complex 3-D impinging jet flowfields [10] and more detailed investigations of single, round, turbulent, impinging jets [11]. In parallel with this last work we have been carrying out detailed experimental studies of impinging jets, to provide CFD validation data. This experimental work has been described in references [12-14] but will be outlined here for completeness.

In parallel with the work described here, our collaborators in R.U. Bochum have been investigating a range of low-Reynolds number impinging jet configurations designed to enhance heat transfer [15,16]. Detailed computational and experimental studies of turbulent, round, impinging jet flowfields have been carried out recently by teams at U.M.I.S.T., Manchester [7,17,18] and K.U. Leuven [19]. Because of the computational challenges posed by the impinging jet flowfield (see Cooper et al. [7]) it has formed the basis of two recent ERCOFTAC/IAHR turbulence modelling workshops [20,21]. The present paper will propose an extension of these earlier test cases to a truly three-dimensional condition - a round jet impinging on a moving surface.

EXPERIMENTAL AND NUMERICAL METHODS

Our experimental set-up is described in ref. [14]. Dried, compressed air is supplied to the laboratory, at pressures of up to 4bar, via a $34m^3$ receiver and then ducted to a small, rectangular plenum chamber. A simple, conical nozzle (D_n=12.7mm) is mounted below the plenum chamber, projecting $10D_n$ below its lower surface. Below the nozzle is a flat board onto which the free jet impinges. The height of the nozzle and plenum chamber assembly above this board can be varied.

Initial measurements were made in the wall jet using Pitot and static probe rakes at various radial positions. Subsequent measurements were made in the free jet using a single-wire hot wire anemometer (HWA) and these revealed a skewed velocity distribution across the nozzle exit. This was corrected by fitting baffles inside the plenum chamber and all further HWA measurements were taken with this configuration (Figs 2 & 3). Axisymmetry checks without these baffles revealed that the Pitot-static probe measurements in the wall jet were not affected by the skewed nozzle exit conditions. Figs 2 & 3 show mean and turbulent velocity profiles in the free jet close to nozzle exit and at various downstream locations. As shown in Fig. 3, the nozzle exit turbulence intensity for these tests is about 5%.

The latest work using this rig has concentrated on cross-wire measurements in the wall jet at a fixed, low nozzle pressure ratio (NPR) of 1.05 and a range of nozzle heights and probe radial positions. The early Pitot-static probe measurements were aimed at determining the effects on the wall jet growth of various parameters, in particular NPR and nozzle height. The next phase of the work used cross-wires to investigate turbulence characteristics of the wall jet. The latest work is again using cross-wires but with longer record lengths being stored so that the turbulence data are even more reliable and time-accurate results can be obtained. The original parametric trends are also being compared with these new data. The results discussed here are from the first two stages of this work.

We have carried out numerical modelling using the commercially-available PHOENICS finite-volume code with turbulence modelling based on the k-ε closure (see [11] and [13]). This is known to overpredict the spreading rate of round jets and underpredict the growth rate of radial wall jets. To address these problems various modifications to the model have been proposed in the past. That due to Rodi [22] was designed to reduce the predicted spreading rate of a round jet, whilst that due to Malin [23] was aimed at increasing the spreading rate of the radial wall jet. Our implementation of these corrections was described in ref. [13].

We have used these models to predict the spreading rate of a wall jet due to a low pressure ratio (NPR = 1.079) impinging round jet, with the nozzle at $7.5D_n$ above the ground. This is the case for which Poreh et al. [3] present experimental data. Calculations were performed using: the basic k-ε model with standard values for the coefficients (see [9] and [11]); the Rodi correction for the free jet with the standard k-ε model for the wall jet; Rodi for the free jet and the Malin correction for the wall jet; and Malin for the wall jet with the standard k-ε model for the free jet. The standard k-ε model was always used for the region above the nozzle exit plane (in the ambient fluid) and inside a box $1.5D_n$ square around the impingement point. Further calculations have been made with the standard k-ε model for a range of nozzle heights and nozzle pressure ratios.

The boundary conditions used were the same for all the above calculations. Nozzle exit conditions consisted of a specified velocity and a fixed pressure. For the results presented here only uniform exit velocities are considered (consistent with our own experiments), the effects of non-uniform nozzle velocities and varying turbulence intensities are presented in ref. [11]. Turbulence production and dissipation rates at nozzle exit were calculated from the specified turbulence intensity using:
$k = (W_{jet} \times T_{int})^2$ and $\varepsilon = 0.009 \times (k^2)/(50 \times \mu)$
The nozzle itself projected down below the upper computational boundary but no internal flow was calculated. The outside of the nozzle was treated with a law-of-the-wall. The free jet had symmetry assumed on its centre-line; the wall jet boundary used the logarithmic law-of-the-wall. The permeable upper and outflow boundaries had fixed ambient pressure on them and were free to pass mass flow in or out but with a damping applied to this flow to prevent computational instabilities. On these boundaries k and ε were fixed to the values calculated at the adjoining, internal cells.

RESULTS AND DISCUSSION

As discussed by Knowles and Myszko [13], the mean flow measurements have revealed that wall jet thickness varies approximately linearly with radial position but at a given radius the thickness is dependent on nozzle height and (inversely) on nozzle pressure ratio. These factors have been included in a new correlation for wall jet growth. These measurements also show that momentum flux (relative to nozzle exit momentum flux) increases as nozzle height is increased. The single-wire measurements show that increasing nozzle height leads to a fall in wall jet peak turbulence intensity (at a fixed radial position) relative to the free jet. In fact the turbulence intensity at entry to the wall jet is fairly constant and it is the turbulence level before the impingement which increases with increasing nozzle height. Cross-wire measurements in the wall jet have revealed that normal stress-driven turbulence is little affected by nozzle height, the main effect seems to be on shear-driven turbulence production which decreases with increasing nozzle height (Fig 4). Our cross-wire measurements also show that the peak levels of turbulence production in the wall jet are at about $r = 2D_n$ [14].

Much of our computational work has been aimed at reproducing the parametric trends discussed above. First, though, we investigated the performance of the k-ε model for these flows. As expected, the basic model overpredicts the mixing rate in the free jet and underpredicts the spreading of the wall jet. The Rodi correction significantly improved the free jet calculation [11] but gives too thin an initial wall jet. The Malin correction slightly increased the spreading of the wall jet but the combination of these two still significantly underpredicts the wall jet thickness. This can be seen in the mean velocity profiles of Fig. 5, where all the turbulence models produce a wall jet with a peak velocity too close to the wall. This suggests that the short-coming is in the near-wall treatment.

We have carried out our parametric investigations using the standard k-ε model. The effect of nozzle height, for instance, is shown in Fig 6. It can be seen that the model correctly captures the change in initial wall jet thickness but at a given radius away from impingement the predicted change in wall jet thickness with changing nozzle height is far too dramatic. The effect of NPR in both the experiments [13] and the calculations [12] is less dramatic but the model correctly captures the trend, at least for subsonic conditions.

We are currently investigating the Reynolds Stress Transport turbulence model for these flows but previous experience does not suggest that this will greatly improve the calculations on its own [20, 21]. An alternative near wall treatment may prove more productive and in connection with this we are proposing a new CFD test case. In many industrial applications impinging jets are used on moving surfaces. Our own earlier experimental work on impinging jets in cross-flow [8] has implied a dramatic effect of a moving surface on wall jet development. No detailed experimental data are apparently currently available for such a case (without cross-flow). At the same time computation of this is a 3-D task, since the jet is axisymmetric but the ground movement is planar.

REFERENCES

[1] Glauert, M., " The Wall Jet", *J. Fluid Mechanics*, 1, 6, 625-643, 1956.
[2] Bakke, B., "An Experimental Investigation of the Wall Jet", *J. Fluid Mechanics*, 2, 467-472, 1955.

[3] Poreh, M., Tseui, Y.G. and Cermak, J.E., "Investigation of a Turbulent Radial Wall Jet", *Trans. ASME J. of App. Mech.*, **89**, 457-463, 1967.

[4] Hrycak, P., Lee, D., Gauntner, J. and Livingood, J., "Experimental Flow Characteristics of a Single Turbulent Jet Impinging on a Flat Plate", *NASA TN D-5690*, 1970.

[5] Donaldson, C. du P. and Snedeker, R.S., "A Study of Free Jet Impingement. Part 1. Mean Properties of Free and Impinging Jets", *J. Fluid Mechanics*, **45**, 2, 281-319, 1971.

[6] Miller, P. and Wilson, M., "Wall Jets Created by Single and Twin High Pressure Jet Impingement", *Aeronautical J. of the Royal Aero. Soc.*, **97**, 963, 87-100, March 1993.

[7] Cooper, D., Jackson, D.C., Launder, B.E. and Liao, G.X., "Impinging Jet Studies for Turbulence Model Assessment - I. Flow-field Experiments", *Int. J. Heat and Mass Transfer*, **36**, 10, 2675-2684, 1993.

[8] Knowles, K. and Bray, D., "Ground Vortex Formed by Impinging Jets in Cross-flow", *AIAA J. of Aircraft*, **30**, 6, 872-878, 1993.

[9] Knowles, K. and Bray, D. "High Mach number impinging jets in cross-flow - comparison of computation with experiment", in: *"Numerical Methods in Laminar and Turbulent Flow"* vol 6, ed. Taylor, C et al., pp 1389-1398. Pub. Pineridge Press, 1989.

[10] Bray, D., Matson, D. and Knowles, K. "3-D numerical modelling of impinging jets in cross-flows", *AIAA 12th Applied Aerodynamics Conference*; Colorado Springs, Colorado, 20-23 June 1994. Paper no. AIAA-94-1807.

[11] Knowles, K., "Computational Studies of Impinging Jets Using k-ε Turbulence Models", *Int. J. Numerical Methods in Fluids* (in print).

[12] Myszko, M. and Knowles, K., "Development of a Wall Jet From an Impinging, Round, Turbulent, Compressible Jet", *AIAA 25th Fluid Dynamics Conference*, Colorado Springs, CO, 20-23 June 1994. Paper no. AIAA 94-2327.

[13] Knowles, K. and Myszko, M., "Studies of Impinging Jet Flows and Radial Wall Jets", *Proc. Int. Symp. on Turbulence, Heat and Mass Transfer*, Lisbon, Portugal, 9-12 August 1994, Vol. 1, pp 2.1.1-2.1.6. (Extended version in *"Monograph on Turbulence, Heat and Mass Transfer"*, to be published by Begell House Inc., New York.)

[14] Knowles, K. and Myszko, M., "Radial Wall Jets Formed by High Speed Impinging Jets", *Proc. ASME/JSME Forum on High Speed Jet Flows*, Hilton Head, S. Carolina, 13-18 August 1995. ASME FED Vol. 214, eds. Raman, Kaji and Freitas.

[15] Potthast, F., Laschefski, H., Mitra, N.K. and Biswas, G., "Numerical Investigation of Flow Structure and Mixed Convection Heat Transfer of Impinging Radial and Axial Jets", *Numerical Heat Transfer, Part A*, **26**, 123-140, 1994.

[16] Laschefski, H., Braess, D., Haneke, H. and Mitra, N.K., "Numerical Investigation of Radial Jet Reattachment Flows", *Int. J. Numerical Methods in Fluids*, **18**, 629-646, 1994.

[17] Craft, T.J., Graham, L.W.G. and Launder, B.E., "Impinging Jet Studies for Turbulence Model Assessment - II. A Comparison of the Performance of Four Turbulence Models", *Int. J. Heat and Mass Transfer*, **36**, 2685-2697, 1993.

[18] Leschziner, M.A. and Ince, N.Z., "Computational Modelling of Three-Dimensional Impinging Jets With and Without Cross-Flow Using Second-Moment Closure", *Computers and Fluids* (in print).

[19] DeClerq, J. and Dutré, W., "Simulation of Impinging Jet With a Reynolds-Stress Turbulence Model and Turbulent Heat Fluxes"", *Proc. Int. Symp. on Turbulence, Heat and Mass Transfer*, Lisbon, Portugal, 9-12 August 1994, Vol. 2, pp P.I.7.1-P.I.7.4.

[20] Brison, J.F. and Brun, G. (eds.), "Round Normally Impinging Turbulent Jets", *Proc. 15th Meeting IAHR Working Group on Refined Flow Modelling*, Lab. Méc.Fluides, ECL, Lyon, France, October 1991.

[21] Leschziner, M.A. and Launder, B.E. (eds.), "Round Normally Impinging Turbulent Jet and Turbulent Flow Through Tube Bank Sub-channel", *Proc. 2nd ERCOFTAC/IAHR Workshop on Refined Flow Modelling*, UMIST, Manchester, U.K., June 1993.

[22] Rodi, W., "Turbulence Models and their Application in Hydraulics - A State of the Art Review", IAHR, Delft, 1980.

[23] Malin, M.R., "Prediction of Radially Spreading Turbulent Jets", *AIAA Journal*, **26**, 750-752, 1988.

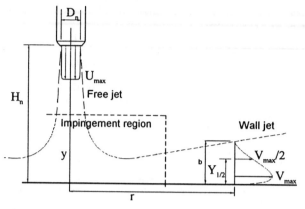

Fig 1 Characteristics of a round impinging jet

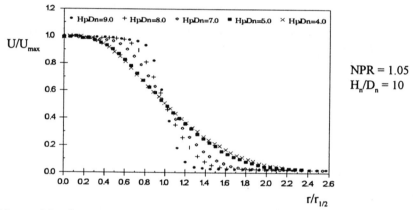

NPR = 1.05
$H_n/D_n = 10$

Fig 2 Measured free jet mean (axial) velocity profiles

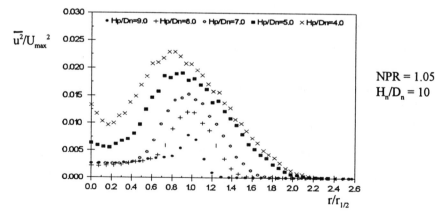

NPR = 1.05
$H_n/D_n = 10$

Fig 3 Measured free jet fluctuating (axial) velocity profiles

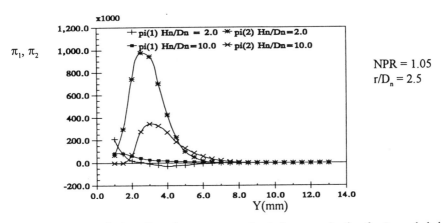

Fig 4 Measured vertical profiles of components of turbulence production for 2 nozzle heights

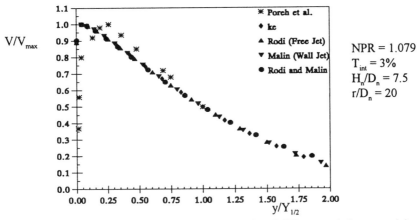

Fig 5 Normalised wall jet mean velocity profiles: results of different turbulence models compared with Poreh et al.

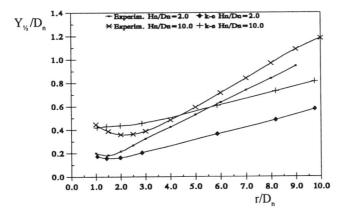

Fig 6 Effect of nozzle height on wall jet growth: computation and cross-wire measurement

NUMERICAL MODELING OF COMBUSTIBLE DUSTY GAS MIXTURE FLOW

V.P. KOROBEINIKOV
Institute for Computer Aided Design, Moscow 123065, Russia

V.V. MARKOV
Mathematical Institute, Russian Academy of Sciences, Moscow 117966, Russia

SUMMARY

The numerical study of combustion and detonation processes in dusty gases is considered in this paper. The moving combustible medium is considered as two-phase, two-velocity, two-temperature continuum including interaction of gas phase and solid phase, combustion, diffusion, turbulence and radiation. The description of the models used, numerical methods and results of the numerical solutions of ignition, combustion and detonation problems is presented in the paper.

INTRODUCTION

The investigation of the unsteady processes of ignition, combustion and detonation is very important for understanding of dusty gas mixture explosion nature and prevention of accidental industrial blasts. The physical processes of grain and coal dusts ignition and combustion are numerically studied using different types of flow descriptions.

It is supposed in the paper that solid particles are rigid spheres, collision among the particles are neglected and the volume of particles is negligible. The perfect gas model is used for gaseous phase. The moving medium is considered as two-temperature continuum including interaction of gas phase and solid phase. One-step overall chemical reaction with induction periods is also included into mathematical models. The following models are used: a) Eulerian two-phase equation system [1]; b) Navier-Stokes description for laminar flow of the medium; c) K-ε approach for the turbulent flow case. The developed numerical methods based on multi-step splitting technique [2] and finite difference scheme including both explicit and implicit parts of numerical algorithm. The method was used for solutions to different problems. Among them there are: ignition of two-phase mixture, combustion and detonation behind shock waves, cellular structure of detonation, transition of slow burning process to detonation. Coal dust particles as well as corn starch dust case were taken for the calculations. The numerically obtained physical effects are briefly discussed.

GOVERNING EQUATIONS

Let us introduce the notations:

v_k for vector velocity of k-phase, $k = 1, 2$; $\rho = \rho_1 + \rho_2$ is a density of mixture; ρ_2 is the density of solid phase; ρ_1 is density of gas the phase and

$$\rho_1 = \rho_g + \rho_v = \rho_{g1} + \rho_{g2} + \rho_v.$$

ρ_{g1} is the density of inert gas; ρ_{g2} is the density of oxidant, ρ_v is volatiles density; ρ_{v2} gas density inside the particles; μ viscosity, p gas pressure, $æ$, D conductivity of heat and diffusion coefficient; $d = 2a$ diameter of the spherical particles, f is the force of phase interactions, F body force (gravity).

Conservation of mass for the gas phase:

$$\frac{\partial \rho_l}{\partial t} + \frac{\partial \rho_i v_{li}}{\partial x_i} = \frac{\partial}{\partial x_i}\left(\rho_k D_l \frac{\partial}{\partial x_i}\left(\frac{\rho_l}{\rho_k}\right)\right) + A_l \dot{K} + B_l \dot{R} + C_l \dot{W}. \tag{1}$$

With the following table of parameters:

l (k)	1 ($k=1$)	2 ($k=2$)	v ($k=2$)	g_2 ($k=1$)	v_2 ($k=2$)
D_l	0	0	D_v	D_0	0
A_l	1	-1	0	0	0
B_l	0	0	-1	$-B_{g2}$	0
C_l	1	0	1	C_{g2}	-1

$l = 1, 2, V_2, g_2, v_2$; B_{g2}, C_{g2} are constants.

Here $\dot{K}, \dot{R}$ are reaction rates for particles and volatiles burning and $\dot{W}$ is the rate of volatiles deposit. Particle volume is negligible. The balance equations for momentum of phases are:

$$\frac{\partial \rho_1 v_{1i}}{\partial t} + \frac{\partial}{\partial x_j}(\rho_1 v_{1j} v_{1i}) + \frac{\partial p}{\partial x_i} = F_i - [f_i + (\dot{W} + \dot{K}) v_{1i}] + \frac{\partial}{\partial x_i} \tau_{ij} \tag{2}$$

$$\hat{\tau} = \tau_{ij} = \mu\left(\frac{\partial v_{1i}}{\partial x_j} + \frac{\partial v_{1j}}{\partial x_i}\right) - 2/3\, \mu \left(\frac{\partial}{\partial x_j} v_{1j}\right) \delta_{ij} \tag{3}$$

$f_i = \theta/3d\, C_d\, \rho_1^0\, (v_{1i} - v_{2i})\, |v_1 - v_2|; \qquad \theta = \pi d^3/6$

$i, j = 1, 2, 3; \qquad \dfrac{\partial}{\partial x_j} a_j \equiv \dfrac{\partial a_1}{\partial x_1} + \dfrac{\partial a_2}{\partial x_2} + \dfrac{\partial a_3}{\partial x_3}$

$$\frac{\partial \rho_2 v_{2i}}{\partial t} + \frac{\partial}{\partial x_j}(\rho_2 v_{2j} v_{2i}) = F_{2i} + [f_i + (\dot{W} + \dot{K}) v_{1i}]. \tag{4}$$

Energy equation for particles phases:

$$\frac{\partial \rho_2 e_2}{\partial t} + \frac{\partial}{\partial x_j}(\rho_2 e_2 v_{2j} + q_{R2j}) = \frac{\rho_2 q}{\theta} - (\dot{W} + \dot{K}) e_2 + \alpha \dot{K} Q_s. \tag{5}$$

Energy equations for the whole mixture:

$$\sum_{k=1}^{2} \left(\frac{\partial \rho_k E_k}{\partial t} + \frac{\partial}{\partial x_j} (\rho_k E_k v_j) \right) + \frac{\partial}{\partial x_j} p v_{1j} - \frac{\partial}{\partial x_j} \left((\hat{\tau} v)_j + \alpha \frac{\partial T_1}{\partial x_j} + q_{Rj} \right)$$

$$= F_j (v_{1j} - v_{2j}) + \dot{K} Q_s + \dot{R} Q_v + \frac{p_2}{\theta} q_{1e} \qquad (6)$$

where $\hat{\tau}$ is the tensor of « viscous » stresses, K_s, Q_v are combustion heats, q_R radiation flux,

$E_k = e_k + v_k^2/2$. $\quad e_k = C_k T + \text{Const}$, $\quad (k = 1, 2); p = R_1 \rho_1 T$.

Radiation transfer:

$$\frac{\partial q_{R1i}}{\partial x_i} = 4 \sigma_B K_p T_1^4, \qquad q_{R2i} = \frac{16 \sigma_B T^3}{3 K_2} \frac{\partial T_2}{\partial x_i}, \qquad (7)$$

K_p is the Planck absorption coefficient for the gas phase, K_2 the Rosseland approximation for the solid phase coefficient ($K_2 = 0.01 \div 1$ [1/cm]); $q = q_T + q_R$, where q_T is the exchange of flow energy between phase (heat conductivity and convection) and q_R is the radiation part of the energy, $q_R = \sigma_B \varepsilon_g T_1^4 + q_{1e}$, ε_g is an empirical coefficient, q_{1e} is the radiant flux due to external sources (hot walls, sparks, etc), if $T_1 \leq 1\,000$ K, q_R is negligible.

REACTION RATES

After an induction period time, t_i, the chemical reaction rate $\dot{R}$ is described by Arrenius law (for laminar case). The rate $\dot{K}$ corresponds to the relation

$$\dot{K} = A_s (1/R_s + 1/R_d)^{-1} \qquad (8)$$

where R_s surface particles chemical reaction rate, R_d the rate of diffusion reaction. A_s is a form factor.

The simple variant for devolatalization process is:

$$\dot{W} = \frac{dm}{dt} = \omega (m_\infty - m), \qquad (9)$$

$\omega = B \exp(-E/RT)$,

$T_0 < T < T_{end}$.

m_∞ is the maximal mass for volatiles, B, E are empirical constants, T_{end} is the temperature for the end of the process.

TURBULENCE INFLUENCE

Turbulence of the gas flow is calculated from the well-known K-ε model, when the viscosity coefficient is replaced in (2), (3) and (6) by $\mu_{eff} = \mu + \mu_t$, etc.

The conservation equations for the determination of K, ε have the form:

$$\frac{\partial \rho \tilde{T}_k}{\partial t} + \frac{\partial}{\partial x_i}(\rho T_k v_{1j}) = \frac{\partial}{\partial x_j}\left(\frac{\mu_{\text{eff}}}{\sigma_k}\frac{\partial \tilde{T}_k}{\partial x_j}\right) + z_{k_1} G + z_{k_2} \rho \varepsilon + W^k_{gp} \qquad (10)$$

$k = 1, 2; \tilde{T}_1 = K, \tilde{T}_2 = \varepsilon.$

G is the generation rate of turbulence, W^k_{pg} are function of particle influence and z_{k1}, z_{k2} are known coefficients.

Influence of Turbulence on Combustion

If the characteristic time of chemical reaction is less than that of turbulence, then we use the above mentioned reactions.

For the case when turbulence characteristic is small, the combustion rate is supposed to be proportional to the rate of dissipation of kinetic energy of turbulence and the Magnus-Hjertager [3, 4] model was used.

NUMERICAL ALGORITHMS

As it was shown above the physical laws of mass and energy conservation, diffusion with chemical reaction and momentum variation for two phases give us the system of equations (1) – (10) of the type:

$$\frac{\partial U}{\partial t} = A(U) + B(U) + C(U) + D(U) \qquad (11)$$

where U is the vector of unknown functions, A is matrix differential operator for the gas phase including gradients, B is a matrix operator with gradients for the solid particle phase, C is for chemical reactions, energy sources and radiation transfer terms of the system, D is the algebraic matrix operator containing the members of mechanical, thermal interactions and mass exchanging between phases as well as turbulence sources.

The developed methods include an implicit part and explicit schemes [5-7] and are based on a splitting technique.

This method was discussed in our work [1, 2] and briefly discussed below. Let us have a vector U known at time t and ΔU the increment

$$\Delta U = U(t + \Delta t) - U(t),$$

from (11) we find:

$$\Delta U = (A + B + C + D)\Delta t + O(\Delta t)\Delta t; O(\Delta t) \to 0, \Delta t \to 0. \qquad (12)$$

The presentation of ΔU by equation (12) shows that the system (1) – (11) can be split on the base of physical meaning of the system terms.

The operators A, B, C are three-dimensional for the general case and they can be also split by two-dimensional and one dimensional algorithms or into three one-dimensional ones with approximating finite-difference schemes for each operator.

Thus we can use for each time step Δt as well splitting by physical principle as geometrical splitting (or alternating direction approach). It is supposed that the computing grid, a moving system and cells sizes are variable over the grid covered computing domain. Initial and boundary conditions were also established for the problem.

EXAMPLES OF NUMERICAL SIMULATIONS

Ignition of Coal Dust/Air Mixture near the Closed End of a Tube

The problem of ignition is important for the determination of combustion beginning. Firstly we studied the problem by the method of mathematical catastrophe theory. It was found analytically the temperature of ignition the medium near closed end of the tube (Fig.1), and time moment for steep gradient of the temperature.

The temperature growth was studied also numerically using laminar Navier-Stokes model (1) – (9), ($n = 1, n = 2$). The results of the calculation are presented in Fig.2 (coal particles in air with 40 % volatiles).

Lifting of Dust Layer and Its Combustion Ignition

Three dimensional problem of dust layer lifting were studied using the Euler model with viscous boundary layer asymptotic. The result is shown in Fig.3.

Two Dimensional Detonation

The unsteady planar detonation in a channel was numerically studied. Let the combustible mixture at time $t = 0$ enter into the tube (channel) with velocity v_0 at coordinate $x_1 = x = 0$. The tube is filled by oxygen/cornstarch mixture with solid phase density, $\rho_{20} = 5.6 \cdot 10^{-3} \text{g/cm}^3$; total heat realizing is $Q = 5$ MJ/kg.

There is a small non uniformity in density, ρ_{20}, distribution along the tube. We supposed that t_i is characteristic time of detonation process and the height of tube $H > D^* t_i^*$ (D^* is Ch.-J. velocity for equilibrium state). It was found numerically, that after several oscillations of the combustion media the cellular structure of detonation is established (see Fig. 4).

CONCLUSION

The numerical investigation of combustion and detonation processes in tube is good remedy for the understanding of the burning phenomenon of two-phase mixture. Including into mathematical models different physical processes can lead to better understanding of combustion nature.

Acknowledgment

The authors are grateful to Prof. P. Wolansky and Dr. R. Klemens from Warsaw University of Technology for useful discussions and the practical help.

References
1. V.P. KOROBEINIKOV, V.V. MARKOV, G.B. SIZYKH, Dokl. Akad. Nauk SSSR, v. 316 p. 1077 (1991).
2. V.P. KOROBEINIKOV, Proc. 5th Int. Symp. on CFD, v.2, Sendai, p. 76, 1993.

3. B.F. HAGNUSSEN, B.H. HJERTAGER, Sixteenth Symp. (Intern.) on Combustion Proc., p. 719, Combustion Inst., Pittsburgh, 1969.
4. J. CHOMIAK, *Combustion*, Gordon and Breach, N.Y., 1990.
5. O.M. BELOTSERKOVSKII, YU.M. DAVYIDOV, *The method of large particles in gas dynamics* (in Russian), Nauka, Moskow, 1982
6. S.K. GODUNOV, A.V. ZABRODIN, G.L. PROKOPOV, Zh. Vychisl. Math. i. Math. F (in Russian) N° 1, p. 1020, 1961.
7. CH. MADER, Numerical modeling of detonation, University of California, 1979.

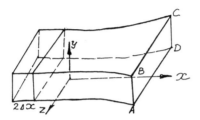

Fig.1 Domain of computation and the zone near the tube end ($x = x_1$, $y = x_2$, $z = x_3$).

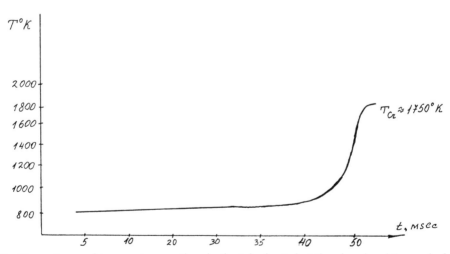

Fig.2 Dependence of temperature on time in the tube heated at the closed end (numerical result), $T_0 = 800$ K.

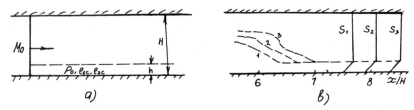

Fig.3 Lifting of dust layer behind the shock wave:
  (a) initial position ($t = 0$);
  (b) upper boundary of dust layer after corresponding shocks.

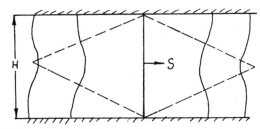

Fig.4 Detonation cell (dashed lines) in cornstarch-oxygen mixture, s is the shock wave position.

Computation of the 3D turbulent Flow surrounding a high-speed train mock-up including the inter-car gap and the bogie, and comparisons with LDV data

LE DEVEHAT Eric, GREGOIRE Rémi
CRESPI Pierre, KESSLER Antoine
SNCF - Direction de la Recherche - Département RP
45 rue de Londres - PARIS - FRANCE -Tel : +33-1-40-08-91-76 / +33-1-42-85-64-53

SUMMARY

This paper presents an important part of the validation process of the French National Railway CFD methodology to compute complex 3D external airflows around trains. A reduced-scale test was first conducted in an open wind tunnel with a 1/7 scale TGV train set mock-up to produce a Laser Doppler Database related to the viscous turbulent air flow. Two components of the instantaneous velocity vector were measured by LDV, along the train mock-up, within the boundary layer, under the train, upstream and downstream the inter-car bogie. Then a complete Navier-Stokes calculation was run, based on this test, and the numerical results were compared with the experimental data. Comparisons were based on the mean velocity and turbulent fields. Quite good agreements were obtained and predictive actions can be today engaged to solve, by the mean of the CFD, technical problems in the field of stationary flows around trains.

INTRODUCTION

The CFD simulation takes today a growing importance in the railway research with the increasing power of computers and the integration of recent fundamental research results in industrial software. Most of the numerical studies in the railway field [1,3] are focused on validation aspects and use a simplified description of train shapes. Complex problems such as drag reduction, aeroacoustics and thermal effects, require a fine modelling of the bogie elements. The purpose of this work, which is the final study of a validation program [4], is to simulate accurately the viscous turbulent flow developed in the inter-car gap of a reduced-scale TGV model tested in an open wind tunnel [6].

I LDV MEASUREMENTS IN THE OPEN WIND-TUNNEL

I.1 The facility and the experimental conditions

A 1/7 scale, 2.5 meter long, mock-up (figure 1) representing two coaches of the TGV "Atlantique" was tested in the Sardou S.A. Wind Tunnel (MSWT, Saint Soupplets, France) [6]. The inter-car gap and the inter-car bogie were well detailed. As the power car was not modelled, a long rounded fairing was set in front of the first coach. It was supposed not to introduce any uncontrolled flow structure in the working section where the air speed was maintained at 150 kph (~ 42 m/s) with an inlet turbulent rate of about 1%. The mock-up was mounted over a moving belt for reproducing the ground effect. The distance between the belt and the wheels corresponded to the height of the rail.

I.2 The measurements

Laser Doppler Velocimetry (LDV) measurements were made within the boundary layers, along the sides of the mock-up, into the inter-car gap, under the train upstream and downstream the inter-car bogie. Both streamwise (X) and vertical (Z) components of the instantaneous velocity vector were measured by the two-component LDV apparatus [5].

Five cross-section planes were investigated, corresponding to five stations along the X axis (stations 1 to 5), from upstream to downstream ; x = -450, -225, 0, 225, 450 mm. For each cross-section plane, six profiles, issued from points on the coach wall, were used in this paper to make comparisons with numerical results (figure 2): three horizontal profiles on the coach side -points H,K,M, a vertical one on the roof - point A, two vertical profiles under the coach - point P,M. In the comparisons, the mean velocity components and the turbulent intensities were divided by a reference velocity periodically measured at the point R (figure 2).

The measurement uncertainty on the velocity component comes from several sources. The four main contributions are the following : fringe spacing estimation, rotation of the beams around the optical axis, perpendicularity between the optical axis and the U-W measurement plane, location of the measurement probe. Globally the uncertainty on the streamwise component U measured is below 1% and the position uncertainty is about 2 mm in the Y direction and 0.1 mm in the X and Z directions corresponding to the dimensions of the probe volume obtained by the crossing of the laser beams.

II THE CFD CALCULATION

II.1 The meshed model and the calculation domain

The CAD model, generated with the CAD system ICEM-DDN, represents the half of the mock-up and half the plane corresponding to the upper surface of the rolling belt. Since the intake and the diffuser of the open wind-tunnel are not modelled, the jet effect is supposed to be weak, so that an infinite upstream flow field at a constant motion rate, the train velocity, is assumed.

As the flow in the bogie cavity is nowadays of great interest for SNCF, a high level of details is brought to the bogie modelling. All structures playing a significant action in the drag force or in the aeroacoustic noise generation are represented as shown on figure 3 : the wheels, the brake systems, the pneumatic suspension, the anti-yawing dampers (coach to coach damper and coach to bogie damper) .

In order to apply the most realistic boundary conditions and to reduce numerical instabilities, the calculation domain (figure 4) is extended about five times the coach height above and beside the model, and about three times in front of the mock-up nose. On the contrary, the mesh domain stops precisely at the rear section of the mock-up, involving an assumption on the boundary conditions to apply, as explained farther. A 3D structured mesh is generated with the software ICEM-CFD using multi-block techniques. Due to the complexity for meshing the bogie and applying wall functions with success -this constraint lead to a 2 mm high cell at the wall- the resulting mesh topology is composed of 270 blocks while the mesh contains 750 000 cells, hexahedrons and prisms. The 2D computational grid on the half mock-up surfaces contains about 45 000 shell cells, 27 500 of them being located in the bogie cavity and the inter-car gap.

II.2 The boundary conditions

A no-slip wall condition is applied on the coach surfaces where a very fine mesh was necessary to satisfy the algebraic " laws of the wall ". In the bogie cavity, the boundary layer development along the different structures has little influence on the flow behaviour compared to blockage effects, impact and separation phenomena. Therefore, in order to reduce the mesh size, slip conditions are applied on the bogie surfaces.

For the calculation, the wheels are motionless while the ground is modelled by a moving wall with a viscous driving, in accordance with the experimental conditions.

Symmetry conditions are applied on the three planes bounding the calculation domain : the symmetry plane of the mock-up, above and beside. Since the planes located above and beside are far away from the mock-up, the channel effect induced by these symmetry conditions is weak. The blockage ratio is about 2%.

At the inlet section of the domain, Dirichlet conditions are imposed according to the experimental conditions encountered in the open-wind tunnel : 42 m/s for the inlet mean velocity and 1% of free stream turbulence rate. At last, at the outlet section, homogeneous Neumann conditions are applied which suppose that the flow is at most slow-varying along the direction normal to the outlet plane. This assumption can be made considering the variations of the physical quantities in the boundary layer are low along this direction. Replacing this condition by a prescribed pressure condition leads to inaccurate results under the coach and strongly reduces the under-coach flow.

II.3 The numerical simulation

For numerically solving the turbulent viscous flow surrounding the mock-up, the fully implicit finite volume based software STAR-CD is used [7]. The flow is assumed to be steady, incompressible and turbulent and the RNG two-equation model is used. The SIMPLE solution algorithm is employed in conjunction with a first order upwind differencing scheme.

The calculation takes 35 hours CPU on a HP 9000/735 workstation to reach convergence. It needs about 250 Mbytes of central memory.

III SIMULATION RESULTS ANALYSIS AND COMPARISONS

Relatively to the 2D measurements carried out by LDV, comparisons can only be made on the mean streamwise component (X) and the vertical one (Z). Comparisons on turbulence are based on the measured RMS value of the streamwise component, since the total turbulent kinetic energy can not be correctly estimated with 2D measurements.

III.1 Boundary layers on the coach

As shown on figure 5, the boundary layer along the coach is thicker in the bottom region downstream the bogie cavity. The ground effect, which introduces a 3D behaviour of the boundary layer, joined to the bogie blockage effect, are the main factors which contribute to this phenomenon.

Figure 6 shows the evolution of the boundary layer thickness and compares computed results with LDV data for cross-sections 1 to 5. The global behaviour of the boundary layer is well reproduced and a very good agreement is observed between numerical and experimental data, except in the lower part of the coach downstream the bogie (stations 4,5). In this area, the computed boundary layer is thicker than the measured one, probably because the numerical turbulent viscosity becomes too strong when the air-flow gets through the bogie-cavity.

On figures 7 and 8, comparisons are made on the mean streamwise component U along three profiles normal to the wall (points H, K and M) and one profile normal to the roof (point A). The velocity component is divided by a reference velocity measured or calculated at the point R (see figure 2).

The mean velocity profiles are globally well predicted and the accuracy on the gradient near the coach wall is good. Added to the experimental uncertainty mentioned above, a numerical one is introduced on the coach surface location because the mesh does not follow exactly the real surface curvature. At least, the uncertainty on the location of one point of comparison is about ± 2 mm along the Y axis.

The velocity profiles issued from the points K and M at stations 2, 3 and 4 (x = -225 mm, 0 mm, 225 mm) are close to the bogie cavity. Therefore, ywall = 0 represents, for these three stations, an extrapolation from the coach wall. While the evolutions of the experimental profiles issued from the point K are well reproduced by the simulation, the comparison based on the profile issued from the point M, just placed under the dampers, reveals an important discrepancy. The secondary flow coming out of the bogie-cavity is probably reproduced not finely enough. This secondary flow would transfer momentum energy to the boundary layer of the coach which would be thinner than the calculated one (figure 6).

The comparisons on the turbulent quantities bring to the fore the limitation of the LDV two-component measurement which does not allow an accurate estimation of the turbulent kinetic energy K. Indeed, in the near wall region on the coach side, the anisotropy of the turbulent flow does not permit to extrapolate the normal component from the tangential ones in order to estimate the turbulent kinetic energy. Consequently, the comparison shown on figure 9 is only based on the root mean square value (RMS) of the fluctuating streamwise component. Numerically, this quantity is estimated by the classical Boussinesq assumption :

$$U_{RMS} = \sqrt{\overline{u'^2}} \; ; \; \overline{u'^2} = -2\nu_T \frac{\partial \overline{U}}{\partial x} + \frac{2}{3} K \quad \left(\nu_T = C_\mu \frac{K^2}{\varepsilon} : \text{turbulent viscosity} \right).$$

Although the predicted turbulent field intensity is lower than the real one, the evolutive trend of the RMS profile, from upstream to downstream, is well reproduced. Particularly, an increase of the turbulent level is observed in the downstream part of the bogie region. This increase is probably related to the wake of the coach-bogie damper, as shown by the calculation presented on figure 10. This turbulent production is one of the identified main sources of the aeroacoustic noise generated by the TGV "Atlantique".

III.2 Flow under the train

On figure 11 are presented the evolutions of two mean streamwise component profiles under the coach at two Y stations (points P and M) upstream and downstream the bogie cavity. A discrepancy is observed between numerical and experimental results, particularly at the station M. This gap both involves an uncertainty on the integrated air discharge under the coach and the velocity gradient near the bottom part of the coach, particularly upstream the bogie. In a first analysis, it seems that a 3D behaviour of the under-coach flow, involving a lateral secondary leakage flow, is not taken into account by the calculation.

Otherwise, the shape modification of these mean velocity profiles caused by the bogie is well reproduced by the simulation. From a flat profile upstream the bogie, a Couette type flow with a linear behaviour between the coach and the ground is obtained downstream the bogie.

CONCLUSION

The analysis of the calculation was essential to perform the validation of the STAR-CD software on a 3D reduced-scale test-case representative of the complex external flows encountered in railway aerodynamic problems. Beyond the validation aspect, this study was very useful to develop a predictive CFD methodology for the train aerodynamics. From these first results, applied research works are today pursued on the drag and on aerodynamic noise reduction.

Acknowledgements

The authors wish to thank the SNCF Research Department for permission to present this work at the IMACS-COST conference. They also wish to acknowledge their colleagues of the SNCF Rolling Stock Department (MEP and MCS) for the significant contribution that they brought all along the experimental program and for the preparation of the numerical work.

REFERENCES

[1] J. CLEMENTSON, A.P. GAYLARD , T. JOHNSON & R. GREGOIRE : **Critical Appraisal of 5 Unstructured CFD codes applied to the flow around a single railway vehicle**, ERRI, Technical Document, Draft Version, Nov. 1994, Utrecht, Netherlands.

[2] F. MASBERNAT, Y. WOLFHUGEL, J.C. DUMAS, S. AITA, A. TABBAL, E. MESTREAU, N. MONTMAYEUR, **CFD Aerodynamics of the French High-Speed Train**, GEC-ALSTHOM technical review, vol. 11, jan. 1993.

[3] T. KAIDEN, S. HOSAKA, T. MAEDE, **A Validation of Numerical Simulation with Field Testing of JR Maglev Vehicle**, International conference on speedup technology for railway and maglev vehicles, 1993.

[4] E. LE DEVEHAT, P. CRESPI, R. GREGOIRE : **Validation of a CFD Software for the Aerodynamics of High Speed Trains and Comparison with a Wind-Tunnel Experiment Using LDV**, World Congress on Railway Research, Nov. 1994, Paris, France.

[5] P. CRESPI, R. GREGOIRE & P. VINSON : **Laser Doppler Velocimetry Measurements and Boundary Layer Survey On-board the TGV High Speed Train**, World Congress on Railway Research, Nov. 1994, Paris, France.

[6] A. KESSLER, R. GREGOIRE, P. CRESPI, **Rapport d'essai de vélocimétrie laser en soufflerie sur maquette TGV-A à l'échelle 1/7** , SNCF technical report, 1995.

[7] STAR-CD Version 2.2 Manuals, Computational Dynamics Limited, 1993.

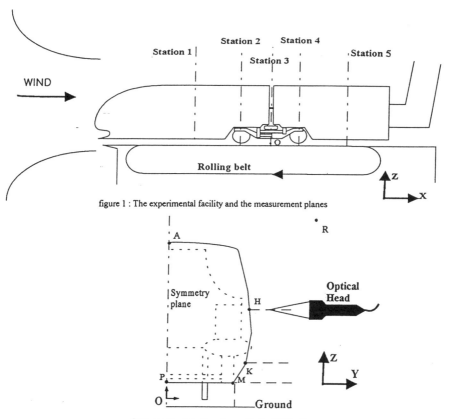

figure 1 : The experimental facility and the measurement planes

figure 2 : Cross section plane with the LDV optic

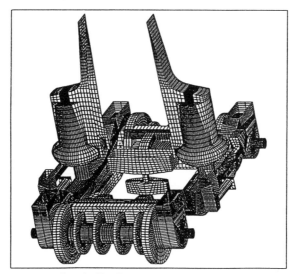

figure 3 : The 2D mesh on the bogie surfaces

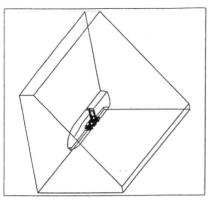

figure 4 : The calculation domain

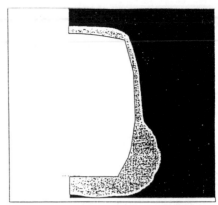

figure 5 : The boundary layer around the coach in a cross-section downstream the bogie

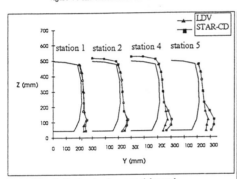

figure 6 : The boundary layer around the coach

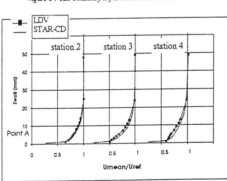

figure 8 : Profiles of the streamwise component of the mean velocity on the roof at the point A

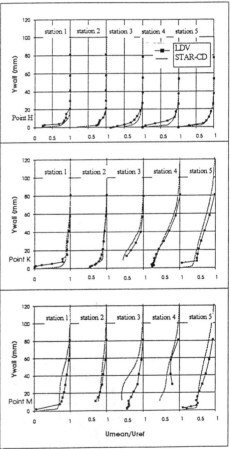

figure 7 : Profiles of the streamwise component of the mean velocity on the coach side at three heights

142

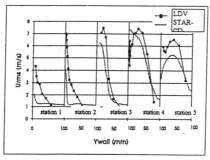

figure 9 : Root Mean square value on the streamwise component at the point K

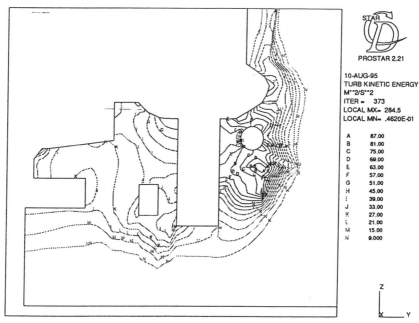

figure 10 : Turbulent kinetic energy contours in the middle section of the inter-car gap (station 3)

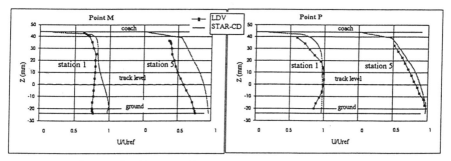

figure 11 : Profiles of the mean streamwise component under the coach at the points P and M

143

COMPUTATIONAL MODELLING OF COMPLEX 3D FLOWS WITH SECOND-MOMENT CLOSURE COUPLED TO LOW-Re NEAR-WALL MODELS

M.A. Leschziner, F.S. Lien, N. Ince and C.A. Lin
UMIST, Mechanical Engineering Department, UMIST, PO Box 88,
Manchester M60 1QD, UK

SUMMARY

The paper exposes some facets of computational research undertaken at UMIST over the past few years, directed towards the use of second-moment closure for predicting physically and geometrically complex 3D flows. The modelling foundation and the numerical framework are summarised, and a selection of solutions is presented for several 3D flows, one of which is transonic. The solutions demonstrate that turbulence modelling is an influential element in the prediction of complex 3D flows and that reverting from eddy-viscosity models to second-moment closure can have a significant impact, at modest additional costs, not only on turbulence quantities but also on mean-flow features.

INTRODUCTION

There is a wealth of evidence to demonstrate that flows featuring complex strain arising from separation, curvature, swirl and impingement cannot be adequately computed with eddy-viscosity models. Such formulations, evolved principally for boundary-layer flows, are known to be afflicted by a whole range of weaknesses, most of which rooted in the rigid linear linkage between stress and strain, and the assumption of normal-stress isotropy implicit in the eddy-viscosity concept. In contrast, wide-ranging validation studies for complex 2D flows (Leschziner, 1990, 1994, 1995) have shown that second-moment is able to capture the influential interaction between complex strain features and turbulence transport, mainly because the generation terms of all stresses are retained in their exact form.

Despite the above evidence, the reality is that the large majority of 3D computations have been and are still being performed with relatively simple eddy-viscosity models. Among the more important reasons for this apparent contradiction are uncertainty on model choice, complexity of implementation, the high sensitivity of numerical stability and robustness to implementation practices, high computational costs, the need for greater numerical accuracy, and difficulties in validating advanced modelling practices in the face of extreme scarcity of sufficiently accurate and comprehensive experimental data.

This paper reports on the outcome of recent studies which investigate the predictive performance of second-moment closure, interfaced with low-Re eddy-viscosity near-wall models or log-law-based "wall functions", when applied to geometrically and physically complex 3D flows. These have been computed with high-accuracy, bounded schemes over meshes of up to 550,000 nodes using strategies which are based on general curved-orthogonal or non-orthogonal finite-volume formulations. The paper summarises salient modelling and computational issues and introduces, drawing on earlier publications, solutions for a combustor flow with multiple radial injection, an internal flow in a circular-to-rectangular transition duct, a external separated flow around a streamlined body at high incidence and multiple impinging jets, both incompressible and transonic. The results included demonstrate that the accurate representation of turbulence transport in the above flows is of considerable importance, even to gross flow parameters. As a result of adopting special model-implementation practices, second-moment-closure solutions can now be obtained at computational costs only 40-50% higher than those required for conventional eddy-viscosity models.

TURBULENCE MODELLING

Outside the thin semi-viscous near-wall layer, turbulence effects are here described by the transport equations for all Reynolds-stress components. In terms of Cartesian tensor notation, the set may be written, formally, as follows:

$$\frac{D\overline{u_i u_j}}{Dt} = P_{ij} + d_{ij} + \Phi_{ij} - \varepsilon_{ij} \tag{1}$$

in which the production terms P_{ij} are given exactly by:

$$P_{ij} \equiv -\overline{u_i u_k}\frac{\partial U_j}{\partial x_k} - \overline{u_j u_k}\frac{\partial U_i}{\partial x_k} \tag{2}$$

and are retained without approximation, viscous dissipation is held to be isotropic, an assumption reflected by the replacement:

$$\varepsilon_{ij} = \frac{2}{3}\varepsilon\delta_{ij} \tag{3}$$

where ε is the rate of dissipation of turbulence energy, and diffusion of the Reynolds-stresses, d_{ij}, is approximated by Daly & Harlow's (1970) *generalised gradient diffusion hypothesis* (GGDH):

$$d_{ij} = \frac{\partial}{\partial x_k}(c_s \overline{u_k u_l}\frac{k}{\varepsilon}\frac{\partial \overline{u_i u_j}}{\partial x_l}). \tag{4}$$

For the pressure-strain terms Φ_{ij} the model of Gibson & Launder (1978) is adopted in most computations to be presented. Variations from this standard practice have entailed the use, for 3D jets, of the cubic pressure-strain model of Fu et al (1987) and the model of Craft & Launder (1992) for the wall-reflection correction associated with the *rapid* fragment of the pressure-strain approximation. No results arising from these variations are included in the present paper, however.

The dissipation rate ε is governed by its own transport equation:

$$\frac{D\varepsilon}{Dt} = \frac{\partial}{\partial x_k}(c_t \frac{\overline{u_k u_l}}{\varepsilon} k \frac{\partial \varepsilon}{\partial x_l}) + 0.5 \frac{\varepsilon}{k} c_{\varepsilon 1} P_{kk} - c_{\varepsilon 2}\frac{\varepsilon^2}{k} \tag{5}$$

which is, formally, identical to that used in the high-Re variant of the k-ε model, except that the diffusion process is represented by the GGDH approximation. In some cases, a proposal of Haroutunian et al (1988),

$$c_{\varepsilon 2} = \frac{1.92}{(1 + 0.6A\sqrt{A_2})} \tag{6}$$

$$A_2 = a_{ij}a_{ji}, \quad A_3 = a_{ij}a_{jk}a_{ki}, \quad A = 1 - \frac{9}{8}(A_2 - A_3), \quad a_{ij} = \frac{\overline{u_i u_j}}{k} - \frac{2}{3}\delta_{ij}$$

has been used to sensitize the dissipation rate to anisotropy.

Close to walls, below $y^+ = O(50)$, turbulence is significantly affected by fluid viscosity, and special practices are needed to account for this interaction. One route is to modify the above Reynolds-stress equations in the manner undertaken, for example, by Hanjalic (1993) and Launder & Tselepidakis (1991). Here, in contrast - and for reasons explained in the introduction, turbulence effects in the semi-viscous sublayer are determined from a low-Re eddy-viscosity model to which the Reynolds-stress model is coupled at $y^+ = O(50)$. The particular model used here is the k-ε variant of Lien & Leschziner (1994a), which has been derived subject to the constraint that the turbulent length scale the model returns close to the wall should asymptote to that prescribed in Wolfshtein's one-equation model (1969). The precise details of the model are of marginal relevance to the main theme of the paper and are not, therefore, included herein.

NUMERICAL FRAMEWORK

The above Reynolds-stress equations have been implanted into two 3D numerical algorithms, one based on a general curved-orthogonal finite volume strategy with staggered storage arrangement for pressure, velocity and Reynolds stresses (Lin & Leschziner, 1989), and the other using a general non-orthogonal framework with a fully-collocated storage for all variables (Lien & Leschziner, 1994b).

With the former strategy, the velocity vector is decomposed into coordinate-directed components. The practice of staggering of velocity as well as stress components is shown by Huang & Leschziner (1985) to be of decisive benefit to stability but necessitates the use of 7 finite-volume sets - 3 for velocity components one for pressure and normal stresses, and 3 for the shear stresses. Discretisation of convection is second-order accurate and effected either with the QUICK scheme of Leonard (1979) or the bounded variant LODA of Zhu & Leschziner (1988).

Most computational solutions have been obtained with a more recent single/multi-block collocated-variable algorithm STREAM of Lien & Leschziner (1994b). Within the non-orthogonal finite-volume system, the velocity vector is decomposed into its orthogonal components to which the momentum equations relate. These components need not retain an invariant orientation, but may be adapted to a user-defined line - say, the centre-line of a curved duct. Convection fluxes are approximated with the UMIST scheme (Lien & Leschziner, 1994c) - a TVD implementation of Leonard's QUICK scheme (1979). Mass conservation is enforced via a pressure-correction equation. This approach is conventional in the case of incompressible flow, but may also be adapted to apply to transonic and supersonic flow. This adapted scheme has been used to predict 2D as well 3D shock-boundary-layer interaction (Lien & Leschziner, 1993a, 1995a).

A fully collocated variable storage, in combination with central differencing for pressure, is known to provoke checkerboard oscillations, reflecting pressure-velocity decoupling. To avoid this, the widely used interpolation practice of Rhie & Chow (1983) has been adopted to interpolate the cell-face velocities from the nodal values. With second-moment closure, an analogous form of numerical instability can arise from the decoupling between velocities and Reynolds-stresses. The manner in which this instability is suppressed is one key facet of a successful implementation of the Reynolds-stress equations into the present numerical framework, and this is considered in some detail in Lien & Leschziner (1995b)

PREDICTIVE PERFORMANCE

Over the past few years, about a dozen different 3D flows have been computed at UMIST with second-moment closure. At the 'simple' end of the range are attached flows in bends of circular and square cross-sectional area, while the most complex flows included twin-impinging transonic jets and multiple vortical separation from streamlined bodies at high incidence. Space constraints do not allow more than a few examples to be introduced by way of representative results extracted from more detailed expositions to which reference is made.

Several configurations involving either single or multiple jets injected into a quiescent environment or into cross-flow, some involving impingement and stand-off shocks, have been computed at UMIST (Ince & Leschziner [1993a,b,c], Leschziner & Ince [1995]). All studies pertain, principally, to VSTOL applications. Fig. 1 relates to a three-dimensional twin-jet configuration (Saripalli [1987]) in which impingement gives rise to a strong fountain originating from the collision of the wall jets formed after impingement. The greatest sensitivity to turbulence modelling is observed in the fountain, and computational results included here relate to the fountain half-width and fountain velocity. As seen, the Reynolds-stress model yields a clearly superior resolution, held to reflect the destabilising influence of curvature in the near-wall shear layers which rise from the wall at the base of the fountain as the wall jets collide.

Results for the transonic twin-impinging jet at nozzle pressure ratio 3.3 are given in Fig. 2. Computed Mach contours reveal both the shock-cell structure and the impingement shock. These agree well with the corresponding features seen in the hand-drawn abstraction of a Schlieren photograph obtained by Abbot & White (1989), with both k-ε and Reynolds-stress models providing a comparable agreement with experiment. The variation of impingement-plate pressure, not included here, is also found to be rather insensitive to turbulence modelling. However, a strong and important response to turbulence modelling is revealed by the comparison of predicted centre-line pressure variations given included in Fig. 2. As seen, the second-moment model returns a much sharper impingement shock than does the k-ε model. This difference reflects the fact that the latter model responds, in contrast to the second-moment form, to irrotational strain within the shock cells preceding the impingement shock by generating a high level of turbulence energy. This is then convected towards the impingement region, strongly elevating turbulent diffusion and hence shock erosion. A further consequence of the considerable lower level of diffusive transport returned by the Reynolds-stress model is the prediction of periodic shock oscillations; the pressure results reported in Fig. 2 are those recorded at half period. Whether such oscillations were present in the experiment is not known.

The interaction of jets with a cross-flow is also of central importance in the application shown in Fig. 3, for which computations have been obtained by Lin (1990). This geometry is part of a combustor model, examined experimentally by Koutmos (1984). A swirling flow enters the chamber from the left-hand inlet section, and two rows of dilution jets are injected radially inwards. These jets entrain swirl momentum and transport this momentum towards the chamber centreline, thus provoking a rapidly swirling vortex in the centre region which significantly distorts the axial flow. The streamwise- and swirl-velocity profiles included in

Fig. 3 demonstrate that the eddy-viscosity model seriously overestimates the entrainment of swirl momentum, thus leading to an excessively strong centreline vortex and unrealistic distortions in the streamwise velocity, the latter arising from the strong axial pressure gradient that goes with steep streamwise variations of swirl.

Amongst the most complex aerodynamic process investigated at UMIST with second-moment closure is open flow separation from the leeward side of a three-dimensional prolate spheroid inclined at 10° and 30° to a uniform on-coming stream, as shown in Fig. 4 and 5. Extensive hot-wire data reported by Kreplin et al (1985) make this a particularly attractive and informative test case. Corresponding calculations have been performed by Lien & Leschziner (1995 b,c) using full Reynolds-stress closure coupled to a low-Re 'k-ε' eddy-viscosity model in the semi-viscous near-wall region. Fig. 4 shows, for 10° incidence, surface streaklines (over half of the prolate spheroid and mapped onto a square), azimuthal profiles of skin friction magnitude and direction at an axial position close to the tail end, and profiles of azimuthal velocity component in the leeward region of the spheroid in which the flow is most complex. While it is evident that the Reynolds-stress model does not return an entirely satisfactory representation of the separation process, the predicted behaviour may be said to be encouraging and is certainly superior to that achieved with the eddy-viscosity model. Fig. 5 gives results for 30° incidence. As is seen from the surface streaklines, there are at least 4 separation lines on the prolate surface. Indeed, the azimuthal variations of skin-friction direction in the rear of the spheroid suggest that there might be six vortices, although only four have been resolved by the computation. Results for skin friction contained in Fig. 5 have arisen from computations with a linear low-Re k-ε model, a non-linear variant (see Lien & Leschziner, 1995c) and the Reynolds-stress-transport closure. The latter two return similar predictive accuracy which is clearly superior to that of the linear k-ε model. Consistently, these two models also predict well the structure of the detached vortex, identified by secondary velocity fields at x/2a=0.917.

The final fluid-dynamic example considered here is the flow in a round-to-rectangular transition duct (experiments by Davis & Gessner (1990); computations by Lien & Leschziner [1995b]). This flow is attached, but involves strong transverse motion which are induced by the lateral squeezing and the streamwise realignment of the vortex lines in the boundary layer. The convective transport associated with the transverse motion creates a strong distortion in the streamwise velocity which persists far downstream. The ability of three models to return the streamwise vortices downstream of transition is conveyed in Fig. 6. Two models are eddy-viscosity variants, one for high-Re conditions only and the other a low-Re variant. The third model is a full second-moment closure (coupled to a low-Re k-ε model in the viscous sub-layer). The results indicate - and this is supported by solutions for pressure, friction factor and stress fields - that the most important process to resolve is the skewness of the flow very close to the wall, arising from the transverse motion reaching its peak just outside the viscous sub-layer. In this flow, then, second-moment closure offers only minor benefits relative to those arising from a detailed resolution of the near-wall mean-flow structure. It is noted, however, that only this closure is able to return a credible representation of the normal-stress fields, especially in respect of anisotropy.

CONCLUDING REMARKS

The paper has argued the merits of using of second-moment closure or the computation of 3D complex flows that feature strong curvature, separation, and swirl. The numerical implementation and performance of one such model, operating in conjunction with a low-Re k-ε model for the viscous near-wall region, has been exposed. This 'zonal' approach is numerically stable, economical and algorithmically simple, especially in terms of the specification of wall boundary condition. Moreover, it allows near-wall processes to be captured in detail, a prerequisite for a credible representation of the skin friction and heat transfer rate.

Reynolds-stress models offer no panacea to all ills, and research continues unabated in efforts to improve these closures, particularly in respect of modelling redistributive and dissipative processes. Moreover, other modelling routes, such as 'RNG' and non-linear eddy-viscosity approaches, are being vigorously pursued. It is arguable, however, that second-moment closure provides the most secure and fundamentally sound basis for a modelling framework required to be applicable to a wide range of complex flow conditions.

REFERENCES

ABBOTT, W.A., AND WHITE, D.R., "The effect of nozzle pressure ratio on the fountain formed between two impinging jets", RAE Technical Memorandum P1166, (1989).

CRAFT, T.J. AND LAUNDER, B.E., AIAA Journal, 30 (1992), p. 2970.

DALY, B.J. AND HARLOW, F.H., Phys. of Fluids, 33 (1970), p. 1.

DAVIS, D.O. AND GESSNER, F., "Experimental investigation of turbulent flow through a circular-to-rectangular duct", AIAA Paper 90-1505, (1990).

FU, S., LAUNDER, B.E. AND TSELEPIDAKIS, D.P., "Accommodating the effects of high strain rates in modelling the pressure-strain correlation", Report TFD/87/5, UMIST, Dept. Mech. Eng. Thermofluids Div., 1987.

GIBSON, M.M. AND LAUNDER, B.E., J. Fluid Mech., 86 (1978), p. 491.

HAROUTUNIAN, V., INCE, N. AND LAUNDER, B.E., "A new proposal for the ε-equation", Proc. 3rd Colloquium on CFD, UMIST, Manchester (1988), p. 1.3.

HANJALIC, K., "Advanced turbulence closure models - A view of current status and future prospects - Report LSTM 378/T/93, University of Erlangen-Nuernberg, 1993.

HUANG, P.G. AND LESCHZINER, M.A., "Stabilization of recirculating flow computations performed with second-moment closure and third-order discretization", Proc. 5th Symp. on Turbulent Shear Flows, Cornel University, (1985), p. 20.7.

INCE, N.Z. AND LESCHZINER, M.A., "Calculation of single and multiple jets in cross-flow with and without impingement using Reynolds-stress transport closure", Paper 23, AGARD Symp. on Computation and Experimental Assessment of jets in Cross-Flow, Winchester (1993a).

INCE, N.Z. AND LESCHZINER, M.A., "Second-moment modelling of incompressible impinging twin-jets with and without cross-flow", Proc. 5th Int. IAHR Symp. on Refined Flow Modelling and Turbulent Measurements, Paris (1993b), p. 39.

INCE, N.Z., PAGE, G.I. AND M.A. LESCHZINER, "Second-moment modelling of subsonic and transonic impinging twin jets", Proc. 1993 European Forum on Recent Developments and Applications in Aeronautical CFD (1993c), p. 18.1.

LESCHZINER M.A. AND INCE, N.Z. "Computational Modelling of three-dimensional jets with and without cross-flow using second-moment closure", Computers and Fluids, 1995, (in press).

LAUNDER, B.E. AND TSELEPIDAKIS, D., "Directions in second-moment modelling of near-wall turbulence", Paper AIAA 91-0219, 29th Aerospace Sciences Meeting, 1991.

LEONARD, B.P., Comp. Meth. Appl. Mech. Eng., 19, (1979), p. 59.

KOUTMOS, P., "An isothermal study of gas turbine flows", PhD Thesis, Unviersity of London (1985).

KREPLIN, H.-P., Vollmers, H. Meier, H.U. and Kühn, A., "Measured mean velocity field around a 1:6 prolate spheroid at various cross sections", DLR Report (1985).

LESCHZINER, M.A., J. of Wind Engineering and Industrial Aerodynamics, 35 (1990), p. 21.

LESCHZINER, M.A., "Refined turbulence modelling for engineering flow", in Computational Fluid Dynamics, Wiley, (1994), p. 33.

LESCHZINER, M.A., "Computation of aerodynamic flows with turbulence-transport models based on second-moment closure", to be published in Computers and Fluids (1995).

LIEN, F-S. AND LESCHZINER, M.A., ASME J. Fluids Engg., 115 (1993a), p. 717.

LIEN, F.S. AND LESCHZINER, M.A., "Modelling the flow in a transition duct with a non-orthogonal FV procedure and low-Re turbulence-transport models", ASME Summer Meeting, Symposium on Advances in CMFD (1994a).

LIEN, F.S. AND LESCHZINER, M.A., Comp. Meth. Appl. Mech. Eng., 114 (1994b), p. 123 and p. 149.

LIEN, F.S. AND LESCHZINER, M.A., Int. J. Num. Meth. Fluids, 19 (1994c), p. 527.

LIEN, F.S. AND LESCHZINER, M.A., "Computational modelling of turbulent aerodynamic flows within the ECARP Programme using anisotropy-resolving closures", Report TFD/95/7, Dept. Mechanical Engineering, UMIST (1995a).

LIEN, F.S. AND LESCHZINER, M.A., "Second-moment closure for three-dimensional turbulent flow around and within complex geometries", Report TFD/95/01, Dept. Mech. Eng., UMIST, Accepted for publication in Computers and Fluids (1995b).

LIEN, F.S. AND LESCHZINER, M.A. "Computational modelling of multiple vortical separation from streamlined bodies at high incidence", Proc. Symp. on Turbulent Shear Flows, Pennstate (1995c), p. 4-19.

LIN, C.A., "Computation of three-dimensional jet-injection processes with second-moment closure", PhD. Thesis, University of Manchester (1990).

LIN, C.A. AND LESCHZINER, M.A. "Computation of three-dimensional injection into swirling flow with second-moment closure", Proc. 6th Int Conf. on Numerical Methods in Laminar and Turbulent Flows, Swansea (1989), p. 1711.

MEIER, H.U., KREPLIN, H.P., LANDHAUSER, A. AND BAUMGARTEN, D., "Mean velocity distribution in 3D boundary layers developing on a 1:6 prolate spheroid with artificial transition", DFVLR Report IB 222-84 A11 (1984)

RHIE, C.M. AND CHOW, W.L., J. AIAA, 21 (1983), p. 1525.

SARIPALLI, K.R., Turbulent Shear Flows 5, Springer (1987), p. 146.

WOLFSHTEIN, M., "The velocity and temperature distribution in one-dimensional flow with turbulence augmentation and pressure gradient", Int. Journal of Heat and Mass Transfer, **12**, 1969, pp. 301-312.

ZHU, J. AND LESCHZINER. M.A., Comp. Meths. Appl. Mech. Engng., 67, (1988), p. 355.

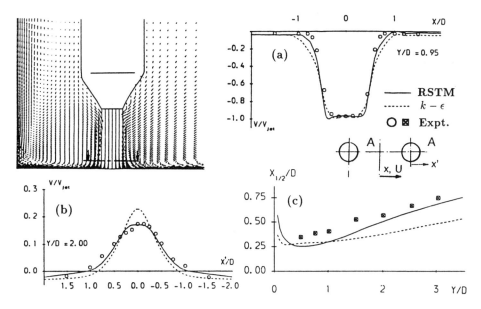

Fig 1   Twin-jet impingement on a flat plate: (a) velocity field across centre bisecting both jets; (b) velocity profiles across fountain; (c) spreading rate of fountain

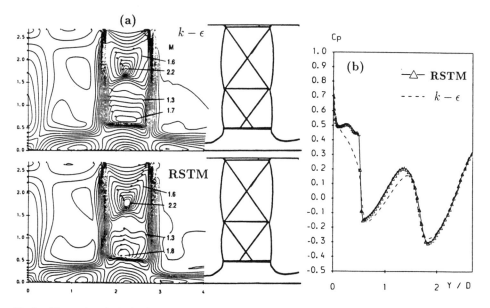

Fig 2   Transonic twin-jet impingement on a flat plate at NPR=3.3: (a) Mach contours; (b) pressure variation along jet axis.

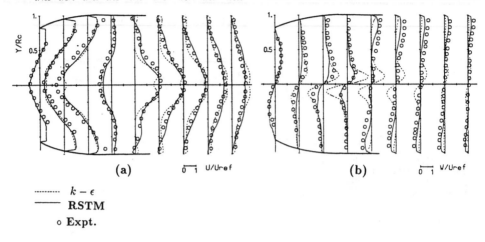

----- $k - \epsilon$
——— RSTM
o Expt.

Fig. 3 Flow in model combustor with multiple injection of dilution jets: (a) profiles of streamwise velocity; (b) profiles of swirl velocity

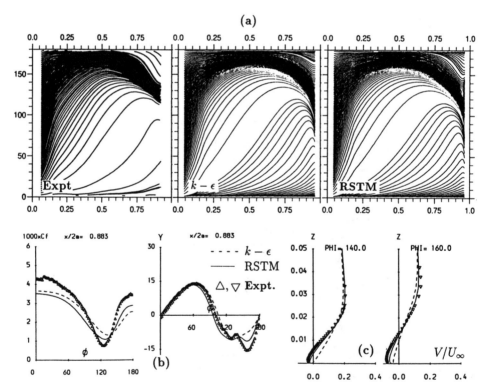

Fig. 4 Flow around prolate spheroid at 10° incidence: (a) surface streaklines; (b) skin-friction magnitude and direction; (c) azimuthal velocity profiles.

152

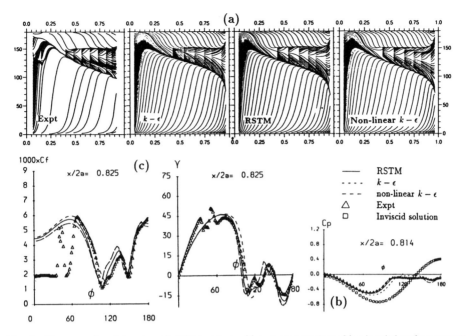

Fig. 5 Flow around prolate spheroid at 30° incidence: (a) surface streaklines; (b) azimuthal surface-pressure coefficient; (c) azimuthal skin-friction direction and magnitude.

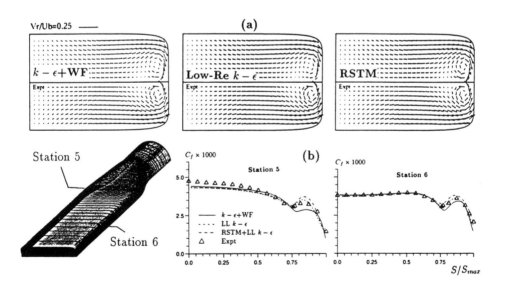

Fig. 6 Flow in a transition duct: (a) transverse velocity fields; (b) skin-friction variations.

Interaction of Two Coaxial Weakly Perturbed Vortex Rings

R. C. K. Leung and N. W. M. Ko
Department of Mechanical Engineering
The University of Hong Kong
Pokfulam Road, Hong Kong

Abstract

A numerical study of interaction of two coaxial vortex rings with initial weak perturbations is reported. The vortex rings are modelled by thin vortex filaments and their evolution is solved by means of three-dimensional vortex blob method. Numerical results reveal the growth of the weak perturbations and hindrance of the repeated leapfrogging, causing the merging of the two vortices. The acoustic radiation before the vortex merging is also presented.

1 Introduction

Circular jet has been the topic of numerous analytical, experimental and numerical studies in past decades. Considerable research efforts were devoted to investigate the jet flow structures, their interactions and sound generation. The laminar boundary layer emanating from the nozzle exit rolls up, forming train of coaxial vortex rings, which undergo vortex interaction and eventual breakdown further downstream [1, 2, 3, 4]. The modelling of vortex rings with core boundary outside which vorticity vanishes, allowed the analytical study of Möhring [5] and numerical study of Kambe and Minota [6] on the sound generated by leapfrogging rings. Recently, Tang and Ko [7] established the importance of acceleration and deceleration of the leapfrogging vortex rings in the production of sound and showed that the rate of change of radial velocity of the leapfrogging vortex rings is crucial to sound generation. As the vortices during interaction are further complicated by the occurrence of weak perturbation waves along its periphery [1, 2], the above studies do not model the real situation. By adopting the perturbation waves proposed by Widnall and Sullivan [8], the present study aims to understand the flow dynamics of two weakly perturbed vortex rings under leapfrogging and the sound generation.

2 Vortex Blob Method

Each perturbed vortex ring is modelled by a vortex filament in unbounded space and the three-dimensional sinusoidal perturbation applied takes the form, $\mathbf{y} = (R + \rho_r e^{im\theta})\hat{\mathbf{e}}_r + (z + \rho_z e^{im\theta})\hat{\mathbf{e}}_z$, where $\mathbf{y}$ is a point on the vortex filament, $\hat{\mathbf{e}}_r$ and $\hat{\mathbf{e}}_z$ the unit vectors in the radial and axial directions, m the wavenumber, θ the azimuthal angle, R the unperturbed ring radius and z the initial axial distance from a frame of reference (Figure 1). The motion of vortex filaments is approximated by means of vortex blob method [9, 10]. The initial vorticity field is discretized into a collection of vortex vector elements [9], also called vortex blobs, which are chosen to lie on the vortex filaments, in the form $\omega(\mathbf{y},t) = \sum_\alpha \Gamma_\alpha \delta \mathbf{s}_\alpha \zeta_\sigma(\mathbf{y}-\mathbf{y}_\alpha(t))$, where Γ_α is the circulation, $\mathbf{y}_\alpha$ the blob centre and $\delta \mathbf{s}_\alpha$ the vector length of vortex blob. The spherically symmetric smoothing function ζ_σ, characterized by a smoothing radius σ, is constructed in the form $\zeta_\sigma(\mathbf{y}) = \zeta(|\mathbf{y}|/\sigma)/\sigma^3$, where $\zeta(\eta) = 15/(8\pi(\eta^2+1)^{7/2})$ with the three dimensional normalization $4\pi \int_0^\infty \zeta(\eta)\eta^2 d\eta = 1$. The smoothing radius is a numerical parameter introduced to inhibit the singular behaviour of Biot-Savart Integral and should be of the same order of magnitude as the physical vortex core radius. In this study the constant smoothing radius was set equal to the physical vortex core radius. Hence, the discretized Biot-Savart velocity field is given by

$$\mathbf{u}_\alpha(\mathbf{y},t) = \frac{1}{4\pi} \sum_{\beta \neq \alpha} \Gamma_\beta \frac{\delta \mathbf{s}_\beta \times (\mathbf{y}_\alpha - \mathbf{y}_\beta)}{|\mathbf{y}_\alpha - \mathbf{y}_\beta|^3} \chi(\frac{|\mathbf{y}_\alpha - \mathbf{y}_\beta|}{\sigma_\beta}), \qquad (2.1)$$

where $\chi(r) = 4\pi \int_0^r \zeta(\eta)\eta^2 d\eta$. The vortex filaments are tilted and stretched by local velocity gradient $\nabla \mathbf{u}$ and these effects are accounted for in the change of vector length of the vortex blob, i.e., where $\mathbf{r}_{\alpha\beta} = \mathbf{y}_\alpha - \mathbf{y}_\beta$,

$$\frac{D\delta \mathbf{s}_\alpha}{Dt} = \frac{1}{4\pi} \sum_{\beta \neq \alpha} \Gamma_\alpha \Gamma_\beta \{ \frac{|\mathbf{r}_{\alpha\beta}|^2 + \frac{5}{2}\sigma_\beta^2}{(|\mathbf{r}_{\alpha\beta}|^2 + \sigma_\beta^2)^{5/2}} (\delta \mathbf{s}_\beta \times \delta \mathbf{s}_\alpha)$$
$$-3(\delta \mathbf{s}_\alpha . \mathbf{r}_{\alpha\beta}) \frac{|\mathbf{r}_{\alpha\beta}|^2 + \frac{7}{2}\sigma_\beta^2}{(|\mathbf{r}_{\alpha\beta}|^2 + \sigma_\beta^2)^{7/2}} (\delta \mathbf{s}_\beta \times \mathbf{r}_{\alpha\beta}) \}. \qquad (2.2)$$

The vortex blob representation of Möhring's vortex sound formula could be similarly derived as

$$p(\mathbf{x}) = \frac{\rho_0}{12\pi c_0^2 |\mathbf{x}|^3} \frac{\partial^3}{\partial t^3} \left\{ \sum_\alpha \Gamma_\alpha(\mathbf{x}.\mathbf{y}_\alpha)\mathbf{x}.(\mathbf{y}_\alpha \times \delta \mathbf{s}_\alpha) - \frac{1}{3}\sigma^2 \mathcal{E} \mathbf{x}. \mathbf{\Omega} \times \mathbf{x} \right\}, \qquad (2.3)$$

where $\mathcal{E} = 4\pi \int_0^\infty \zeta(r) r^4 dr = 3/2$ and the total vorticity, $\mathbf{\Omega} = \sum_\alpha \Gamma \delta \mathbf{s}_\alpha$, vanishes in unbounded potential flow.

The convergence of the vortex blob method was proved by Beale [11] and Cottet [12]. According to their convergence proofs, the appropriate error norm for the discretized vorticity and velocity fields reduces to zero as the number of vortex blobs increases and the smoothing radius decreases, provided that the vortex blobs overlap, i.e. $\sigma/h > 1$, where h is a typical separation between vortex blobs. This constraint was revealed by a test on the translating velocity of a vortex ring (Figure 2). In order to maintain the proper extent of overlapping, a vortex blob was split into two, whenever it had been elongated, due to vorticity stretching, to twice of its original value. Likewise, a vortex blob was fused to its nearest neighbour, as its length was shorter than half of its original value so as to avoid velocity blow-up. Pedrizzetti's divergence filtering was employed to maintain the vorticity field divergence-free during the simulation [13]. At each computation time step δt, the vector length δs_α was modified by, $\delta s_{\alpha,new} = (1 - \mathcal{K}_F)\delta s_\alpha + \mathcal{K}_F |\delta s_\alpha| \hat{\omega}_\alpha$, where $\mathcal{K}_F = \delta t/T_F$ and T_F is the filter time scale, which was adjusted with respect to the time scale of the physical phenomena. The time integration was effected by the classical fourth order Runge-Kutta scheme. All the lengths and time were normalized by the initial values of R_T and R_T^2/Γ_T. Physically, the accuracy of the numerical computation was governed by the linear vorticity impulse $\mathbf{I} = \sum_\alpha \Gamma_\alpha (\mathbf{y}_\alpha \times \delta \mathbf{s}_\alpha)$, an invariant of potential vortex motion. The allowable relative error was kept within one percent. By trial and error, the range of filter time scale T_F was from 0 to 0.5.

3 Results and Discussions

Figure (1) shows two perturbed vortex rings of unity circulation ($\Gamma_T = \Gamma_L = 1.0$) at the start of simulation. The ratio of vortex core radius to unperturbed ring radius was chosen to be 0.2. Unless otherwise stated, the quantities with suffixes L and T denote, respectively, the leading and trailing vortex rings and by solid and dotted lines. All quantities were normalized by the combination of ring radius R_T and circulation Γ_T of the trailing vortex ring. In this study, the wavenumber m was set to one. Initially both the radial and axial amplitudes of weak perturbation waves were 0.02 times the unperturbed ring radius and were in phase. The initial separation, λ, between two vortices equals the unperturbed ring radius of the trailing vortex. The normalized time step, $\delta t/(R_T^2/\Gamma_T)$, was selected as 0.01. Figure (3) depicts the temporal variations of linear vorticity impulse, which preserve their values well throughout the course of computation, indicating the computation had attained good accuracy.

Due to the invariance of the linear vorticity impulse, the circular axis of a perturbed vortex ring at a particular instant could be defined as a circular vortex ring having identical vorticity impulse at that instant. The radius ρ and axial position a of a circular axis were constructed in accordance with the formulae given by Saffman [14] and Batchelor [15], i.e. $(0,0,a) = \frac{\Gamma}{2} \int \frac{\mathbf{y}(s) \times \partial \mathbf{y}(s)/\partial s \cdot \hat{\mathbf{I}}}{|\mathbf{I}|} ds$ and $\rho = \sqrt{|\mathbf{I}|/\pi \Gamma}$. In the case of an ordinary vortex ring, ρ and a reduce to the radius and axial position of the vortex. For comparison the interaction of identical vortex rings without perturbation was simulated and the radial accelerations are depicted in Figure 4. Two sharp peaks occur at the instants when two vortex rings are on the same radial plane; the first one by trailing vortex at $t' \approx 3.7$ and the second one by leading vortex at $t' \approx 11.15$, agreeing with the previous findings [7]. For the perturbed vortices under consideration, the evolution of the radii of their circular axes also gives two distinct slip-throughs; the first one occurs at $t' \approx 3.7$ but the second one occurs at $t' \approx 11.3$ (Figure 5) and their corresponding radial accelerations are shown in Figure 6. It is found that the first radial acceleration peak is almost identical to the unperturbed interaction but the second peak is delayed to $t' \approx 12$ with weaker amplitude. This reveals the perturbation waves affect the vortex dynamics of second slip-through. Motion of circular axis, which definition averages out the details of perturbation development, is informative as long as the amplitude of wave is not comparable to the ring radius and the two vortices do not merge. The detailed motion of the two vortex rings is shown in Figure (8). Initially, both vortex rings tilt towards the negative x-domain. Owing to their mutual induction the trailing vortex ring accelerates and shrinks with angle of tilting increasing, while the leading vortex ring decelerates and widens with reducing angle of tilting. During $3.2 \leq t' \leq 4.0$, trailing ring rushes through the almost planar leading ring (Figures 8a and 8b). After the first slip-through the portion of the leading vortex ring on the negative x-domain is closer to the trailing vortex ring. This results in stronger induced velocity on that portion and causes tilting of the leading vortex ring changes. Then, the two vortex rings exchange their roles and undergo second slip-through during $10.2 \leq t' \leq 12$ (Figures 8c and 8d). Their tilting angles are opposite and increase with time. During the approach to the third slip-through, the nearest portion of the trailing vortex ring to the leading ring on the positive x-domain causes the latter to stretch rather than to change its tilting. This may account for the appearance of ripples on the radial accelerations of the circular radii (Figure 6) and linear vorticity impulse (Figure 3) at $t' > 16$. The leading vortex ring elongates and tries to align with the trailing vortex ring (Figure 8e). Since two vortex rings are in close proximity, the induced velocity and straining fields are so intense that the aligned portions of the vortices stretch severely and rapidly entangle with one another (Figure 8f). Such partial merging may be responsible for the earlier observation of vortex coalescence [2].

The sound pressure fluctuation at $120R_T$ on the x-axis reveals that two distinct pressure pulses occur exactly when the radial accelerations of the circular axes peak (Figure 7). The findings are similar to that of Tang and Ko [7]. Before the first slip-through occurs the amplitudes of perturbation waves are small as compared with vortex radii and two perturbed vortices interact in a similar way to their unperturbed counterpart. As a result, an almost identical sound pressure fluctuation, indicating the leapfrogging is the major contribution of sound generation at the early stage. However, at its later stage, sound signature is superimposed by high frequency components of increasing amplitude. Before the second slip-through instant, the high frequency contribution of this oscillation is no longer negligible and distorts the sound pressure fluctuations. The occurrence of second pressure pulse is also delayed to $t' \approx 12$. As the vortices are severely stretched and merge, the sound signature is heavily contaminated with the high frequency resulting in significant numerical errors.

4 Conclusions

The interaction of two vortex rings with weak perturbations and its sound radiation were studied by means of vortex blob method. Numerical results show that in the presence of weak perturbation waves, only two slip-throughs are possible and the interaction ends up with vortex merging. The sound generation is well predicted before the vortex merging commences. Based on the numerical results, leapfrogging is the sole contribution to sound generation at the first slip-through but the effect of perturbation becomes significant at the second slip-through.

References

[1] Yule, A. J.: "Large-scale structure in the mixing layer of a round jet", J. Fluid Mech. 89, (1978), pp. 413–432.

[2] Paschereit, C. O., Oster, D., Long, T., Fiedler, H. E., Wygnanski, I.: "Flow visualization of interactions among large coherent structures in an axisymmetric jet", Expt. in Fluids 12, (1992), pp. 189–199.

[3] Anderson, A. B. C.: "Vortex-ring structure-transition in a jet emitting discrete acoustic frequencies", J. Acoust. Soc. Am. 28(5), (1956) pp. 914–921.

[4] Laufer, J., Yen, T. C.: "Noise generation by a low Mach number jet", J. Fluid Mech. 134, (1983) pp. 1–31.

[5] Möhring, W.: "On vortex sound at low Mach number", J. Fluid Mech. 85, (1978) pp. 685–691.

[6] Kambe, O. M., Minota T.: "Sound radiation from vortex system", J. Sound Vib. 74, (1981) pp. 61–72.

[7] Tang, S. K., Ko, N. W. M.: "A study on the noise generation mechanism in a circular air jet", Trans. ASME J. Fluids Eng. 115, (1993) pp. 425–435.

[8] Widnall, S. E., Sullivan J. P.: "On the stability of vortex rings", Proc. R. Soc. A 332, (1973) pp. 335–353.

[9] Knio, O. M., Ghoniem, A. F.: "Numerical study of a three-dimensional vortex method", J. Comp. Phys. 86, (1990) pp. 75–106.

[10] Winckelmans, G., Leonard, A.: "Contributions to vortex particle methods for the computation of three-dimensional incompressible unsteady flows", J. Comp. Phys. 109, (1993) pp. 247–273.

[11] Beale, J. T.: "A convergent 3-D vortex method with grid-free stretching", Math. Comp. 46, (1986) pp. 401–424.

[12] Cottet, G. H.: "A new approach for the analysis of vortex methods in two and three dimensions", Ann. Inst. Henri Poincaré 5, (1988) pp. 227-285.

[13] Pedrizzetti, G.: "Insight into singular vortex flows", Fluid Dyn. Res. 10, (1992) pp. 101-115.

[14] Saffman, P. G.: "Vortex dynamics", Cambridge University Press, 1992.

[15] Batchelor, G. K.: "An introduction to fluid dynamics", Cambridge University Press, 1967.

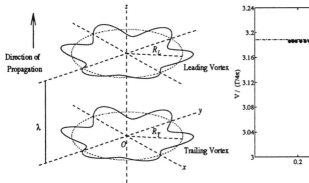

Figure 1 Two perturbed vortex rings at the start of computation.

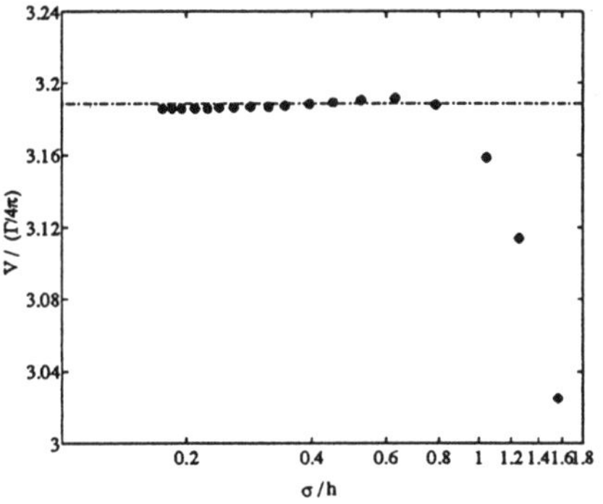

Figure 2 Effect of σ/h on the convergence of velocity of single vortex. $-\cdot-\cdot-$, Saffman [14].

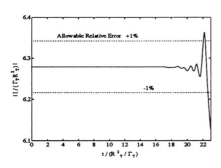

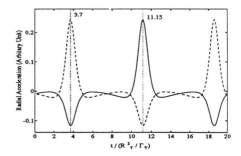

Figure 3 Temporal variation of linear vorticity impulse of the vortex system.

Figure 4 Temporal variation of radial accelerations of unperturbed vortex rings. ———, Trailing Vortex, $-\cdot-\cdot-$, Leading Vortex.

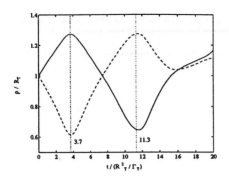

Figure 5 Temporal variation of radii of circular axes. ———, Trailing Vortex, — · — · — · —, Leading Vortex.

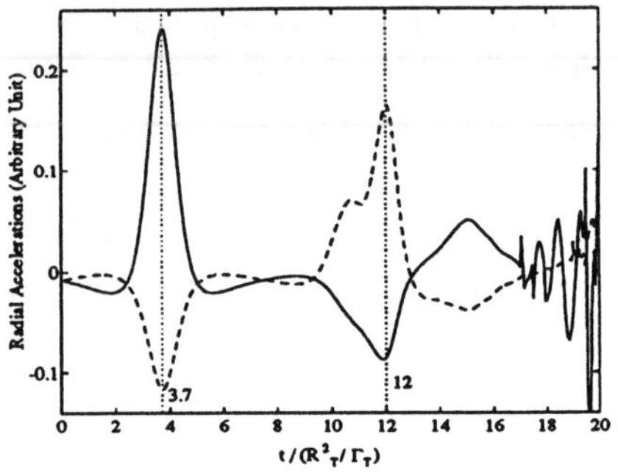

Figure 6 Temporal variation of radial accelerations of circular radii. ———, Trailing Vortex, — · — · — · —, Leading Vortex.

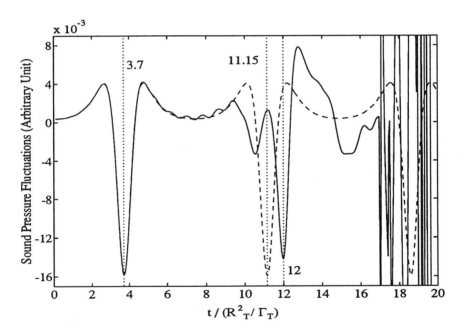

Figure 7 Sound pressure fluctuations observed at $120R_T$ on the x-axis. — · — · — · —, Sound radiated from two unperturbed coaxial vortex rings.

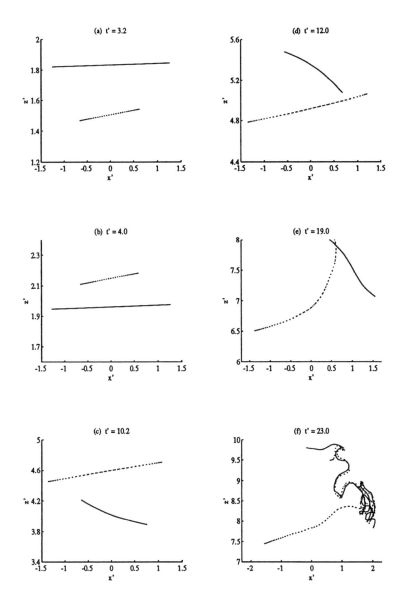

Figure 8 Topology of two interacting weakly perturbed vortex rings. ...., Trailing Vortex, — · — · — · —, Leading Vortex, $t' = t/(R_T^2/\Gamma_T), x' = x/R_T, z' = z/R_T$.

AERODYNAMICS OF COMBUSTION CHAMBERS FOR AERONAUTICAL ENGINES

V. Michelassi, F. Martelli, F. Pigari
Energetics Department, University of Florence
Via S.Marta 3, 50139 Florence, Italy

SUMMARY

The aerodynamics of combustion chambers for aeronautical engines is studied by means of a compressible Navier-Stokes solver without chemical reactions. The three-dimensional (3-D) code is designed for the simulation of complex geometries by introducing non orthogonal curvilinear coordinates. The implicit solver, which is based on the scalar approximate factorization method, accounts for turbulence effects by using a zero-equation turbulence model and solves in terms of primitive variables. Particular care is devoted to boundary conditions to account properly for swirl velocities and cooling air lateral holes. The code showed good flexibility and design capability for 3-D complex flow fields. The tests show that the presence of circumferential and toroidal recirculation zones can be controlled by either the inlet swirl velocity or the lateral jet holes.

INTRODUCTION

The accurate prediction of the flow field inside combustion chambers is a crucial task for the design and development of gas turbine engines for both aeronautical and power plants applications [1,2]. Aeronautical engines have additional constraints due to the necessity of reducing size and weight of the apparatus, so that an advanced design is compulsory for the components optimization. Although the chemical reactions are of primary importance for the flow development inside the combustion chamber, due to the large density and temperature gradients, the prediction of the cold aerodynamic response of a combustor might already give some qualitative informations in the design step. Modern combustors are normally divided into three main areas; first the fuel injection area, the secondary area where additional fresh air is injected, and the third diluition zone where the combustion is stopped by injecting further air before the turbine first stage.

Air is injected inside combustors by axial swirlers in proximity to the fuel injectors. Additional lateral holes are used to control the fuel-to-air ratio and increase the turbulence level for a better combustion. Further, film cooling holes are drilled to prevent damages to the combustion chamber side walls. The inlet swirl gives a sub- or supercritical flow depending on the ratio of the axial to angular inlet momentums. Most of the gas turbine combustors are designed to have a supercritical flow immediately downstream the inlet section, with a subsequent toroidal recirculation downstream of which the flow reverts to subcritical [3,4]. The sudden increase in cross section facilitates the vortex breakdown phenomena, as spotted by Benjamin [5]. The first effect of the inlet swirl, with the growth of the toroidal recirculation, is an increase of the so called "mean residence time" since it is possible to have high velocities with a small axial component. A high swirl number is normally obtained by injecting the cool air through a bladed section, or by using a rotating inlet section (to obtain a "solid body rotation" effect). Lateral holes in the first and second sections are normally intended to increase the turbulence level and gradually increase the amount of fresh air to control combustion. A large swirl might induce large total pressure losses with a detrimental effect on the overall performances of the engine, so that alternative approaches are often investigated. This paper is mainly devoted to the investigation of the effect of lateral jets on the aerodynamics of the combustion chamber and to the interaction between lateral jets and inlet swirl for the development of lateral and toroidal recirculation zones.

MATHEMATICAL FORMULATION: THE FLOS3D CODE

The FLOS3D program [6,7] is based on a scalar implicit algorithm [8]. The equations, written in compact vector form, are discretized by centered finite differences in a curvilinear non orthogonal coordinate system. The program solves three-dimensional compressible flows with complex boundaries in inviscid, laminar and turbulent regime. In order to make the approximate factorization method less computational costly Pulliam and Chaussée [8] proposed a scalar form which drops the viscous contribution to the implicit operator which was found weak for internal turbulent flows. To cure this problem the implicit side is modified to account for an approximate expression of the viscous eigenvalues. The modified algorithm reads:

$$T_\xi \cdot \left[I + \Delta t\left(\delta_\xi \Lambda_\xi - \delta_\xi^2 \Lambda_\xi^v\right)\right] \cdot N \cdot \left[I + \Delta t\left(\delta_\eta \Lambda_\eta - \delta_\eta^2 \Lambda_\eta^v\right)\right] \cdot P \cdot \left[I + \Delta t\left(\delta_\zeta \Lambda_\zeta - \delta_\zeta^2 \Lambda_\zeta^v\right)\right] \cdot T_\zeta^{-1} \cdot \Delta Q = \text{RHS} \quad (1)$$

in which Λ and T are the eigenvalues and eigenvectors respectively, and $\Delta Q = Q^{n+1} - Q^n$, the matrices $N = T_\xi^{-1} T_\eta$ and $P = T_\eta^{-1} T_\zeta$ are solution independent. The vector of unknowns is $Q = [\rho, \rho u_1, \rho u_2, \rho u_3, e]^T$, and the eigenvalue vectors are:

$$\Lambda_l = \left(U_l, U_l, U_l, U_l + a\sqrt{l_{x_1}^2 + l_{x_2}^2 + l_{x_3}^2}, U_l - a\sqrt{l_{x_1}^2 + l_{x_2}^2 + l_{x_3}^2}\right)^t$$

$$\Lambda_l^v = \left(0, \frac{\rho \mu_{\mathit{eff}}\left(l_{x_1}^2 + l_{x_2}^2 + l_{x_3}^2\right)}{Re}, \frac{\rho \mu_{\mathit{eff}}\left(l_{x_1}^2 + l_{x_2}^2 + l_{x_3}^2\right)}{Re}, \frac{\rho \mu_{\mathit{eff}}\left(l_{x_1}^2 + l_{x_2}^2 + l_{x_3}^2\right)}{Re}, \frac{\gamma \rho \mu_{\mathit{eff}}\left(l_{x_1}^2 + l_{x_2}^2 + l_{x_3}^2\right)}{Re}\right)^t$$

in which $l = \xi, \eta, \zeta$, and U_l is the contravariant velocity defined as $U_l = l_{x_1} u_1 + l_{x_2} u_2 + l_{x_3} u_3$. Further details about the eigenvector matrices may be found in Pulliam and Chausshée [8] and in Michelassi et al. [6,7]. Artificial damping terms are introduced in both the implicit and the explicit sides of the operator in equation (1). The non-linear artificial second plus fourth order damping formulation is the one by Jameson et al.[9]. Equation (1) is modified as follows:

$$T_\xi \left\{I + \theta \Delta t \left[\delta_\xi \Lambda_\xi - \left(\delta_\xi^2 \Lambda_\xi^v + \delta_\xi\left(\Omega^\xi \omega_\xi^2 \delta_\xi(J)\right)\right) + \delta_\xi\left(\Omega^\xi \omega_\xi^4 \delta_\xi^3(J)\right)\right]\right\} \cdot$$
$$N \left\{I + \theta \Delta t \left[\delta_\eta \Lambda_\eta - \left(\delta_\eta^2 \Lambda_\eta^v + \delta_\eta\left(\Omega^\eta \omega_\eta^2 \delta_\eta(J)\right)\right) + \delta_\eta\left(\Omega^\eta \omega_\eta^4 \delta_\eta^3(J)\right)\right]\right\} \cdot \quad (2)$$
$$P \cdot \left\{I + \theta \Delta t \left[\delta_\zeta \Lambda_\zeta - \left(\delta_\zeta^2 \Lambda_\zeta^v + \delta_\zeta\left(\Omega^\zeta \omega_\zeta^2 \delta_\zeta(J)\right)\right) + \delta_\zeta\left(\Omega^\zeta \omega_\zeta^4 \delta_\zeta^3(J)\right)\right]\right\} \cdot T_\zeta^{-1} \cdot \Delta Q =$$

$$= \Delta t \left(-\frac{\partial F_k}{\partial x_k} + \frac{\partial F_k^v}{\partial x_k} + D_k^2 - D_k^4\right)^n$$

The terms D_k^2 and D_k^4 are second and fourth order differences of the transported quantities given by vector Q. In the i-direction these extra terms are discretized as follows:

$$D_i^2 = \nabla_i\left(\Omega^\xi_{i,j,k}\omega^2_{i,j,k}\right)\Delta_i(J\, Q_{i,j,k}) \quad, \quad D_i^4 = \nabla_i\left(\Omega^\xi_{i,j,k}\omega^4_{i,j,k}\right)\Delta_i\nabla_i\Delta_i(J\, Q_{i,j,k})$$

in which Δ and ∇ are forward and backward difference operators respectively, and the ω^2, ω^4 coefficients are the artificial terms weights computed as suggested by Jameson et al.[9]. The Ω terms represent a directional scaling as a function of the directional spectral radius:

$$\Omega^\xi = \lambda_\xi + \left(\lambda^\sigma_\eta + \lambda^\sigma_\zeta\right)\cdot\lambda_\xi^{(1-\sigma)} \;;\; \Omega^\eta = \lambda_\eta + \left(\lambda^\sigma_\xi + \lambda^\sigma_\zeta\right)\cdot\lambda_\eta^{(1-\sigma)} \;;\; \Omega^\zeta = \lambda_\zeta + \left(\lambda^\sigma_\xi + \lambda^\sigma_\eta\right)\cdot\lambda_\zeta^{(1-\sigma)}$$

in which:

$$\lambda_l = |U_l| + a\sqrt{l^2_{x_1} + l^2_{x_2} + l^2_{x_3}}\,.$$

For the local time step, a wide number of numerical tests showed that in case of turbulent viscosity fields with large spatial variations, the introduction of the viscous contribution into the time step formula gave the best convergence rate. The expression adopted in the code is:

$$\Delta t = \frac{CFL}{1/\Delta t_{conv} + 1/\Delta t_{diff}}$$

in which the convective and diffusive time steps, Δt_{conv} and Δt_{diff}, are defined as:

$$\Delta t_{conv} = \frac{1}{|U_\xi| + |U_\eta| + |U_\zeta| + a\sqrt{\xi^2_{x_1} + \xi^2_{x_2} + \xi^2_{x_3} + \eta^2_{x_1} + \eta^2_{x_2} + \eta^2_{x_3} + \zeta^2_{x_1} + \zeta^2_{x_2} + \zeta^2_{x_3}}}$$

$$\Delta t_{diff} = \frac{1}{C_d \frac{\mu_{eff}}{Re}\left(\xi^2_{x_1} + \xi^2_{x_2} + \xi^2_{x_3} + \eta^2_{x_1} + \eta^2_{x_2} + \eta^2_{x_3} + \zeta^2_{x_1} + \zeta^2_{x_2} + \zeta^2_{x_3}\right)}$$

C_d allows weighting the viscous term contribution, "a" is the sound speed. Best results are obtained with C_d ranging between 1. and 2.5. The *CFL* number ranges between 7 to 15.

Inlet and lateral jet holes are modelled by fixing the total pressure and temperature, and specifying the flow angle. Details about the periodic boundary treatment my be found in Michelassi et alt. [10].

RESULTS

The FLOS3D code has been developed and validated for the flow simulation of aeronautical and aerospace propulsion system. The particularly complex geometry and flow field indicated the combustor designed by ENEL [11] for gas turbine applications as an interesting test. The scaled geometry was gridded by using 131×25×25 nodes, which cover only a 60-deg portion of the full combustor due to the cylindrical symmetry of the overall geometry, including the lateral holes. Figure 1 shows a perspective view of the grid utilized for the calculations. The computations were carried out so as to have an exit Mach number around 0.1-0.2 which matches the IGV inlet Mach number. The inlet velocity is computed so as to maintain the given exit Mach number, while the inlet flow angle is adjusted to have the desired swirl number S. Only the two rows of lateral holes are modelled, while the film cooling holes are neglected regarding their influence on the overall flow development as negligable. Runs have been performed varying the inlet swirl in the range 0 to 1.5, while the lateral jet strenght has been adjusted by changing the ratio between the main inlet to lateral jet total pressures. The flow was kept at constant total temperature and the Reynolds number is approximately 5×10^5.

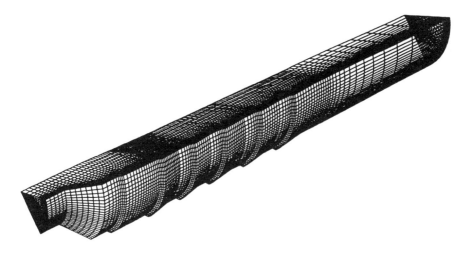

Figure 1. Combustion chamber grid 131×25×25 (60-deg sector shown).

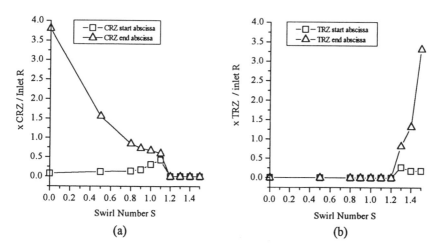

Figure 2. Corner(CRZ) and Toroidal(TRZ) recirculation separation and reattachment points.

Although a simple algebraic turbulence model may be inadequate for the simulation of such a complex flow field (see references [12,13]), the Baldwin-Lomax model [14] was adopted for all the calculations. This choice is motivated by the simplicity and relative fast convergence of the code in practically all the test runs. Due to lack of space only few results will be reported here. The main aim of the tests was to discover when and how the penetration of the lateral jets could provoke a toroidal recirculation at low swirl number, so that the usual vortex breakdown and the transition to a supercritical phase was not induced by an increase in the inlet swirl number. Figure 2 summarizes the large number of tests carried out without lateral flow injection. Figure 2.a shows that the corner recirculation is gradually reduced as the inlet swirl number S increases. Notably, at S=1.2 there is no corner recirculation. For the same S, there is no toroidal recirculation, as shown in figure 2.b. For larger values of S the TRZ size gradually increases. Apparently the S=1.2 case is free of recirculations and appears to be approximately in the critical regime since a small increase in the inlet swirl produces a dramatic change in the flow field.

The introduction of two rows of lateral jets modifies the flow pattern. Figure 3 shows two typical cross sections across the first and second row of the lateral air flow injection. In figure 3.a the velocity vectors show that for S=0.5 and $P_{0\text{-jet}}=1.3 \times P_{0\text{-inlet}}$ there is flow recirculation on both sides of the jet, whereas the upstream recirculation disappears for the second row of holes in which $P_{0\text{-jet}}=1.1 \times P_{0\text{-inlet}}$. While the second row of lateral holes does not substantially alter the flow field, the first row has the abity of producing the toroidal recirculation at low swirl number S. Figure 4 shows the particle traces of the lateral jets for S=0.5 and $P_{0\text{-jet}}=1.3 \times P_{0\text{-inlet}}$ for the first row and $P_{0\text{-jet}}=1.1 \times P_{0\text{-inlet}}$ for the second row. The plot shows no TRZ. Figure 5 shows the same flow visualization for S=0.5 and $P_{0\text{-jet}}=1.5 \times P_{0\text{-inlet}}$ for the first row and $P_{0\text{-jet}}=1.1 \times P_{0\text{-inlet}}$ for the second row. The flow clearly experiences a toroidal recirculation which takes place at a swirl number in which it was absent without lateral flow injection. This phenomena is caused by the lateral jet penetration which is enough to produce the TRZ only when its total pressure is approximately 150% of the inlet P_0. The bifurcation of the lateral jet stream takes place in proximity of the symmetry axis. This indicates that the jet needs a large momentum to act as barrier to the axial flow for the onset of the TRZ. Observe that the TRZ separation point does not move further upstream increasing the lateral jet momentum.

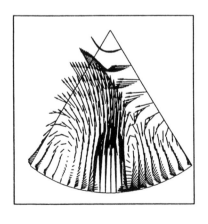   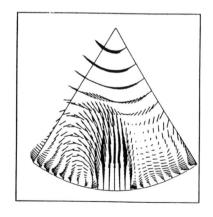

(a)   (b)

Figure 3. Cross sections

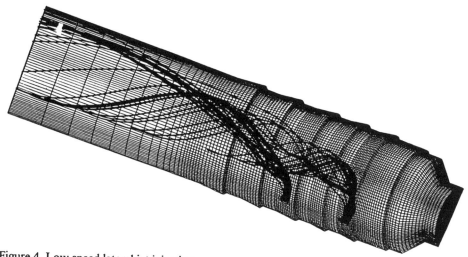

Figure 4. Low speed lateral jet injection.

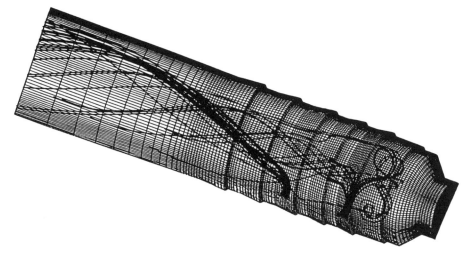

Figure 5. High speed lateral jet injection.

CONCLUSIONS

The FLOS3D code proved able to predict the complex flow field inside a gas turbine combustor in absence of chemical reactions. The large set of calculations carried out on the cylindrical combustor clearly evidentiated a vortex breakdown phenomena at S=1.2 in absence of lateral jet holes. The presence of lateral jet holes produce a dramatic change in the flow field, provided that the jet momentum is enough for a significant penetration in the main stream. Here the toroidal recirculation starts at S=0.5. The recirculation size is enough to produce a large turbulence level and increase the mean residence time with inlet tangential momentum values far below those required in absence of lateral air injection. This phenomena is strictly linked to the relative values of the cooling air and tangential momentum, but the large number of important parameters still require further calculations by using more refined grids. Futher calculations will indicate the relation between inlet angular momentum and lateral jet momentum which provokes the toroidal recirculation. The effect of the geometry has to be carefully investigated as well. The computations proved that even with a simple turbulence model it is possible to obtain overall informations about the flow field without combustion processes.

ACKNOWLEDGMENTS

The authors would like to gratefully acknowledge CINECA for granting computer time on a CRAY C-90 supercomputer, ENEL for providing the geometrical data of the combustor, and ENEL and ASI for partially supporting the research.

REFERENCES

[1] Wassel, A.B., "Combustion Problems in Gas Turbine Applications: The Design and Development of High Performance Combustors", Von Karman Institute for Fluid Dynamics, January 10-14 1977.
[2] McGuirk, J.J., J.M. Palma, "The Flow Inside a Model Gas Turbine Combustor: Calculations", Transaction of the ASME, Vol. 115, July 1993.
[3] Ahmed, S.A., Nejad, A.S., "Swirl Effects on Confined Flows in Axisymmetric Geometries", Journal of Propulsion and Power, Vol.8, No. 2, March-April 1992.
[4] Escudier, M.P., Keller, J.J., "Recirculation in Swirling Flow: A Manifestation of Vortex Breakdown", AIAA JOURNAL, Vol. 23, No. 1, January 1985.
[5] Benjamin, T.B., "Significance of the Vortex Breakdown Phenomenon", Transaction of the ASME, June 1965.
[6] Michelassi, V., Liou, M.-S., Povinelli, L.A., Martelli, F., 1990, "Implicit Solution of Three-Dimensional Internal Turbulent Flows", NASA CP-10045.
[7] Michelassi, V., Martelli, F., 1993, "3-D Implicit Navier-Stokes Solver for Internal Turbulent Compressible Flows", Journal de Phys. III, France 3, pp. 223-235.
[8] Pulliam,T.H., Chaussee, D.S., 1981, "A Diagonal Form of an Implicit Approximate-Factorization Algorithm". Journal of Computational Physics, N. 39.
[9] Jameson, A., Schmidt, W., Turkel, E., 1981 : "Numerical Solutions of the Euler Equations by Finite Volume Methods Using Runge-Kutta Time-Stepping Schemes", AIAA 81-1259.
[10] Michelassi, V., Theodoridis, G.S, Papanicolaou, E.L., "Low Speed Turbine Computation by Pressure-Correction and Time-Marching Methods", Proceedings, ASME 1995 International Mechanical Engineering Congress & Exposition (IMECHE), 12-17 November 1995, San Francisco (CA), USA.
[11] Benelli, G., Private Communication, ENEL-Italy, 1994.
[12] Hogg, S., Leschziner, M.A., "Computation of Highly Swirling Confined Flow with a Reynolds Stress Turbulence Model", AIAA JOURNAL,Vol. 27, No. 1, January 1989.
[13] Fu, S., Huang, P.G., Launder, B.E., Leschziner, M.A., "A Comparison of Algebraic and Differential Second-Moment Closures for Axiaimmetric Turbulent Shear Flows Whith and Without Swirl", Transaction of the ASME, Vol. 110, June 1988.
[14] Baldwin, B.S., Lomax, H., "Thin Layer Approximation and Algebraic Model for Separated Turbulent Flows", AIAA 16TH AEROSPACE SCIENCES MEETING Huntsville, Alabama January 16-18, 1978.

TIME-DEPENDENT 3D NUMERICAL SIMULATION OF OLDROYD-B FLUID USING FINITE VOLUME METHOD

G. Mompean and M. O. Deville
IMHEF-DGM, EPFL, Switzerland 1015

Abstract

A *three-dimensional* numerical simulation of an Oldroyd-B fluid through a 4:1 planar contraction was performed using a time-dependent finite volume method on a staggered grid. The forward Euler scheme was employed to march in time. The non-linear terms in the momentum equations were discretized using a quadratic upstream interpolation. The program was validated on the start-up Couette flow at Reynolds number 1 and different Deborah numbers from 0.5 to 900, and the 2D 4:1 planar contraction of a Newtonian fluid. The time-dependent nature of these flows is analysed taking the solution at various times, from the initial condition to the steady-state, as frames in a motion picture. Results for vortex formation and growth are presented, then the differences between the simulated 2D and 3D planar contractions are discussed. To our knowledge, this is the first time-dependent numerical simulation of a viscoelastic fluid through a *three-dimensional* 4:1 contraction.

1 INTRODUCTION

The 4:1 planar contraction has been used to test the numerical methods as well as the constitutive equations of the fluids. The flow through this geometry is exposed to shearing and extensional effects at the same time. The existence of a stress singularity at the corner has been a hard test case for numerical methods. The flow of a viscoelastic fluid through contractions has been studied by many researchers. The 4:1 planar contraction for an Oldroyd-B fluid has been simulated numerically for the *two-dimensional* case by several authors. Yoo and Na [1] use a finite volume technique to simulate a steady-state 2D flow of an Oldroyd-B fluid. Marchal and Crochet [2] show some results of a 2D numerical calculation using rectangular finite elements. A time-dependent two-dimensional numerical simulation has been made by Bodart and Crochet [3].

The prediction of flows for a high Deborah number has been a challenge for several numerical methods. As pointed out by Boger [4], it is clear that the three-dimensional and the time dependence characteristics of the flow have to be taken into account.

The aim of the present paper is to show some results of a 3D time-dependent numerical simulation using a finite volume method attaining high Deborah numbers. The conservation equations are discretized using a staggered grid, this method is derived from SOLA [5], a technique rarely used in non-Newtonian numerical works. Recently, Resh and Kuster [6] have presented briefly the design of an algorithm in 2D for time-dependence. The paper of Bschorer and Brunn [7] has shown some results for a 2D 4:1 planar contraction flow of an Oldroyd-B fluid using the SIMPLE algorithm [8]. In this study all inertia terms are taken into account, even though we solve a small Reynolds number (creeping) flow. In section 2 we present the transport equations. The numerical method and the boundary conditions are shown in section 3. In section 4, we present the results for a time-dependent start-up Couette flow as a validation case, and then the 2D and 3D 4:1 planar contraction for the Newtonian and Oldroy-B fluids.

2 TRANSPORT EQUATIONS

In this paper, the flow of a viscoelastic fluid is considered. The equations governing the velocity components (U_i) and pressure (P) field are obtained from the mass and momentum equations. The mass conservation imposes the velocity field to be divergence-free. The momentum equation is:

$$\rho \left(\frac{\partial U_i}{\partial t} + \frac{\partial}{\partial x_j}(U_j\, U_i) \right) = -\frac{\partial P}{\partial x_i} + \frac{\partial}{\partial x_j}(\eta_o \frac{\partial U_i}{\partial x_j} + T_{ij}), \qquad (2.1)$$

where η_o is the Newtonian viscosity, ρ the fluid density and T_{ij} the viscoelastic extra-stress components. For an Oldroyd-B fluid the following constitutive relation is assumed [9]:

$$T_{ij} + \lambda(\frac{\partial T_{ij}}{\partial t} + \frac{\partial}{\partial x_m}(U_m T_{ij}) - \frac{\partial U_i}{\partial x_k}T_{kj} - \frac{\partial U_j}{\partial x_k}T_{ki}) = \eta_1(\frac{\partial U_i}{\partial x_j} + \frac{\partial U_j}{\partial x_i}). \qquad (2.2)$$

Here, λ is the relaxation time and η_1 the constant viscosity of the viscoelastic extra-stress tensor. The Reynolds number is defined as $Re = \rho <u_x> D/\eta$, where $\eta = \eta_o + \eta_1$. For the start-up flow, the reference $<u_x>$ is the velocity of the plate, and for the contraction, it is the average velocity in the small channel defined as $<u_x> = Q/hD$, where Q is the volumetric flow rate, h the channel width and D the channel height.

The dimensionless number describing the elastic effects is called Deborah number and is given by $De = \lambda \dot\gamma$, where $\dot\gamma$ is the characteristic shear rate for the flow. For the planar contraction $\dot\gamma$ will be evaluated as $\dot\gamma = 2<u_x>/D$

3 NUMERICAL METHOD

The numerical method used to solve the set of equations is a classical finite volume technique. The conservation equations are integrated over a control volume, and then the Gauss theorem is used to transform the volume integrals into surface integrals. The method uses a staggered mesh; the pressure and the normal extra-stress components are treated at the center of the control volume; the velocities are computed in the center of the faces and the cross components of the extra-stress tensor are attached to nodes located at the mid-edges. Figure 1 shows the position of the variables. The non linear terms in the momentum equation are discretized using a quadratic upstream interpolation proposed by Leonard [10], namely QUICK. The transient solution is obtained by a time-marching algorithm. The extra-stress components are treated by an explicit Euler scheme. The decoupling procedure for the pressure is derived from a SOLA type algorithm, Hirt et al. [5]. Below is shown the time discretization used for the momentum conservation equations:

$$\frac{U^{n+1} - U^n}{\Delta t} + [\nabla.(UU) - \nabla.(\nu_o \nabla U - T)]^n + \frac{1}{\rho}\nabla P^{n+1} = 0; \quad (3.3)$$

or in another form:

$$U^{n+1} = \Delta t (SU^n - \frac{\nabla P^{n+1}}{\rho}), \quad (3.4)$$

where SU^n is the explicit term of Eq. (3.3). The time step is denoted by Δt. The method of solution consists in substituting the equation for the velocity at time $n+1$ (Eq. (3.4)) into the discretized equation for mass conservation at the same time level. This yields a Poisson pressure equation. The matrix of the pressure system is positive definite and symmetric. It can be solved either by a Cholesky factorization or a preconditioned conjugate gradient method. Afterwards, the velocities are calculated from Eq. (3.4). The solution results in a velocity and pressure field that enforce the conservation equations.

The choice of the time step was made using the Courant-Friedrich-Lewy (CFL) condition. At this stage no attempt was made to take into account the limit that the equations (2.2) of the extra-stress tensor may impose. We note that for the worst case where the time step imposed restrictions for the convergence, the problem was solved using the global time step divided by two.

For the simulation of the planar contraction, the boundary conditions used for the extra-stress tensor were the usual ones. In the *entry section*, far from the step (see section RESULTS), all the values were set to zero. In this section (*entry*), a constant value for the pressure was given, which has permitted to obtain a calculated velocity profile.

In the *outlet section* the same procedure was used for the pressure, setting it to zero. For the extra-stress components the Neumann homogeneous condition was used. At all the walls the no-slip condition was employed.

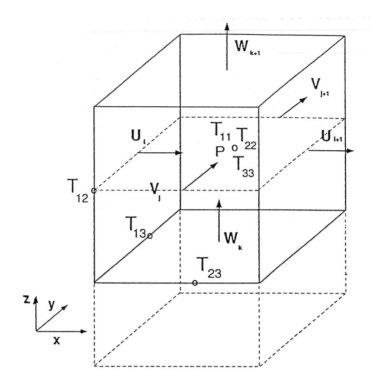

Figure 1 Variables position.

4 RESULTS

For all the cases, we have used the classical value of the relation between the Newtonian and the viscoelastic viscosity for the Oldroyd B, respectively $\eta_o = 1/9$ and $\eta_1 = 8/9$.

4.1 Time-dependent start-up Couette flow

For this simulated shear flow, the Reynolds number was set to 1.0. The calculations were compared with the analytical solution [11]. The upper plate is moved from rest to the velocity U instantaneously; this is known as the start-up Couette flow. The results obtained for a Newtonian fluid are in excellent agreement with the analytical solution. We note that the velocity profile reached a linear distribution after a short time (T=0.4s), the evolution between T=0.s and T=0.4s shows the monotonic increase of the velocity.

For the viscoelastic fluid we carried out the validation varying the Deborah number ($\lambda U/L$, where L is the distance between the plates) from 0.5 to 900. We obtained an excellent agreement between analytical and numerical solutions, even for a very high Deborah number (De=900).

4.2 Two dimensional flow through a 4:1 planar contraction

Newtonian fluid

To validate the code, the creeping flow of a Newtonian fluid through a two-dimensional 4:1 planar contraction was simulated (Re = 0.071). Several simulations were made changing the entry lengths, and no significant difference was observed in the vortex size as previously observed by Yoo and Na [1]. The independence of the inlet length was verified using D_u and $D_u/2$ as the dimension for the inlet length; for both cases the dimensionless reattachment length X (see Figure 2), was the same. The dimensionless reattachment length $X = 0.19$ was obtained for the corner vortex. The results are in very good agreement with previous steady-state calculations made by Keunings and Crochet [12], Yoo and Na [1] and measurements [4]. To verify the grid independence for the solution two different grids were employed: a) $100(x) \times 80(z)$ and b) $80(x) \times 40(z)$, using the symmetry condition at the center, where (x) and (z) represent respectively the streamwise and the normal directions. The solutions were similar, and the difference for the reattachment length obtained with the two grids was less than 0.5 percent.

Oldroyd-B fluid

Several cases were simulated for different Deborah numbers. The same test of grid independence employed for the Newtonian case described above was carried out. All results show that for the Oldroyd-B fluid the steady-state solution was also similar to the Newtonian case, giving exactly the same reattachment length. The comparison between Newtonian and viscoelastic fluids was made using the same mesh configuration. For the same boundary conditions, imposed pressure gradient between the inlet and the outlet, the Reynolds numbers of the flow change from 0.08 (Newtonian) to 0.138 (viscoelastic, De = 5.2) due to the inclusion of the viscoelastic viscosity. These Reynolds numbers are inside the range that characterizes creeping flows.

The major difference is in the transient state of the solution. Visual observations of the time evolution for the streamlines near the corner show that the transient solution for the Oldroyd-B fluid differs substantially from the Newtonian transient solution. The visual analysis was made possible animating the solutions for different time steps. We have used about fifty images to describe

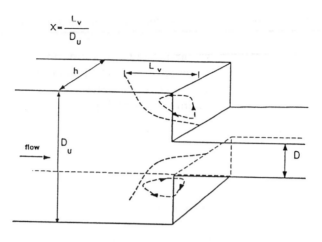

Figure 2 Geometry for the 3D 4:1 planar contraction.

the time dependent nature of the solution from the beginning - initial conditions equal zero for all fields (velocity, pressure and extra-stress) - up to the steady state.

4.3 Three dimensional flow through a 4:1 planar contraction

In Figure 2 the configuration used for the simulation is shown. The ratio $D_u/h = 1$ was chosen to observe the three-dimensional effects.

The grid independence was verified with two different grids: *a)* 50x14x30 and *b)* 100x20x40. For both grids, a greater concentration of points near the corner was used. The results presented and commented here were obtained with the grid *b*. Only one quadrant of the whole contraction was simulated, due to symmetry considerations. The mesh used and the streamlines obtained for a case with De = 27.3 are shown in Figure 3.

The first calculations used an entry length of 2.5 D_u while the downstream length was 10 D. To save memory space and computing time, we have reduced the dimensions of the domain, because we were only interested in what happened near the corner. We observed that the solution did not change until an entry length of 1.5 D_u and a corresponding downstream length of 2.5 D were employed.

The comparison between the streamlines obtained for the 2D and 3D problems, shows that the size of the center vortex (in the symmetry plane) for the 3D case is always smaller than the 2D one.

The time-dependence of this flow was observed in a film made from several frames of the solutions at different time-steps, for Deborah numbers (3.2, 7.5 and 27.3). In Figure 3 are shown the streamlines for one intermediate time for

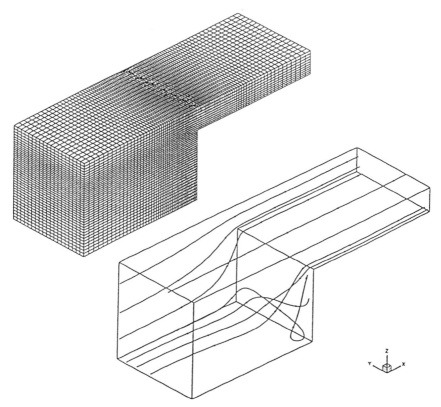

Figure 3 Mesh used 100x20x40 (x,y,z) and streamlines for the Oldroy-B fluid with De = 27.3.

De = 27.3 . The visual observation shows a very complex vortex formation and growth near the corner.

We have calculated the eigenvalues of the tensor $\mathbf{T_A}$, defined by:

$$\mathbf{T_A} = \mathbf{T} + (\frac{\eta_1}{\lambda})\mathbf{I} . \tag{4.5}$$

During the transient and the steady-state solutions we have found negative values for the eigenvalues of $\mathbf{T_A}$ in the region near the corner, but no problem with the convergence was observed. The positive-definiteness of the tenseur $\mathbf{T_A}$ was not required for the Oldroyd-B fluid in order to converge the algorithm.

5 CONCLUSIONS

The code was validated using the start-up Couette flow (Re = 1.0), and for all the cases (We = 0.5, 2.0, 5.0 up to 900) the algorithm converged. The vortex size for the 4:1 2D contraction of a Newtonian fluid agrees very well with previous computations for creeping flow. A converged solution was always obtained for the Oldroy-B fluid ($\eta_o = 1/9$ and $\eta_1 = 8/9$), through the 4:1 3D contraction.

The algorithm was tested up to De = 157. A steady state solution was always found for all cases. From the observation of the streamlines we saw that the 3D results differ substantially from the 2D results. The code runs at 72 MFlops. The CPU time for the 4:1 3D contraction (De = 27.3), from the initial state to the steady-state, using the grid (100x20x40) was about 10 hours in a CRAY-YMP. This is a reasonable time for a 3D viscoelastic time-dependent calculation, that became possible using finite volume as the numerical method.

Future simulations will be extended to another choice of constitutive equations beyond the Oldroyd-B fluid. The next choice should incorporate the extensional viscosity of the test fluid.

References

[1] J. Y. Yoo and Y. Na, "A numerical study of the planar contraction flow of a viscoelastic fluid using the SIMPLER algorithm," J. Non-Newtonian Fluid Mech. **39**, 89 (1991).

[2] J. M. Marchal and M. J. Crochet, "Hermitian finite elements for calculating viscoelastic flow," J. Non-Newtonian Fluid Mech. **20**, 187 (1986).

[3] Ch. Bodart and M. J. Crochet, "Time-Dependent Numerical Simulation of Viscoelastic Flow and Stability", Theor. and Comput. Fluid Dynamics **5**, 57 (1993).

[4] D. V. Boger, "Viscoelastic flows through contractions", Annu. Rev. Fluid Mech. **19**, 157 (1987).

[5] C. W. Hirt, B. D. Nichols and N. C. Romero, "SOLA - Numerical solution algorithm for transient fluid flow," Los Alamos Laboratory, Report LA-5852 (1975).

[6] M. Resch and U. Kuster, "On viscoelastic fluid flow simulation using finite volume method," Computational Fluid Dynamics '94, Eds. S. Wagner, E. H. Hirschel, J. Périaux, R. Piva, 989-993 (1994).

[7] S. Bschorer and P. O. Brunn, "Finite volume method for simulation of non-Newtonian fluid flow," Computational Fluid Dynamics '94, Eds. S. Wagner, E. H. Hirschel, J. Périaux, R. Piva, 989-993 (1994).

[8] V. S. Patankar, *Numerical heat transfer and fluid flow*, (Hemisphere Publishing Corporation, New York, 1980).

[9] D. D. Joseph, *Fluid Dynamics of Viscoelastic Liquids*, (Springer-Verlag, New York, 1984).

[10] B. P. Leonard, "A stable accurate convective modelling procedure based on quadratic upstream interpolation," Comp. Meth. Appl. Mech. Eng., **19**, 59-88 (1979).

[11] J. J. Van Schaftingen, "Méthodes d'éléments finis pour les écoulements viscoélastiques", Thèse de Doctorat, Université Catholique de Louvain, Louvain-la-Neuve (1985).

[12] R. Keunings and M. J. Crochet, "Numerical simulation of the flow of a viscoelastic fluid through an abrupt contraction," J. Non-Newtonian Fluid Mech. **14**, 279 (1984).

A re-analysis of Stanford's 30 degree bend experiment

A.Nakkasyan

IMHEF-DGM, Ecole Polytechnique Fédérale de Lausanne, Switzerland 1015

ABSTRACT

Our two-step data analysis technique is applied to re analyze the second- and third-order velocity correlations measured by Walter R.Schwarz and Peter Bradshaw [1], [2], [3] at Stanford. First, the experimental eigenvalues and eigenvectors of the measured Reynolds-stress tensors are obtained by local matrix diagonalisations. Next, the probability density functions (p.d.f.s) of velocity fluctuations are investigated with a Monte Carlo (MC-) method, assuming three independent (trial) 'perturbed centered Gaussian' distributions along the eigenvectors. Sample-averaged correlations along the measurement axes are compared with corresponding 16 measured profiles. Our results based on statistical ideas compare favorably with the predictions of the eight models discussed by the authors [2]. Measured triple correlations are projected along the experimental eigenvectors, for the first time.

1 INTRODUCTION

In the stationary three-dimensional turbulent boundary layer (3DTBL) of this experiment [1], [2], [3], relatively important cross-stream pressure gradients were generated by forcing a 30 degree bend in the horizontal plane (Figure 1). A total of 22 stations labeled by the authors as 0,1,2,.. 21 are situated along the centerline where the streamwise pressure gradients are quasi-null and the crossflow pressure gradients are negative with a minima near station 8 which is located at x=1875 mm and z=0 mm -where x and z are the 'centerline-coordinates' of the set-up. Accordingly, the potential flow near centerline is accelerating 'more and more' down to station 8 and accelerating 'less and less' beyond. Some of these stations (1,4,7,10,13,18,20) form 'triplets' with two other stations 'A' and 'B' located at distances z=+127 mm and z=−127 mm from their center-line station. At all 36 stations, the measured data on data summary tape or DST [1] include the time-averaged double and triple correlations ($\overline{u_i u_j}_{mes}$ and $\overline{u_i u_j u_k}_{mes}$; with i,j,k=1,2,3 and '$_{mes}$' for measured) along the -more conventional- coordinate axes ($\hat{x}$', $\hat{z}$', $\hat{y}$') fixed at the upstream test section where 2DTBL conditions are more or less satisfied. Symbol/station identifications are avoided by plotting our profiles for stations 0,1,2,3,... with the letters A,B,C,D,..., respectively. Consequently, enclosed figures for the authors' categories "crossflow developments (stations 0 to 12)" and " crossflow decay (stations 12 to 20)" are plotted with symbols 'A to M' and 'A; M to U' (A to remind the upstream characteristics; U for station 20)'. Note that the *slopes* of crossflow Cp gradients (where Cp is the the pressure-coefficient) are near zero and positive at stations 9 to 12 (I to M). Profiles involving 'triplet' stations 'A' and 'B' are presented with the symbols '+' and '-', respectively. Unfortunately, the measured Cp gradients were not precise enough to establish the existence or the absence of one-to-one correlations between the observed intrinsic properties (inside the 3DTBL) and the properties of the *local* driving forces (the first- and second-order derivatives of Cp).

Our data analysis technique [4], [5] was developed, starting 1988, with the purpose of identifying the gradual changes in the statistical properties of homogeneous turbulence [$\overline{u_1 u_1} = \overline{u_2 u_2} = \overline{u_3 u_3} = \overline{q^2}/3 \equiv (\overline{u_1 u_1} + \overline{u_2 u_2} + \overline{u_3 u_3})/3$]; first, put in contact with a smooth boundary (a wall); and next, submitted to pressure gradients. It is optimized to detect the existence of privileged directions and to determine departures from centered-Gaussian distributions inside 2D and 3D TBLs, in view of studying data from the point of view of statistical physics.

2 DATA ANALYSIS METHOD

A stationary, ergodic flow is required and the method is applied, in two steps, at each measurement position of a measurement station.

Part I involves the local diagonalisation of a fully measured 3x3 symmetric Reynolds-stress tensor $(\overline{u_i u_j})_{mes}$, to obtain the experimental orthonormal eigenvectors $(\hat{d}_1, \hat{d}_2, \hat{d}_3)$ and the corresponding eigenvalues $(\overline{U_1 U_1}, \overline{U_2 U_2}, \overline{U_3 U_3})$ -subscripts '$_{exp}$' are implied whenever subscripts are omitted. By definition, $(\overline{U_1 U_1} > \overline{U_2 U_2} > \overline{U_3 U_3})$ and U_i are statistically-independent variables, since $\overline{U_i U_j} = 0$ for $i \neq j$.

To avoid diagonalisation problems, we use *right-handed* local 'measurement systems' $(\hat{x}_1, \hat{x}_2, \hat{x}_3)$ with axes parallel to a unique axial system, in general. For Stanford's 30 degree bend experiment, our *right-handed* systems $(\hat{x}_1, \hat{x}_2, \hat{x}_3)$ were defined as $(\hat{x}', \hat{z}', -\hat{y}')$ - i.e., $(u_1, u_2, u_3) \equiv (u, w, -v)_{mes}$ along $(\hat{x}_1, \hat{x}_2, \hat{x}_3)$.

At any time t we can write
$$\vec{q}' = u_1 \hat{x}_1 + u_2 \hat{x}_2 + u_3 \hat{x}_3 = U_1 \hat{d}_1 + U_2 \hat{d}_2 + U_3 \hat{d}_3$$
and use the 3x3 coordinate transformation matrix (d_{ij}) whose column j contains the matrix elements (d_{1j}, d_{2j}, d_{3j}) which are the cosine-directors of $\hat{d}_j$ along $(\hat{x}_1, \hat{x}_2, \hat{x}_3)$. To calculate u_i as a function of U_j (i,j=1,2,3) we apply:

$$\begin{pmatrix} u_1 \\ u_2 \\ u_3 \end{pmatrix} = \begin{pmatrix} d_{11} & d_{12} & d_{13} \\ d_{21} & d_{22} & d_{23} \\ d_{31} & d_{32} & d_{33} \end{pmatrix} \begin{pmatrix} U_1 \\ U_2 \\ U_3 \end{pmatrix} . \tag{2.1}$$

We visualize the evolution of $\hat{d}_1, \hat{d}_2, \hat{d}_3$ in two different way. A global view is obtained with 9 sub-figures (d_{ij} vs y [mm]; with y$\leq \delta$, where δ is the boundary layer thickness), disposed like the elements of a 3x3 matrix. Plus, the depth-wise evolution of the tip-positions of eigenvectors are traced on the surface of a sphere which is station-wise fixed and has radius 1. Then, we find the approximate coordinate-transformation matrix $(d_{ij})_{app}$:

$$\begin{pmatrix} d_{11} & d_{12} & d_{13} \\ d_{21} & d_{22} & d_{23} \\ d_{31} & d_{32} & d_{33} \end{pmatrix} \approx \begin{pmatrix} cos\varphi & -sin\varphi & 0. \\ sin\varphi & cos\varphi & 0. \\ 0. & 0. & 1. \end{pmatrix} \begin{pmatrix} cos\theta & 0. & -sin\theta \\ 0. & 1. & 0. \\ sin\theta & 0. & cos\theta \end{pmatrix}. \quad (2.2)$$

The above approximation indicates that, at the 2DTBL limit (when $\varphi \to 0$), the privileged directions $(\hat{d}_1,\hat{d}_3)$ remain in the vertical plane of symmetry $(\hat{x}_1,\hat{x}_3)$ and a rotation about the horizontal $\hat{x}_2 \equiv \hat{d}_2$ axis takes place with an angle θ. Correspondingly, the local p.d.f.s look like an 'ellipsoid'.

Because we require an ergodic flow, time-averaged correlations along the measurement axes ($\overline{u_i u_j}$ and $\overline{u_i u_j u_k}$) are equal to the sample-averaged correlations along the same axes ($< u_i u_j >$ and $< u_i u_j u_k >$). Also, we have learned in the past (see analysis part II) that the double correlations are in practice insensitive to the details of the p.d.f.s along the eigenvectors $\hat{d}_i$. And therefore, 'centered Gaussian' distributions are perfectly compatible with measured *double* correlation data. Further, each of these p.d.f.s are perfectly defined if their r.m.s.-values are known. So, we can replace sample-averaged *double* correlations $< u_i u_j >$ by a single operation: the vector dot product of the components $(u_{i,rms} \hat{x}_i; u_{j,rms} \hat{x}_j)$ of the r.m.s- vector:

$\vec{q'}_{rms} = u_{1,rms}\hat{x}_1 + u_{2,rms}\hat{x}_2 + u_{3,rms}\hat{x}_3 = (\overline{U_1^2})^{0.5}\hat{d}_1 + (\overline{U_2^2})^{0.5}\hat{d}_2 + (\overline{U_3^2})^{0.5}\hat{d}_3.$

In practice, we replace (d_{ij}) in (2.1) with $(d_{ij})_{app}$ in (2.2) and replace $\vec{q'}$ above (2.1) with $\vec{q'}_{r.m.s.}$ and obtain, for the general case $(u_1, u_2, u_3) \equiv (u,w,v)$;

$\overline{uv} = cos\varphi \cdot sin\theta \cdot cos\theta \cdot (\overline{U_1^2} - \overline{U_3^2})$
$\overline{wv} = sin\varphi \cdot sin\theta \cdot cos\theta \cdot (\overline{U_1^2} - \overline{U_3^2})$
$\overline{uw} = sin\varphi \cdot cos\varphi \cdot ((cos\theta)^2 \overline{U_1^2} - \overline{U_2^2} + (sin\theta)^2 \overline{U_3^2})$
$\overline{uu} = (cos\varphi \cdot cos\theta)^2 \overline{U_1^2} + (sin\varphi)^2 \overline{U_2^2} + (cos\varphi \cdot sin\theta)^2 \overline{U_3^2}$
$\overline{ww} = (sin\varphi \cdot cos\theta)^2 \overline{U_1^2} + (cos\varphi)^2 \overline{U_2^2} + (sin\varphi \cdot sin\theta)^2 \overline{U_3^2}$
$\overline{vv} = (sin\theta)^2 \cdot \overline{U_1^2} + (cos\theta)^2 \cdot \overline{U_3^2}.$

From the above equations we find $\overline{wv}/\overline{uv} = sin\varphi/cos\varphi$. Thus, φ [defined by $(cos\varphi = d_{22}; sin\varphi = -d_{12})$ in (2.2)] is equal to $tan\gamma_\tau \equiv \overline{wv}/\overline{uv}$.

Note, when 2DTBL conditions are approximately satisfied, we simply require (φ=0; $cos\varphi$=1; $sin\varphi$=0) in the preceding equations.

In part II, the MC-method allows us to investigate the p.d.f.s of the velocity fluctuations along the previously determined experimental eigenvectors. Clearly, 'centered Gaussians' are not valid when $\overline{u_i u_j u_k} = < u_i u_j u_k > \neq 0$. And, velocity distributions along measurement axes are generally joint-p.d.f.s, hence complicated. On the other hand, the p.d.f.s along the three eigenvectors $\hat{d}_i$ can be assumed to be independent functions, at least for the lower-order statistics. Therefore, it seems easier to make a small detour by starting with the statistically-independent variables u_i; using the transformation matrices (d_{ij}) or $(d_{ij})_{app}$; and finally, calculating the sample averaged-correlations in the measurement system.

All MC-calculations start with the (trial) simple assumption that the p.d.f. along an eigenvector $\hat{d}_i$ is a 'shifted centered-Gaussian' distribution:

$$u_{i,MC} = u_{i,c.-G.,MC} + \Delta u_{i,MC} \quad (2.3)$$

[in the above equation, '$_{c.-G.}$' stands for the 'centered Gaussian' whose r.m.s. value $(\overline{u_i u_i})^{0.5}_{MC}$ is equal to the previously determined $(\overline{u_i u_i})^{0.5}$; and, $\Delta u_{i,MC} \neq 0$ for $i = 1, 2$ represents the shift/drift of the '$_{c.-G.}$' distribution as a whole].

Technically, the sample-averaged double and triple correlations (along the measurement axes $(\hat{x}_1, \hat{x}_2, \hat{x}_3)$, at each measurement position) are obtained as follows:

(a) First, we MC-generate the three independent components $(u_{i,c.-G.,MC}; i = 1, 2, 3)$ for the n-th time (n=1,2,3,... N) by making (at least) six calls to a random-number-generator function. We choose N=5000, at least. Also, we use tail-less 'centered Gaussians' satisfying $|u_{i,c.-G.,MC}| \leq 3(\overline{u_i u_i})^{0.5}$.

Next, we add the hypothetical $(\Delta u)_{i,MC}$.

(b) We use the coordinate transformation matrix (d_{ij}) or $(d_{ij})_{app}$.

(c) We calculate (update) $<u_i u_j>_{n,MC}$ and $<u_i u_j u_k>_{n,MC}$

...We repeat (a) to (c), after increasing n by 1. If n=N, we store the last 16 updated correlations and go to the next measurement position where we repeat (a) to (c), starting with n=1.

As soon as all measurement positions (of the same measurement station) are settled, we test the quality of our hypothetical p.d.f. profiles: if the simultaneously sample-averaged 16 profiles ($<u_i u_j>_{N,MC}$ and $<u_i u_j u_k>_{N,MC}$) are compatible with the corresponding time-averaged measurements, the trial p.d.f.-profiles are declared 'realistic' (at that measurement station).

Recently, for the first time to our knowledge, the measured triple correlations $\overline{u_i u_j u_k}_{,mes}$ were projected on local experimental eigenvectors to obtain $(\overline{u_i u_j u_k})$.

From these, we have extracted the stationwise *largest* 'drift velocity (Δu_{12})' whose two components $(\Delta u_1, \Delta u_2)$ are $\Delta u_1 = (\overline{u_1 u_1 u_1})^{1./3.}$ and $\Delta u_2 = (\overline{u_2 u_2 u_2})^{1./3.}$; by assuming, (a) the actual perturbations are quasi-independent along the eigenvectors and (b) $\Delta u_3 = 0$ -according to $\overline{u_3 u_3 u_3} \approx 0$. In spite of the fact that $\Delta u_{1,MC}$ and $\Delta u_{2,MC}$ are related in a semi-empirical way in the present calculations -see later, Figure 8-, we cannot suggest analytical expressions for the 'drift velocities', yet. This is partially due to the fact that $\varphi < 0$ had to be corrected by an angle of $180°$ (for the first time since we apply this method) and we are not quite sure if this is related with the observed (unexpected) jumps in φ-profiles -see next section.

3 RESULTS AND CONCLUSION

In Figure 2 the experimental (d_{ij}) profiles are visualized in such a way that the underlying symmetry properties of this transformation matrix become obvious: $(d_{11} \approx d_{22} \approx d_{33} \approx 1), (d_{12} \approx -d_{21}), (d_{13} \approx -d_{31})$ and $(d_{23} \approx d_{32} \approx 0)$.

In Figure 3, the intersections of three vertical cones (with $10°, 20°, 30°$ opening angles) with the fixed sphere (of radius 1) show that $|\theta|$ close to the wall are typically $20°$ and that eigenvectors for flow accelerating 'more and more' have different characteristics than those accelerating 'less and less'. At station 8, we observed a sort of agitation, which in our opinion, indicates that local effects are more important than 'history effects'. Use of $(d_{ij})_{app}$ had systematic effects on the double correlations in this experiment: correlations which contained w were systematically deteriorated. On the other hand, effects of $(d_{ij})_{app}$ in triple correlations remained ambiguous.

In Figure 4, the $\overline{u_i u_i}/q^2$ -profiles for stations '0 to 12' (at left) and stations '12 to 20' (at right) are compared against the idealized values $1/7, 2/7$ and $4/7$ (for ii=33,22,11, respectively) which we had observed previously in the quasi-absence of pressure gradients. Note, $\overline{u_3 u_3}/q^2$ are the least affected ratios.

In Figures 5 for θ profiles we see a concentration about $-20°$ and a sudden jump to $90°$ -always present, but not always seen. Note, θ was evaluated as $-22.5°$, in our previous analysis of 2DTBLs.

Figure 6 is for angles φ (= phi(d12) in figure) and $\gamma_\tau = tan^{-1}(\overline{wv}/\overline{uv})$ (= phi(tau) in figure). As already explained in the previous section, φ and γ_τ are equal in our $(d_{ij})_{app}$. But, unexpected discontinuities in φ profiles -only- were observed whenever these crossed the zero-line.

In Figures 7, we have plotted -for stations '0 to 12' (at left); and stations '12 to 20'(at right)- the differences of γ_g and γ_τ (= GAMAG and GAMATAU in figure) directly from the DST [1] (where $\gamma_g = tan^{-1}[(dW/dy)/(dU/dy)]$ and U and W are the local mean velocities). Errors, as much as $5°$, are evaluated by the authors. But, at the last two 'triplet' stations (not shown)- the isolated profiles of $(\gamma_g - \gamma_\tau)$ for z=-127 mm meet the $10°$ limit. Because the other intrinsic properties at these two stations show also isolated profiles, we attribute their cause to the slightly particular flow conditions near x=2.948m -see next.

In Figure 8, the measured (dots) and measured+projected (broken lines) triple correlation profiles with three identical indexes (iii=111,222,333) are compared, at a given station. Note, the profiles for iii=333 were practically zero at all stations (i.e, the experimental p.d.f.s along eigenvectors $\hat{d}_3$ -in the vicinity of $\hat{x}_3$- are quasi 'centered-Gaussians' inside Stanford's 3DTBL). Further, the station-wise *largest* 'drift velocities (Δu_{12})' are extracted from the measured+projected triple correlation profiles and plotted as 'maximum drift velocities'. Our first investigations indicate that 'maximum drift velocities' are maybe proportional, both, to local turbulent kinetic energy and to (y/δ). In a complementary study, we compare the directions of the measured Cp gradients, the 'maximum drift velocities' and corresponding φ.

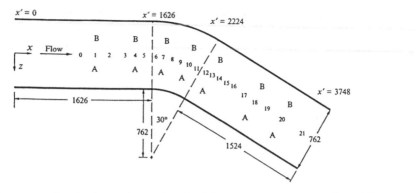

Figure 1 Stanford's 30 degree bend with 36 measurement stations (distances in mm) -Ref. [3].

In Figure 9, finally, we compare our sample-averaged profiles with Stanford measurements at three stations discussed by the authors [2]: station 7 (x=1775 mm, z=0), station 13 (x=2415 mm, z=0) and station 18 (x=2948 mm, z=0). For '7' and '13', partial results are presented with 5000 MC-events. For '18', all 16 profiles are shown with 80000 MC-events.

Double correlations were MC-predicted always with excellent precision whenever d_{ij} was used and their overall quality remained superior to the predictions of the four models (based on pressure strain 'redistribution' [2]). In general, our triple correlation profiles compared favorably with the predictions of the four models (based on turbulent transport 'diffusion' [2]). Still, correlations containing w remain problematic.

In conclusion, this data analysis technique, based on matrix diagonalisations and Monte Carlo calculations, allowed us to extracted new intrinsic information from Stanford's measured statistics, by extending text-book ideas from Kinetic Theory of perfect gas.

References

[1] Schwarz W.R., Bradshaw P., DST (data summary tape) on floppy disk sent kindly by the authors, upon request.

[2] Schwarz W.R., Bradshaw P., 1994, Phys. Fluids vol 6, No. 2, 986-998.

[3] Schwarz W.R., Bradshaw P., 1994, J. Fluid. Mech. vol 272, 183-209.

[4] Nakkasyan A., 1992, Proc. of the ERCOFTAC Workshop, at EPFL-Lausanne, March 1990, p 172-177.

[5] Nakkasyan A., 1993, Appl. Scientific Research vol 51, 115-121. - (N.B. there are printing errors in this publication!)-

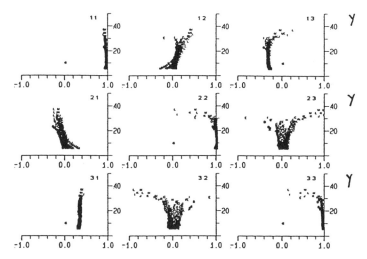

Figure 2 Global view of the nine d_{ij} profiles for $y \leq \delta$, for stations '0 to 12', .

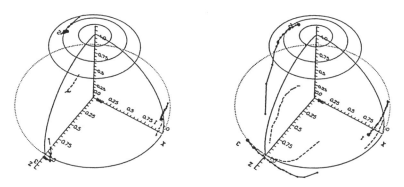

Figure 3 Typical evolution of the tip-positions of $(\hat{d}_1, \hat{d}_2, \hat{d}_3)$ for $y \leq \delta$, for stations '0 to 7' (left) and '9 to 20' (right).

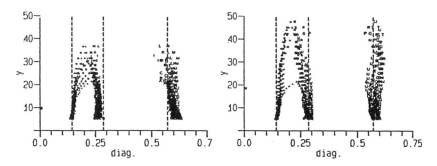

Figure 4 The eigenvalue profiles for stations '0 to 12' (left) and '12 to 20' (right). Profiles remain grouped about their 2DTBL values.

183

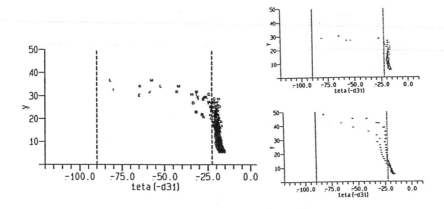

Figure 5 The θ-profiles assuming $(d_{ij})_{app}$; for stations '0 to 12' (left), 'triplet-4' (top) and 'triplet-18' (bottom).

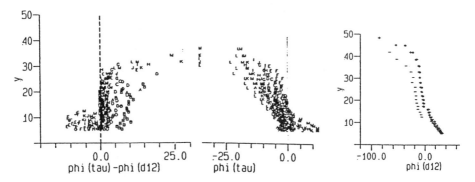

Figure 6 Angles φ with discontinuity at zero are -maybe- indicating a measurement problem.

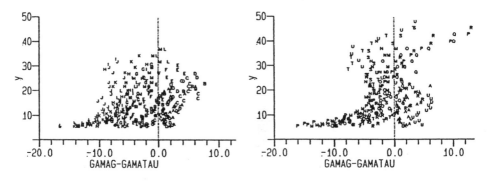

Figure 7 The $(\gamma_g - \gamma_\tau)$ profiles, for stations '0 to 12' (left) and '12 to 20' (right) -from [1].

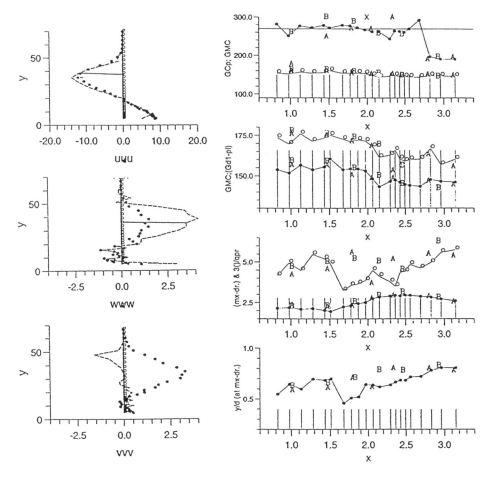

Figure 8 Figure at left: Measured+projected triple correlations with three identical indexes (broken lines) along corresponding eigenvectors, at a particular station. Figures at right: The y/δ ($= y/d$) locations of 'maximum drift velocities' at all stations -dots- (right, bottom). 'Maximum drift velocities' (=mx-dr.) at all stations -dots- (right, level 2). 'Maximum drift velocities' divided by local turbulent kinetic energies -open circles- (right, level 2) look like the (y/δ) curve of the previous figure. Angles (=GMC) between the fixed x' direction and the directions of 'maximum drift velocities' -dots-. The angles between x' and the eigenvectors $\hat{d}_1$ (at 'maximum drift velocities') are larger than 270 degrees -i.e., φ is negative. When we subtract 180 degrees (=Gd1-pi) -see open circles- and compare with (GMC), the difference is found to be about 20 degrees (right, level 3). Angles (=GCp) between x'(?) and the directions of local Cp gradients from DST -dots- (right, top).

185

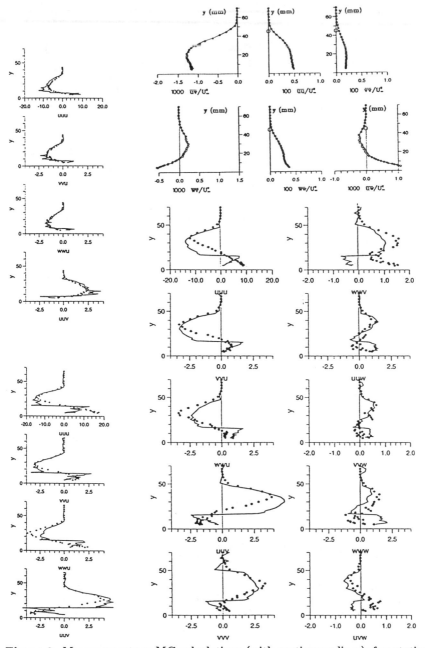

Figure 9 Measurements vs MC-calculations (with continuous lines); for: station 7 at x=1775 mm (top), station 13 at x=2415 mm (bottom) and station 18 at x=2948 mm (right).

Computation on Vortex Interactions behind Two Circular Cylinders

C.W. Ng and N.W.M. Ko
Department of Mechanical Engineering
The University of Hong Kong
Pokfulam Road, H. K.

Abstract

The evolution of vortices behind two circular cylinders of equal diameter in bistable flow at high subcritical Reynolds regime was studied. Impulsively started incompressible flows past two circular cylinders placed side-by-side in a uniform stream were investigated numerically by a discrete-vortex model. Flow development was initiated by translating one of the cylinders a small distance downstream for a short duration and back. Two phases of vortex evolution are identified in the near-wake: the 'symmetric' shedding phase, characterized by two quasi-antiphase vortex streets, and the flip-flopping phase, characterized by biased gap flow, switching at irregular times. For the flip-flopping phase, amalgamation or amalgamation/pairing of vortices, vortex pairing and vortex dipole are found. The initiation and transition from the symmetric to flip-flopping phase are caused by the asymmetry of one of the gap vortices. Flow visualization was carried out to support the findings from the model.

1 Introduction

The study of flows past two circular cylinders is of great importance to the understanding of many engineering problems, such as the flows around buildings and structures [1]. One feature is the occurrence of asymmetric flow patterns at the critical spacings of cylinders of 1.5 <T/d< 2.0 in side-by-side arrangement [2], where T is the transverse distance between cylinder centers and d the cylinder diameter. Within the asymmetric flow regime, narrow and wide wakes are found and are separated by the biased gap flow. The gap flow is bistable and switches at irregular intervals. The present paper describes the results of the computation on the interactions of vortices behind two circular cylinders of equal diameter in bistable flow.

2 Numerical Method

As the present study concentrates on the near wake at high subcritical regime, where vortices are dominant, the inviscid discrete-vortex method used can model many of the features observed experimentally reasonably well [3]. Basically, the method is a Lagrangian grid-free numerical technique in solving the continuity and Euler equations [4]. The rotational or free-space velocity field is evaluated by the Biot-Savart Law [5] and the irrotational flow is considered by the method of successive images [6]. The trajectories of discrete vortices were evaluated according to a first-order Euler scheme [7]. A reduction scheme, based on Kiya et al. [8], on the circulation of all elementary vortices was introduced to allow for the turbulent decay characteristic of primary vortices in real flow. The generation of vorticity near solid boundary of high subcritical flow regime was approximately represented by introducing nascent vortices near laminar separation points [9], which were determined by the method of Thwaite [10]. With the intention of generating asymmetric flow patterns, the first perturbation method [9] was by translating the upper cylinder a streamwise distance $\Delta \bar{x}=0.05$ downstream and back during the time $7 < \tau < 17$, where $\bar{x}=2x/d$, $\tau=2tU/d$ and U the freestream velocity. The present perturbation method is inherently asymmetric and restores the translation of the upper cylinder to its original position. The time interval was based on the minimum rate of circulation in which the vortex spirals are most susceptible to small disturbances [7]. Other perturbation methods were also used for better understanding of the problem.

All computations, based on a time step $2\Delta tU/d$ of 0.0625, were performed serially either on a DECstation 5000/200 PXG with 6 MFLOPS peak speed or on a IBM 9076 SP1 system with 8 RS/6000 processors and 1 GFLOPS aggregate peak speed. Typical run time per time step on the latter system is 240s. The cylinders spacing was fixed at T=1.75d.

3 Results and Discussion

The development of the near wakes behind the two circular cylinders is shown in Fig. 1. In order to better visualize the vortex motion, the frame of reference is moving with a speed of 0.8U downstream. Two phases of vortex evolution are identified: the 'symmetric' shedding phase and the flip-flopping phase. In the 'symmetric' shedding phase two vortex streets are developed behind the cylinders and they are basically antiphased (Fig. 1a). As shown in Fig. 1a, five cycles of these quasi-antiphase vortices are found. The degree of symmetry varies slightly from the first to fifth cycle. The vortices in the fifth cycle are less 'symmetrical', especially the gap vortices, as they develop further downstream (Fig. 1b). This change in symmetry marks the beginning of the change-over to other types of interactions. The development of the two vortex streets agrees with the flow visualization results of two impulsively started circular cylinders in the side-by-side arrangement of slightly different gap spacing [11]. In this shedding phase the deflection angles of the two gap shear layers θ_{gap2} and θ_{gap3} are basically antiphased (Fig. 2). Thus, no deflection of the gap flow is found. The deflection angle θ was determined from the least square line of the first ten discrete vortices, which distances from their respective cylinder were less than 2d. In order to simplify the following presentation, the outer and gap shear layers of the upper cylinder are denoted by -UO1 and +UG2 respectively, while those of the lower one by +LO4 and -LG3.

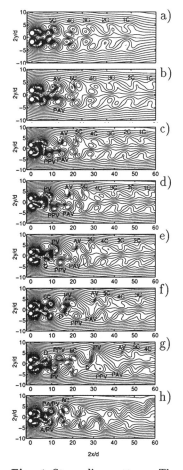

Fig. 1 Streamline patterns. The flow is from left to right. (a) Symmetric shedding phase. $\tau=55.3$. (b) Transition process. $\tau=66.5$. (c) Flip-flopping phase. $\tau=72.1$. (d) $\tau=75.8$. (e) $\tau=79.5$. (f) $\tau=88.9$. (g) $\tau=102.6$ (h) $\tau=131.4$. Symbols same as in Fig. 2.

Table 1 Circulations and their ratios of vortices of cycle, amalgamation, pairing and dipole.* For Cycle: ratio of means of two groups of vortices.

Time τ_D	Process	Circulation $\kappa/\pi Ud$ ($|\kappa/\kappa_{min}|$)*			
		-UO1	+UG2	-LG3	+LO4
15.9	1 Cycle	-1.16	1.48 (1.31:1)	-1.47	1.08
25.3	2 Cycle	-0.91	1.19 (1.34:1)	-1.24	0.91
34.6	3 Cycle	-0.98	1.30 (1.34:1)	-1.28	0.94
44.0	4 Cycle	-0.87	1.19 (1.30:1)	-1.23	0.96
53.4	5 Cycle	-0.91	1.23 (1.32:1)	-1.23	0.94
66.5	Amalgamation	-.074 (1.01)	0.73 (1.0)	-0.9 (1.27)	-
75.8	Pairing	-1.03 (2.51)	0.41 (1.0)	-0.91 (2.22)	-
79.5	Dipole	-	1.10 (1.53)	-0.80 (1.11)	0.72 (1.0)
88.9	6 Cycle	-0.88	1.07 (1.24:1)	-1.02	0.72
98.3	Pairing	-	1.10 (1.90)	-0.58 (1.0)	1.29 (2.22)
102.6	Dipole	-0.77 (1.0)	0.87 (1.13)	-1.10 (1.43)	-
112.0	7 Cycle	-0.74	1.17 (1.36:1)	-1.14	0.96
121.4	8 Cycle	-0.97	1.20 (1.32:1)	-1.10	-0.76
133.3	A/P	-	0.97 (1.29)	-0.75 (1.0)	0.81 (1.08)
137.0	Dipole	-0.86 (1.13)	1.11 (1.46)	-0.76 (1.0)	-
148.2	A/P	-0.76 (1.25)	0.61 (1.0)	-0.83 (1.36)	-
156.4	Pairing	-0.79 (1.41)	0.56 (1.0)	-0.92 (1.64)	-
160.1	Dipole	-	0.76 (1.0)	-0.88 (1.16)	0.82 (1.08)
171.4	9 Cycle (Distorted)	-0.89	1.23 (1.32:1)	-0.91	0.73
179.5	10 Cycle	-0.99	1.10 (1.33:1)	-1.10	0.66
188.9	A/P	-	1.03 (1.14)	-1.05 (1.17)	0.90 (1.0)
196.4	A/P (Not definite)	-0.79 (1.98)	1.03 (2.58)	-0.4 (1.0)	-
202.6	Pairing (Not definite)	-	0.26 (1.44)	-0.18 (1.0)	0.79 (4.39)
204.5	Dipole	-0.87 (1.14)	0.76 (1.0)	-0.89 (1.17)	-
213.9	11 Cycle	-0.64	1.02 (1.27:1)	-0.97	0.93
225.1	Pairing	-0.81 (1.84)	0.44 (1.0)	-1.13 (2.51)	-
231.4	Dipole	-	0.89 (1.10)	-0.81 (1.0)	0.89 (1.10)
240.8	A/P	-0.84 (1.0)	0.97 (1.15)	-0.89 (1.06)	-
250.1	Pairing	-1.30 (2.11)	0.48 (1.0)	-0.92 (1.92)	-
254.5	Dipole	-	-1.2 (1.54)	-0.85 (1.09)	0.78 (1.0)

After the fifth cycle, at $\tau \approx 55.0$, though anti-phased θ_{gap} is still found, θ_{gap2} has reduced to low value and the succeeding θ_{gap2} at $\tau \approx 59.0$ has near zero value (Fig. 2). Coupled with trailing behind of the lower gap vortex (from -LG3) and its greater induction on the adjacent lower outer vortex (from +LO4), the latter convects further downstream, resulting in the 'amalgamation' of the other three succeeding vortices at the upper half at $\tau=66.5$ (Fig. 1b). This indicates that with this change of the two gap deflection angles at $53.4 < \tau < 60.9$, though of lower $|\theta_{gap2}|$, to gap deflection mainly on one side of θ_{gap3}, the vortices change from quasi-antiphase arrangement to that of a single lower outer vortex (named the pre-amalgamation vortex PAV) and three amalgamating vortices (AV). The deflections of the two gap shear layers play an important role on the formation of PAV and the amalgamation of the other three vortices. The smaller θ_{gap2} peak at $\tau \approx 55.0$ and the slightly below zero of θ_{gap2} peak at $\tau \approx 50.0$ indicate the instant of the change of phase of θ_{gap2} (Fig. 2). At $\tau > 59.0$ the two gap deflection angles are basically in-phased till $\tau \approx 75.3$, giving the flip-flopping motion of the gap flow. This suggests that the amalgamation process acts as the transition between the symmetric shedding phase to the flip-flopping phase and the initiation of amalgamation is due to the loss of symmetry of the gap vortices within the two streets. At $\tau=72.1$ the fluids from the outside of the upper cylinder (from -UO1) break away, due to the cessation of the ingestion of boundary layer fluids, forming another negative vorticity vortex at $(\bar{x},\bar{y}) \approx (6,3)$ (Fig. 1c). This negative vortex interacts with the negative cross-wise vortex from -LG3, forming a rotating pair (PV) at $(\bar{x},\bar{y}) \approx (8,2)$ (Fig. 1d). This pairing process is similar to the vortex pairing of successive vortices of the same direction of rotation in free shear layer [12], except that in this case the origins of the two pairing vortices are from the alternate shear layers of the two cylinders.

Another process is observed during the same period of vortex pairing. At $\tau=79.5$, two succeeding gap vortices of opposite vorticity (from +UG2 and -LG3) are found to form another structure at $(\bar{x},\bar{y}) \approx (7,-1)$ (Fig. 1e). Also shown in Fig. 2, these two gap vortices are formed in the last part of the flip-flopping at $68.5 < \tau < 76$, during which the gap flow is deflected downwards and then upwards and the flip-flopping changes over to 'anti-phase' mode. The merged structure, consisted of two contra-rotating gap vortices, resembles a vortex dipole (D) [13, 14]. These two processes of vortex pairing and formation of the vortex dipole found within the flip-flopping phase, are mutually related. The vortices of the pairing process are those just preceding the formation of the asymmetric flow, in which the gap flow with their gap vortices is deflected towards the lower cylinder, resulting in the formation of the vortex dipole in the lower part of the wake. Because of the flip-flopping of the gap flow, the locations of the pre-pairing vortex, the pairing vortices and the vortex dipole are in the lower, upper and lower halves respectively. After these two processes, the four vortices at $\bar{x} \approx 7$ (Fig. 1f) form the sixth cycle of quasi-antiphase vortices.

Table 1 shows the interaction processes and circulations of the individual vortices during amalgamation, pairing and formation of vortex dipole. It is shown in Fig. 1g that another sequence of cycle, pairing-dipole and cyle is found. Also, at 121.4≤ τ_D ≤171.4, where τ_D is the approximate time for the structure to reach $\bar{x}$ ≈8, the sequence of cycle, amalgamation/pairing, dipole, amalgamation/pairing, pairing, dipole, cycle is found (Fig. 2 and Table 2). Fig. 1h shows the the formation of amalgamation/pairing, during which the vortices do not indicate clearly amalgamation or pairing, hence called amalgamation/pairing vortices (A/PV). For the amalgamation and amalgamation/pairing processes, the circulation of the third vortex of opposite vorticity is comparable with those of the other two vortices of the same sense of vorticity (Table 2). The circulation ratios of these three vortices are mainly about 1.30:1. On the other hand, for pairing the circulation ratio of the third vortex of opposite vorticity is significantly lower. Thus, the above findings show the importance of the strength or circulation of the third vortex of opposite vorticity on the processes of pairing-dipole and amalgamation.

In order to establish these four processes under other perturbations, another perturbation with transverse translation $\Delta\bar{y}$=-0.05 at 7< τ <17 of the lower cylinder yields results (not shown here) indicating the occurrences of cycles of quasi-antiphase vortices within the symmetric shedding phase and the amalgamation and pairing-dipole within the flip-flopping phase.

As the above results from the two perturbations indicate the importance of the condition of one of the gap vortices in the initiation of the other interaction processes, other perturbations had also been introduced to study the effect of the relative importance of either the outer or gap vortices. Two series of perturbation were attempted. The sequences of the four interaction processes, thus obtained, are tabulated in Table 2. The introduced asymmetry, by shifting the lower gap vortex, produces significant differences (Table 2b). With increasing τ, the number of cycles of quasi-antiphase vortices reduces from three at $\Delta\bar{x}$=-0.05 to no cycle at $\Delta\bar{x}$ ≥-0.75. Thus, at such large perturbations no quasi-antiphase vortices are found, rather, the direct occurrence of amalgamation. This further supports the above findings that the asymmetry of a gap vortex is the mechanism responsible for the occurrence of other interaction. On the other hand, the perturbation in maintaining the in-phased condition of the two gap vortices results in more stable condition of the four quasi-antiphase vortices and the symmetric shedding phase (Table 2c).

Interactions of the individual vortices from the four shear layers within a water tunnel were also observed. Same experimental setup, as in Lam et al. [15], was used. At the Reynolds number of 860 the amalgamationg vortices (AV) with the pre-amalgamation vortex (PAV) (Fig. 3a), pairing (PV) with the pre-pairing vortex (PPV) (Fig. 3b), the vortex dipole (D) (Fig. 3c) and a cycle of quasi-antiphase vortices (C) (Fig. 3d) are shown. Based on Fig 3, the estimated convection velocities of these interacting vortices are about 0.8U to 0.9U, agreeing with the quantitative results of 0.8U [16]. These convection velocities also agree with those of the present numerical estimation of about 0.8U.

4 Conclusions

Based on a two-dimensional discrete-vortex model, the evolution of vortices behind two circular cylinders of equal diameter in side-by-side arrangement in the bistable regime was investigated. With the present perturbation methods two phases of vortex evolution are identified in the near-wake: the 'symmetric' shedding phase and the flip-flopping phase. The symmetric shedding phase is characterized by two quasi-antiphase vortex streets. The flip-flopping phase is characterized by biased gap flow, switching at irregular times. For the flip-flopping phase, amalgamation or amalgamation/pairing of vortices, vortex pairing, and vortex dipole are found. The occurrence of pairing of two vortices of the same sense of direction and of amalgamation of three vortices depends on the condition of the third vortex of opposite vorticity. Pairing of two vortices is observed when the circulation of the third vortex is comparatively small, while amalgamation is found when its circulation is comparable with those of the other two vortices. Results also indicate that pairing is followed by the formation of vortex dipole, forming a joint process of pairing-dipole. The occurrence of these different types of interaction processes depends on the deflections of the two gap shear layers. The flow visualization results in a water tunnel support the numerical findings of the four interaction processes.

References

[1] ZDRAVKOVICH, M. M.: "Review of flow interference between two circular cylinders in various arrangemnets", ASME J. Fluids Engng., 99 (1977) pp. 618-633.

[2] ZDRAVKOVICH, M. M.: "The effects of interference between circular cylinders in cross flow", J. Fluids and Struct., 1 (1987) pp. 239-260.

[3] OHYA, A., OKAJIMA, A., HAYASHI, M.: "Wake interference and vortex shedding", in Ency. Fluid Mech.,(N.P. Cheremisinoff, ed.) Gulf, Houston (1989).

[4] NG, C. W., KO, N. W. M.: "Flow interaction behind two circular cylinders of equal diameters - a numerical study", J. Wind Engng. Ind. Aero., 54/55 (1995) pp. 277-287.

[5] SARPKAYA, T.: "Computational methods with vortices - the 1988 Freeman scholar lecture", ASME J. Fluids Engng., 111 (1989) pp. 5-52.

[6] DALTON, C., HELFINSTINE, R. A.: "Potential flow past a group of circular cylinders", ASME J. Basic Engng., 93 (1971) pp. 636-642.

[7] SARPKAYA, T., SCHOAFF, R. L.: "Inviscid model of two-dimensional vortex shedding by a circular cylinder", AIAA J., 17 (1979) pp. 1193-1200.

[8] KIYA, K., SASAKI, K., ARIE, M.: "Discrete-vortex simulation of a turbulent separation bubble", J. Fluid Mech., 120 (1982) pp. 219-244.

[9] STANSBY, K.: "A numerical study of vortex shedding from one and two circular cylinders", Aero. Quar., 32 (1981) pp. 48-71.

[10] SCHLICHTING, H.: "Boundary Layer Theory", 6th ed., McGraw-Hill, New York 1968.

[11] KAMEMOTO, K.: "Formation and interaction of two parallel vortex streets", Bull. JSME, 19 (1976) pp. 283-290.

[12] WINANT, D., BROWAND, F. K.: "Vortex pairing: the mechanism of turbulent mixing layer growth at moderate Reynolds number", J. Fluid Mech., 63 (1974) pp. 237-255.

[13] COUDER, Y., BASDEVANT, C.: "Experimental and numerical study of vortex couples in two-dimensional flows", J. Fluid Mech., 173 (1986) pp. 225-251.

[14] VOROPAYEV, S. I., AFANASYEV, YA. D.: "Two-dimensional vortex-dipole interactions in a stratified fluid", J. Fluid Mech., 236 (1992) pp. 665-689.

[15] LAM, K. M., WONG, P. T. Y., KO, N. W. M.: "Interaction of flows behind two circular cylinders of different diameters in side-by-side arrangement", Expt. Ther. Fluid Sc., 7 (1993) pp. 189-201.

[16] KIM, J. H., DURBIN, P. A.: "Investigation of the flow between a pair of circular cylinders in the flopping regime", J. Fluid Mech., 196 (1988) pp. 431-448.

Acknowledgement

This work was partly supported by a research grant from the Research Grants Council, Hong Kong.

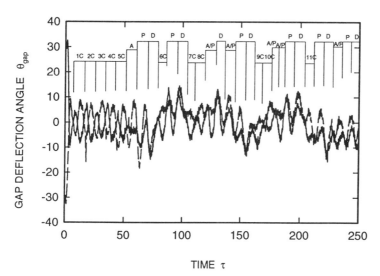

Fig. 2 Time variations of deflection angles of gap shear layers. ——,θ_{gap2};— — —,θ_{gap3}. Symbols: C, cycle of quasi-antiphase; PAV, pre-amalgamation vortex; A, amalgamation; PPV, pre-pairing vortex; P, pairing; D, vortex dipole; A/P, amalgamation/pairing; PA/PV, pre-amalgamation/ pairing vortex.

Table 2 Sequences of interaction processes at different perturbation methods. *: Perturbation; △: asymmetry. Symbols same as in Fig. 2

(a) Upstream streamwise shift and back of lower outer vortex											
$\Delta\bar{x}$	-0.05	C*	C	C△	A/P	A/P	D	A/P	P	D	A	P
	-0.25	C*	C	C	C△	A/P	P	D	A	P	D	C
	-0.5	C*	C	C	C△	A	P	D	A	A/P	A/P	A
	-0.75	C*	C	C	C△	C△	A	P	D	C	C	A
	-1.0	C*	C	C	C	C△	A	P	D	A/P	A/P	A/P
(b) Upstream streamwise shift and back of lower gap vortex											
$\Delta\bar{x}$	-0.05	C*	C	C△	A	P	D	C	A	A	A/P	A/P
	-0.25	C*	C△	A	A/P	A	A	A/P	D	A	A/P	D
	-0.5	C*	C△	A	A/P	A/P	D	C	A	A	A/P	D
	-0.75	A*	A	A	A/P	D	C	C	C	A	A/P	A/P
	-1.0	A*	A/P	A/P	D	A	P	D	C	A	A/P	D
(c) Upstream streamwise shift of both gap vortex											
$\Delta\bar{x}$	-0.05	C*	C	C	C△	C△	A	P	D	A/P	A/P	A/P
	-0.25	C*	C	C	C	C△	C△	C△	A	A/P	D	A/P
	-0.5	C*	C	C	C	C	C△	C△	P	D	A/P	A/P
	-0.75	C*	C	C	C	C△	A	P	D	A	P	D
	-1.0	C*	C	C	C	C	C	C△	C△	A/P	P	D

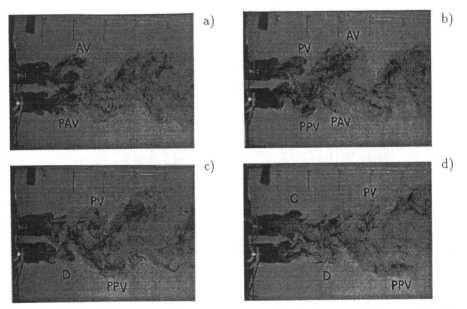

Fig. 3 Flow visualization. Reynolds number = 860. Time $\Delta\tau$ between figures is 4.75. Symbols same as in Fig. 2

THREE-DIMENSIONAL STUDY OF EXTRUSION PROCESSES BY THE STREAM-TUBE METHOD : NEWTONIAN AND VISCOELASTIC RESULTS

Magdeleine NORMANDIN & Jean-Robert CLERMONT
Laboratoire de Rhéologie, Domaine Universitaire
BP 53, 38041 GRENOBLE CEDEX 9, FRANCE

SUMMARY

The stream-tube analysis is considered for the problem of the free surface determination for a 3D duct. The method allows computation of the unknown free surface by only considering a "peripheral stream tube" limited by the wall and the free surface and an inner stream surface, in a mapped computational domain where streamlines are rectilinear. The unknowns are determined by considering the conservation equations together with an optimization criterion, solved by using the Levenberg-Marquardt optimization algorithm. The number of degrees of freedom and the storage area are reduced. Similarly to previous 2-D studies, singularity problems in the vicinity of the junction points of the wall and the free surface are avoided. However, the stress peaks due to the singularity at the exit may still be evaluated. Results are presented, using a Newtonian fluid and a memory-integral constitutive equation, for a duct of square cross-section.

INTRODUCTION

Three-dimensional extrudate swell problems were considered in recent years, at the end of the eighties (e.g. Bush and Phan-Thien [1], Karagiannis et al. (1989) [2], Shiojima et Shimazaki (1990) [3], Wambersie and Crochet (1992) [4], Legat and Marchal (1992) [5]). The authors generally retained finite element methods to compute the relevant unknowns, namely the velocity components (u, v, w) and the pressure p, leading to a significant number of degrees of freedom. Purely viscous constitutive equations were generally considered in these flow simulations. It should be pointed out that 3D numerical flow simulations with memory-integral constitutive equations still remain a difficult task, since the flow pathlines are not planar curves and do not pass through the mesh points (" particle tracking problem").

The stream-tube method, introduced some years ago [6,7], allows flow computation by means of an unknown transformation T, assumed to be one-to-one, between a physical flow domain D and its mapped domain D*, used as computational domain, in which the transformed streamlines are parallel and straight. Since the extrudate swelling flow involves only open streamlines, a one-to-one transformation may be defined, allowing no restriction to apply stream-tube analysis. Previous studies of 2D swell flows with this method were carried out for Differential [7] and integral constitutive equations [8] were used in 2D swell flows with this method, the main features of which may be summarized as follows :

1) Mass conservation is automatically verified by the formulation. In contrast to classic methods, the primary unknowns of the problem are the pressure and one or two mapping functions (f) or (f and g), in the respective 2D or 3D cases.

2) The mesh, very simple for 2D and 3D situations, is built, in the computational domain, on the rectilinear mapped streamlines. Differential and integral operators which may be involved in viscoelastic constitutive equations, may be easily taken into account [7,8].

3) The flow may be computed by considering successive sub-domains of D, the stream tubes,

from the wall to the central flow region, provided the action of the complementary flow domain is taken into account. This property is to be emphasized in the case of the swelling flow problem, for which only consideration of the "peripheral stream tube" involving the wall and the free surface permits determination of the unknown jet surface. Thus, reduction of the number of unknowns, as well as significant lowering of CPU time are expected.

In the present work, we apply the method to extrudate swell flows of fluids obeying memory-integral constitutive equations. Similarly to a recent work on 2D swell flows [7], the analysis presented here avoids to consider explicitly the singularity problems in the vicinity of the junction points of the wall and the free surface. In the case of a die of constant cross-section along the z-axis, the duct length is considered to be large enough to assume a fully-developed flow for sections $z \leq z_1$. A solid flow region is obtained in the free jet at section z_2 (Fig 1).

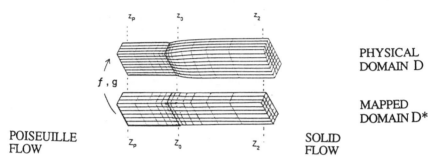

Fig. 1 Physical and mapped domains related to a 3D extrudate swell problem

STREAM-TUBE METHOD IN 3D SWELL PROBLEMS

Physical and mapped domains D and D* corresponding to the transformation $T : D \rightarrow D^*$ are illustrated in the example of Fig 1. Cylindrical coordinates (r, θ, z) are used in the flow domain D. Mapped domain D* is a cylinder the basis of which is identical to the upstream section A_0 of the physical domain D at $z = z_1$, which limits the fully developed flow region. The two domains D and D* are related to the following equations :

$$r = f(R, \Theta, Z) ; \qquad \theta = g(R, \Theta, Z) \quad ; \quad z = Z . \tag{1}$$

f and g denote the mapping functions related to T, to be determined numerically, (R, θ, Z) are the cylindrical coordinates associated with the mapped domain D*, in which the streamlines are rectilinear and parallel to the z-axis. At section z_1, the following relations:

$$r = R = f(R, \varphi, z = z_1) \quad ; \quad \theta = \varphi = g(R, \varphi, z = z_1) \quad ; \quad z_1 = Z_1 \tag{2}$$

may easily be verified. The velocity vector $V = u(r, \theta, z)c_r + v(r, \theta, z)c_\theta + w(r, \theta, z)c_z$ is known at the upstream section z_1. (c_r, c_θ, c_z) denotes the orthonormal frame related to the cylindrical coordinates. We get :

$$V = w(r, \theta, z_1) c_z \quad \text{for} \quad z = z_1 \; ; \; V = w_2 c_z \quad \text{for} \quad z = z_2 \; , \tag{3}$$

where w_2 denotes the solid flow velocity. For a simply-connected die of general shape, we may define a function $\psi(R, \varphi)$ such that :

$$w(R, \varphi, Z_1) = \psi(R, \varphi)/R \quad , R \neq 0.$$

Then, the velocity components may be readily written as

$$u = f'_Z \psi(R, \varphi)/(f\Delta) \; ; \; v = g'_Z \psi(R, \varphi)/\Delta \; ; \; w = \psi(R, \varphi)/(f\Delta), \qquad (4)$$

In equations (6), $\Delta = |\partial(r,\theta,z)/\partial(R,\varphi,Z)| = f'_R g'_\varphi - f'_\varphi g'_R$ denotes the Jacobian, assumed to be non-singular, of the transformation T. The derivative operators $\partial/\partial r$, $\partial/\partial\theta$ and $\partial/\partial z$ relating the sets of variables (r,θ,z) and (R,φ,Z) may be evaluated.

For a simply-connected geometry, the stream tubes considered in the present flow analysis are limited, at the upstream section z_1, by contour values of the velocities w = Constant. The zero-velocity contour curve corresponds to the boundary line at the wall. The positions of stream tubes at the solid flow section z_2 are unknown.

STREAMLINE APPROXIMATION- DISCRETIZATION OF UNKNOWNS

For approximating the free surface in the 3D extrudate swell problem, we generalized the equation given by Batchelor and Horsfall [11], also considered in the literature, which was proved to fit well with swell experimental data of various polymer solutions and melts and adopted by the authors of this paper [7,8], for streamlines in 2D flow domains. Accordingly, a stream surface f related to equation (1) may be written as follows :

$$f(R, \varphi, Z) = C_1(R, \varphi, Z) + A(R, \varphi) \cdot q(Z) \{1 - \exp[-(Z - Z_0) B(\varphi, Z)]\}, \qquad (5)$$

in the tube $(Z_1 \leq Z \leq Z_0)$. In the jet, for $Z_0 \leq Z \leq Z_2$, we assume the following equation

$$f(R, \varphi, Z) = f(R, \varphi, Z_2) - A(R, \varphi) \exp[-(Z - Z_0) B(\varphi, Z)]\}. \qquad (6)$$

Functions $q(Z)$ and $C_1(R, \varphi, Z)$ have the same meanings than those defined in 2D situations [7,8]. Streamline warping curves $\mathcal{L}$ belong to stream surfaces $f(R, \varphi, Z)$. The function g, which must satisfy the boundary condition (2) is computed by point values on the three-dimensional mesh defined in the mapped domain D*. The surfaces f in the peripheral stream tube are close enough to assume that, in the mapped domain D*, the quantities $\| R_0(\varphi, Z) - R_2(\varphi, Z) \|$ are small. Given a section Z of the flow domain, for $R \in [R_2(\varphi, Z) - R_0(\varphi, Z)]$, the functions A, B and the function g may be written as

$$A(R,\varphi) = \alpha_0(\varphi) + \alpha_1(\varphi)[R - R_0(\varphi)] + O((R - R_0(\varphi))^2), \qquad (7)$$

$$B(R,\varphi) = \beta_0(\varphi) + \beta_1(\varphi)[R - R_0(\varphi)] + O((R - R_0(\varphi))^2), \qquad (8)$$

$$g(R, \varphi, Z) = \gamma_0(\varphi, Z) + \gamma_1(\varphi, Z)[R - R_0(\varphi, Z)] + O[(R - R_0(\varphi, Z))^2]. \qquad (9)$$

The functions $\alpha_0(\varphi)$, $\alpha_1(\varphi)$, $\beta_0(\varphi)$, $\beta_1(\varphi)$, $\gamma_0(\varphi, Z)$ and $\gamma_1(\varphi, Z)$ of equations (7-9) are to be determined numerically from the governing equations of the problem.

GOVERNING EQUATIONS FOR THE PERIPHERAL STREAM TUBE

The relevant equations are now considered, under isothermal conditions, with restriction to flow in the peripheral stream tube. Surface tension, inertia and body forces are ignored and no traction on the jet at section z_2 is assumed. Mass conservation is automatically verified from the formulation.

(1) The use of variables (R, φ, Z) of the mapped domain D^* and approximating functions for f leads to express, in terms of the pressure gradient components, the dynamic equations as:

$$\mathcal{A}_1 \partial p/\partial R + \mathcal{B}_1 \partial p/\partial \varphi = \mathcal{E}_1, \qquad (10)$$

$$\mathcal{A}_2 \partial p/\partial R + \mathcal{B}_2 \partial p/\partial \varphi = \mathcal{E}_2, \qquad (11)$$

$$\mathcal{A}_3 \partial p/\partial R + \mathcal{B}_3 \partial p/\partial \varphi + \mathcal{C}_3 \partial p/\partial \varphi = \mathcal{E}_3, \qquad (12)$$

$\mathcal{A}_1, \mathcal{B}_1, \mathcal{A}_2, \mathcal{B}_2, \mathcal{A}_3, \mathcal{B}_3, \mathcal{C}_3, \mathcal{E}_1, \mathcal{E}_2$ and $\mathcal{E}_3$ denote functions of $\{f, g\}$ or, equivalently, of the unknowns $\{\alpha_0(\varphi), \alpha_1(\varphi), \beta_0(\varphi), \beta_1(\varphi), \gamma_0(\varphi, Z), \gamma_1(\varphi, Z)\}$. Equations (10-12) are written on streamlines $\mathcal{L}^*$ of the mapped domain D^*. The pressure p, assumed to be zero at the solid flow section z_2, may be determined by integration, using the following relation, where the integrand $\mathcal{H}(R, \varphi, \zeta)$ is evaluated from equations (10-12):

$$p(R, \varphi, Z) = \int_{z_2}^{Z} \mathcal{H}(R, \varphi, \zeta) \, d\zeta. \qquad (13)$$

(2) The solving of the equations may be carried out on the peripheral stream tube, provided the action of the complementary flow domain Ω of boundary $\partial \Omega$ is taken into account (e.g. [7]). This condition may be expressed, in 3D flows, by the following equations related to respective following resultant and moment vectors $\mathcal{R}$ and $\mathcal{M}_0$:

Resultant vector: $\mathcal{R} = \int_{\partial \Omega} \sigma \, n \, ds = \int_{\partial \Omega} (-p \mathbb{I} + \mathbb{T}) \, n \, ds = 0$

$$\qquad (14)$$

Moment vector: $\mathcal{M}_0 = \int_{\partial \Omega} OM \wedge \sigma(M) \, ds = 0$.

n denotes the outward unit vector normal to the surface $\partial \Omega$. For ducts involving symetries with respect to the z-axis, equations (14) reduce into only one scalar equation.

(3) The zero-stress condition at the free surface Σ_0 is given by the following equation:

$$[\int_{\partial \Sigma_0} |(-p \mathbb{I} + \mathbb{T}) \, n \, ds)| \cdot c_z = 0. \qquad (15)$$

(4) The constitutive equations to be expressed in the context of the mixed formulation concern:

(i) a Newtonian fluid (η_0 is the fluid viscosity and $\mathbb{D}$ the rate-of-strain tensor) :

$$\mathbb{T} = 2 \; \eta_0 \; \mathbb{D} \; , \tag{16}$$

(ii) a K-BKZ viscoelastic memory-integral equation where the stress tensor is written as :

$$\mathbb{T}(t) = \int_{-\infty}^{t} m(t-\tau) \cdot h(I_1, I_2) \; \mathbb{C}^{-1}{}_t (\tau) \; d\tau \tag{17}$$

where $m(t-\tau)$ denotes the memory-function, $\mathbb{C}^{-1}{}_t$ the codeformational Finger tensor. $h(I_1, I_2)$ is a damping function taken on the form proposed by Papanastasiou, Scriven and Macosco (1983) [10], already used in the literature for 2D flow computations.

The use of integral constitutive equations involves determining particle evolution versus time. In the transformed domain D* where the streamlines are rectilinear, the following relation is obtained between the time t related to the particle and a reference time τ_1 :

$$\tau - \tau_1 = \int_{Z_1}^{Z} ds/w(R,\varphi,s) \tag{18}$$

where $Z_1 (= z_1)$ and Z denote the respective positions of a particle at times τ_1 and τ. The time τ_1 corresponds to the position of the particle at section z_1. The procedure of evaluation of codeformational kinematic tensors, applied to the present study, has been given elsewhere [11].

The set of the governing equations must be considered together with the classic "simple" boundary condition equations related to mapping functions f, g.

NUMERICAL PROCEDURE AND SWELL RESULTS FOR EXTRUSION FROM A SQUARE DIE

Governing equations for the peripheral stream tube as also approximating schemes retained for the streamlines are defined such that the number of equations and unknowns are different. As previously done in 2D swell studies [7,8], we use optimization methods in order to solve the equations. These computational approaches have proved their robustness and efficiency for problems involved by the stream-tube analysis, with regard to the significant sensivity of the equations to changes of the mapping functions f and g. The Levenberg-Marquardt algorithm is adopted in the present work.

Newtonian results

The Newtonian liquid swells or retracts depending on the azimutal angle φ. The swell ratio for $\varphi = \pi/4$ is found to be 1.15, which is close to the computed swell value (1.18) given by Wambersie et Crochet [4]. The ratio corresponding to the relative reduction in size of the jet radius for $\varphi = 0$, is found to be 0.848, which is consistent with numerical results in the literature. Results are shown in Fig 2 and Fig 3 and indicate stress peaks in the vicinity of the junction points between the wall and the free surface.

Results with the memory-integral viscoelastic equation

As in 2D swell situations, the K-BKZ fluid was found to exhibit a swell surface depending on flow conditions. An illustrative example of the computed jet and stresses is given in Fig 3 and Fig 4, by considering a peripheral stream tube (2590 equations and 224 unknowns).

DUCT WALL & COMPUTED JET SURFACE

CORRESPONDING STRESS COMPONENT T^{rr} (Wall and free surface)

CORRESPONDING STRESS COMPONENT T^{zz} (Wall and free surface)

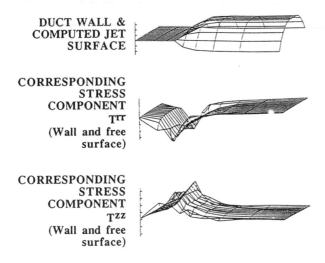

Fig 2 Computed jet and corresponding stresses for a Newtonian fluid (the domain involves one side of the duct the computed corresponding free surface)

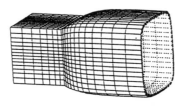

Fig 3 Computed jet for a K-BKZ fluid emerging from a duct of square cross-section

CONCLUDING REMARKS

In this paper, a formulation related to stream-tube analysis has been presented to determine the unknown free surface in the 3D swell problem for a Newtonian fluid and a memory-integral viscoelastic fluid. From the numerical point of view, the use of a simple mapped computational domain leads to define simple and stable discretization schemes. The possibility of the method to compute the free surface and singularity effects by using a peripheral stream tube the corresponding mesh in D* of which remains unchanged, as also the efficiency of the Levenberg-Marquardt algorithm for solving the equations are to be underlined. The results obtained by the present method are consistent with numerical data in the literature. To our knowledge, these results with memory-integral equations appear to be the first in literature.

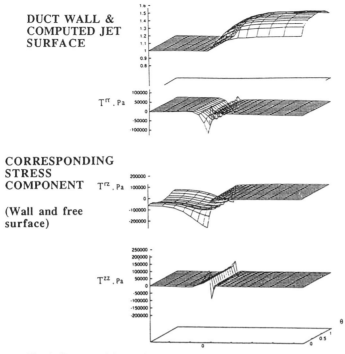

Fig 4 Computed jet and corresponding stresses for a K-BKZ memory-integral fluid (the domain involves one side of the duct and the corresponding free surface)

REFERENCES

[1] M.B. Bush and N. Phan-Thien, "Three-dimensional viscous flows with a free surface : flow out of a long square die", J. Non-Newtonian Fluid Mech. 18 : 211- (1985)
[2] A. Karagiannis, A. N. Hrymak and J. Vlachopoulos, "Three-dimensional extrudate swell of creeping Newtonian jets", AIChE J. **34**, 2088-2094 (1988)
[3] T. Shiojima and Y. Shimazaki, "Three-dimensional finite element method for extrudate swells of a Maxwell fluid", J. Non-Newtonian Fluid Mech. 34 : 269-288 (1990)
[4] O. Wambersie and M.J. Crochet, " Transient finite element method for calculating steady state three-dimensional free surface", Int. J. Num. Meth. Fluids **14**, 343-360 (1992)
[5] V. Legat and J.M. Marchal, "Prediction of three-dimensional general shape extrudates by an implicit iterative scheme", Int. Journal for Num. Meth. Fluids **14**, 609-625 (1992)
[6] J.R. Clermont, "Sur la modélisation numérique d'écoulements plans et méridiens de fluides non-newtoniens incompressibles", C.R. Acad. Sci. Paris, Série II, **297** (1983)
[7] J.R. Clermont and M. Normandin, "Numerical simulation of extrudate swell for Oldroyd-B fluids using the stream-tube analysis and a streamline approximation", J Non-Newtonian Fluid Mech **50**, 193-215 (1993)
[8] P. André and J.R. Clermont, "Experimental and numerical study of the swelling of a viscoelastic liquid using the stream-tube method and a kinematic singularity approximation", J Non- Newtonian Fluid Mech., 38, 1-29 (1990)
[9] J. Batchelor and H. Horsfall, " Die-swell in elastic and viscous fluids", Rubber and Plastics Research Association of Great Britain", Rep. 189, (1971)
[10] A.C. Papanastasiou, L.E. Scriven and C.W. Macosco, "An integral constitutive equation for mixed flows: viscoelastic characterization", J. Rheology 27 (4) 387-410 (1983)
[11] J.R. Clermont, Calculation of kinematic histories in two- and three-dimensional flows using streamline coordinate functions, Rheol. Acta 32 : 82-93 (1993)

NATURAL CONVECTION IN VOLUMETRICALLY HEATED AND SIDE-WALL HEATED MELT POOLS: THREE DIMENSIONAL EFFECTS

R.R. Nourgaliev, T.N. Dinh, and B.R.Sehgal
Division of Nuclear Power Safety, Royal Institute of Technology (KTH)
Brinellvägen 60, 10044 STOCKHOLM, SWEDEN
Fax + (46)(8) 790-76-78, e-mail: robert@ne.kth.se

Abstract

The paper presents calculational results of three-dimensional analysis of flows and heat transfer in internally and side-wall heated naturally convecting melt pools. Both slice-type semicircular cavities and hemispherical pools are considered. Physical aspects of simulation of internal energy sources by sidewall heating are discussed. In case of internal heating it is shown that heat transfer rates to the cooled upper pool surface in the 3D simulations are higher than those obtained from 2D computational modeling. Also, the result indicates some differences in local heat flux distribution on the cooled bottom pool surfaces of 2D semicircular cavities, 3D semicircular cavities with different slice thicknesses, and 3D hemispherical pools. All examined cases ($Ra \simeq 10^{12}$) experience convection induced effects of low fluid Prandtl numbers.
Keywords: natural convection, heat transfer, internal energy source, side-wall heating, effects of fluid Prandtl number, stable and unstable stratification, three-dimensional effects.

1 Introduction and problem formulation

Natural convection is the major mechanism of heat transfer in a liquid (melt) pool with internal energy sources. There are in existence various technical and geophysical processes that experience volumetrically heated natural convection. For instance, hypothetical severe (meltdown) accidents in nuclear reactors might lead to a formation of core melt pools with nuclear decay heat generation. This is also the application motivating the present CFD analysis (for relevant review see [1]). A hemispherical geometry is the most probable configuration of melt pools in reactor situations of interest. Since solidification of core melt occurs at a pool-crust interface, isothermal boundary conditions are used for both the upper and downward (curved) surfaces of the melt pool. Due to their technological advantages, semicircular slice-type cavities have been proposed for use as test sections in experimental investigations of thermal-hydraulic and physico-chemical phenomena. In order to fully represent major physical mechanisms of interest, appropriate slice thicknesses must be selected. The role of the slice thickness has been addressed in the recent work of Dinh et al. [2], in which three-dimensional computations were performed for square slice-type cavities. These computations had been limited to rather low Rayleigh numbers ($Ra \propto 10^7 - 10^{10}$).

Physical similarity to prototypical conditions requires providing the isothermal conditions at the upper and downward surfaces of the semicircular cavity. Furthermore, the nuclear decay heat generation must be simulated. Direct electrical heating (DEH) and side-wall heating (SWH) have been considered as the major candidates for energy supply in high-temperature experiments (2400-3000K). In the first case (DEH), the face and back surfaces of the slice cavity must be thermally isolated. The second case (SWH) employs the face and back surfaces of the cavity for heat supply with heat fluxes defined as $q_{face,back} = q_v \cdot \Delta z/2$, where Δz is the cavity thickness. Thus, cases of the internal heating (IH) and side-wall heating (SWH) are equivalent in terms of energy input. The Rayleigh (Ra) number in the side-wall heating case is defined by means of the equivalent volumetric heat generation rate q_v. Non-prototypical conditions in simulant and real melt pool experiments (including those due to methods of simulating of nuclear energy sources) have recently been disscused by Dinh et al. [2]. In order to support the design of experimental facilities using side-wall heating, natural convection in side-wall heated melt pools must be analysed. Previously, it has been shown that, for $Ra \leq 10^{10}$, Nusselt (Nu) numbers at the vertical and bottom walls of square cavities are similar to the cases of side-wall heating and internal heating [2]. The Nu numbers at the upper surface in side-wall heating cases are, however, as much as two times larger than those of the internal heating cases. In the present work, we focus our analysis on more complex flows in semicircular cavities and under higher Rayleigh number conditions.

Dinh et al. have provided 2D analyses of fluid-Prandtl-number effects on the local heat flux distribution

[2]. From computational results it was found that the fluid Pr number has small effect on the averaged Nu numbers in the convection dominated regions. The decrease of the Pr number causes a decrease of Nu numbers on the top and side walls of cavities up to 30% in the considered Pr number range ($Pr \propto 0.2 \div 7$). In the conduction dominated regions two phenomena, namely $\nu-$ and $\alpha - phenomena$ were identified (see fig.1). Higher heat transfer rates were predicted for lower Pr numbers. These effects depend on the cavity geometry (curvature) and the Ra numbers. Essentially, the question is whether the Pr number effects, predicted by 2D modeling in semicircular cavities and found to be significant for $Ra = 10^{12}$, exist in 3D case and in hemispherical cavity.

Summarizing, the objectives of the paper are *(1) to assess effects of geometry and three-dimensionality in semicircular and hemispherical pools*; [1] *(2) to provide the detailed computational comparative analysis of natural convection flows and heat transfer in internally heated and side-wall heated three-dimensional liquid pools*; and, *(3) to investigate Pr number effects in 3D semicircular and hemispherical internally heated pools*.

2 Features of CFD-analysis

The mathematical model of the problem at hand is based on the set of the 3D governing equations for mass, momentum, and energy conservation. The *continuity equation*:

$$\frac{\partial \rho}{\partial t} + \nabla \cdot (\rho \mathbf{U}) = 0, \qquad (1)$$

The *momentum equation*:

$$\frac{\partial \rho \mathbf{U}}{\partial t} + \nabla \cdot (\rho \mathbf{U} \otimes \mathbf{U}) = \mathbf{B} + \nabla \cdot [-P\delta + \mu_{eff}(\nabla \mathbf{U})] \qquad (2)$$

where $\mathbf{B}$ is a body force (gravitational force $\rho \mathbf{g}$ is the only body force in the present study). The *energy equation*:

$$\frac{\partial \rho h}{\partial t} + \nabla \cdot (\rho \mathbf{U} h) - \nabla \cdot [\lambda_{eff}(\nabla T)] = \frac{\partial P}{\partial t} + q_v, \qquad (3)$$

Since the Ra number of interest (10^{12}) indicates that the natural convection flows in question might be turbulent, turbulence modeling must be considered. In eqs.(2,3) $\mu_{eff} = \mu + \mu_t$ and $\lambda_{eff} = \lambda + \lambda_t$ are the effective viscosity and conductivity, respectively. Here μ_t and λ_t represent turbulence transport properties. Accurate description of such complex flows seems to require at least the second-moment closure level with low-Reynolds-number corrections in order to ensure predictions of non-standard behaviour pertinent to the flow of interest (see Dinh and Nourgaliev, [1]). Therefore, in the present work the low-Reynolds-number $\kappa - \varepsilon$ turbulence model of Launder and Sharma [3] is employed.

In the present work the general-purpose CFDS-FLOW3D code [4] has been employed to solve the governing equations. Body-fitted coordinates are included in CFDS-FLOW3D to allow the treatment of arbitrary complex geometries. Coordinate transformation is given numerically. The basis of the code is a conservative finite-difference method with all variables defined at the centre of control volumes which fill the physical domain being considered. To avoid the chequerboard oscillations in pressure and velocity the improved Rhie-Chow interpolation method is used. The third-order accurate CCCT differencing scheme for advection term treatment, fully implicit backward difference time stepping procedure and SIMPLEC velocity-pressure coupling algorithm, have been utilized. The multi-block approach is used to design a computational mesh with about 10^5 nodes. An unsteady approach is used to obtain the time-dependent flow and temperature fields and heat transfer characteristics. The results are then averaged over time intervals to produce the quasi-steady values of heat transfer characteristics.

Experience in 3D computations of similar *unsteady-state* problems shows that numerical solutions can be obtained on modern workstations with reasonable computer times, if discretization nodes are of order 10^5. In general, both single-block and multi-block approaches can be used to design the computational mesh. However, the single-block grid is hardly an optimal one in describing the physics of fluids in liquid pools with *curved downward walls*. In fact, in order to be capable of modeling heat transfer to the

---

[1] Up to date, no systematic studies of effects of geometry and three-dimensionality in semicircular and hemispherical pools with internal energy sources have been reported.

isothermally cooled walls, very dense grids must be used in near-wall layers. Also, it is known that natural convection flows, in an internally heated liquid pool, involve stably stratified layers at the lower region of the pool and mixing region in its upper part. Complex three-dimensional flows take place in the mixing region, necessitating refined nodalization in order to properly describe momentum and heat transfer in buoyant plumes and thermals generated by the upper cooled pool surface. Furthermore, we are interested in the local heat flux distribution on the curved downward surface of core melt pool. In particular, heat fluxes at the lowermost part of the hemispherical pools are of paramount importance for the application (reactor safety) motivating this work. Description of such effects would require refined grids along the curved wall at the bottom region of the semicircular pools. Computational errors and loss of important information (local heat fluxes at the very bottom) may be caused by using single-block mesh generation, assuming node numbers are limited to 10^5. In the present work, we employ the computational mesh design method which is based on multi-block methodology, implemented in the CFDS-FLOW3D computer code. The grid system used in three-dimensional computations of natural convection flows and heat transfer in semicircular cavities is presented in fig.2.

In order to assess the quality of the computational meshes used to solve the governing equations, grid analyses have been performed. For semicircular geometry, five grids have been used for the $[x, y]$-plane. Both the numbers of blocks and nodes have been varied. Results of 2D the computations employing these computational meshes are presented in Table 1. It can be seen that the G-4/3975 and G-12/3988 grids can be used to describe natural convection flows on the $[x, y]$-plane. These grids are utilized in further 3D computations of semicircular slices in the present work.

A 3D computational mesh has been generated for a hemispherical pool, comprised of 14 blocks and about $0.8 \cdot 10^5$ grid points. Fig.3 depicts the computational mesh in a plane through the symmetry line of the hemispherical pool. Even though there exist differences in design of the computational meshes, nodalization in both the near-wall layer and the core region of the hemispherical pool has been made in the same manner as that of the semicircular cavities. In particular, because of the importance in describing recirculation and stagnation regions of descending boundary flows, a very fine grid has been utilized for the lower pool portion.

3 Results and discussion

3.1 3D effects in internally heated semicircular cavities

Apparently, detailization of computational meshes depends much on the specific features of the transport processes under consideration. In particular, the nodalization on the $[x, z]$-plane of the upper surface is of paramount importance, since heat transfer to the upper surface is governed by the interaction between that surface and ascending plumes and by the generation and departure of cooled (denser) blobs from the isothermal wall. There are no experimental observations available to quantify whether large tongues or small blobs dominate heat transfer to the cooled upper wall. At least there exist a lot of blobs which potentially can fall down to the mixing region. Based on the direct numerical simulation (DNS) of free convection in a fluid layer cooled from the top and heated within, Nourgaliev and Dinh [5] formulated requirements for grid nodalization in horizontal direction as $(\Delta z, \Delta x) \propto H \cdot Ra^{-0.272}$. This scale becomes sub-mm for core melt pools of $H = 0.4m$ and $Ra = 10^{12}$. Such a requirement is out of capability of modern workstations. Quantification of subgrid phenomena is yet to be addressed in future parallel computations. It is assumed here, however, that the heat removed by subgrid blobs is relatively small.

In this paper, time and surface-average Nu numbers are the most important heat transfer characteristics. Unfortunately, no heat transfer data is available for internally heated semicircular cavities and hemispherical pools with $Ra = 10^{12}$. Table 2 summarizes some of the prediction results. It can be seen that 3D computations resulted in *higher (about 30%)* [2] *Nusselt numbers* on the upper pool surface Nu_{up}, and therefore, lower temperature differences driving the pool convection. It is perhaps instructive to note that 2D computations of internally heated pools were the only option available in the past due to computing time and memory requirements of 3D modeling. In many cases the 2D problem formulation looks acceptable, since heat is generated uniformly inside the liquid pool. However, an assumption of zero W-velocity

---

[2]Such a percentage must be considered as a trend rather than a strict numerical conclusion concerning the effect of *three-dimensionality*, until effects of sub-grid phenomena can be quantified.

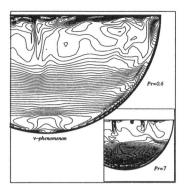

Figure 1: *Temperature fields in the semicircular cavity (Prandtl number effect), 2D computation, $Ra = 10^{12}$, [2].*

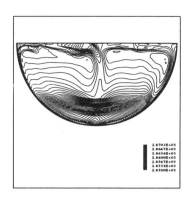

Figure 4: *Temperature fields in the semicircular cavity, slice thickness-to-height ratio: 1:25, $Ra = 10^{12}$, $Pr = 0.6$.*

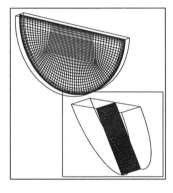

Figure 2: *Computational meshes in [y,x]-plane (2500 nodes ~ 50 × 50) and in [y,z]-plane (50 × 22) of the 3D slice-type semicircular cavity.*

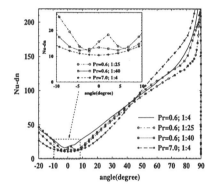

Figure 5: *Slicewise distribution of the depthwisely averaged Nusselt number on the curved surface of the semicircular cavity, $Ra = 10^{12}$.*

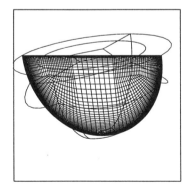

Figure 3: *Computational grid in a plane through the symmetry line of the hemispherical pool.*

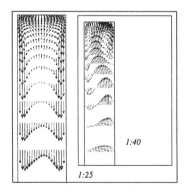

Figure 6: *Effect of slice thickness on the velocity field in the semicircular cavity, $Ra = 10^{12}$, $Pr = 0.6$, $\Delta z/H$ =1:25 and 1:40.*

might reduce the freedom of naturally convecting flows. Most notably, interaction between cooled blobs falling down (thermals) and ascending buoyant plumes is a *three-dimensional phenomenon*. In this sense, the 2D numerical treatment of the flow in question is able to suppress transport and exchange processes in the mixing region and, therefore, reduce heat removal rates to the upper cooled pool surface.

Similarly to [2], the Nu_{up} numbers in cases with $Pr = 7$ are about 15-20% higher than those obtained for $Pr = 0.6$. Temperature and velocity fields obtained in the 3D computational modeling of natural convection flows in semicircular cavities are qualitatively similar to those of 2D ones. Figs.1 and 4 present temperature fields in the semicircular cavity for 2D computation and 3D slice with thickness-to-height ratio 1 : 25. The mixing region in the upper pool portion is larger and the ν-phenomenon is less notable in the 3D cavity. Probably, the better mixing in the upper region is related to enhanced heat transfer to the upper surface. Consequently, the stratified layers are more suppressed, reducing the driving temperature differences for descending boundary flows along the cooled curved walls. Nevertheless, the local Nu numbers in the downward surface of the 2D and 3D semicircular pools behave in a similar way. Most notably, there exists a convection-induced local increase of Nu_{dn} in the lowermost part of the 3D semicircular cavities with $Pr = 0.6$ (see fig.5).

Qualitatively similar flow and temperature patterns were also obtained for the relatively thin slices with slice thickness-to-height ratio in the range 1:40 - 1:15. The surface-averaged Nu_{dn} changed only slightly with slice thickness. A very thin slice might suppress the free convection in the upper pool portion. One can see differences in the velocity field of the slice with slice thickness-to-height ratio 1:25 and that of the 1:40 slice (fig.6). A symmetric pattern of large oscillating vortices is found in the thicker slice, whereas an asymmetrically fixed vortex structure is presented in the thinner slice due to a reduced degree of freedom. In addition, there might be less significant Pr-number effect on Nu_{dn} in a very thin slice (see curve $Nu_{dn} = f(\phi)$ in fig.5 for the 1:40 slice. Perhaps in the cases of small slice thickness, the side walls are capable of damping recirculation flows of the ν-phenomenon.

It must be noted here that most experiments in the past have employed relatively thin slices as test sections when studying natural convection heat transfer in internally heated fluid layers, square and semicircular cavities, [2]. Slice thickness-to-height ratios were typically in the range 0.04-0.15, allowing holographic observations of flow structure and temperature pattern. Even though the face and back surfaces of the slices were thermally isolated, wall effects (primarily on flow structure) might be present, affecting, ultimately, heat transfer on the isothermally cooled surfaces of the slice cavity. Most of the previous computational work analysing natural convection flows in square and semicircular cavities have also been performed in a 2D formulation, presuming no flow effects of the adiabatic slice surfaces and, more importantly, zero W-velocity in the z-direction, [2]. Most of the 2D computations underpredict the Nu number on the upper cooled pool surface for intermediate (say, 10^{12}) and higher Ra numbers ([1]). This fact has been related mainly to the underprediction of turbulence generation in the upper pool portion by the turbulence models employed. The results of the above analysis indicate, however, that validating computational models on data from slice experiments, one must take into consideration the 3D and wall effects of slice test section.

3.2 Hemispherical cavity

3D computations for hemispherical pools have also been performed in the present work. Analysis of the calculational results indicate qualitative similarity of the flow and heat transfer behavior in semicircular and hemispherical geometries. Time and surface average Nu numbers on the upper and curved downward walls of the hemispherical pools are nearly identical to those of semicircular slices with slice thickness-to-height ratios higher than 1:15 (see Table 2). The computations performed for $Pr = 0.6$ indicate that the ν-phenomenon is more notable in the hemispherical geometry (fig.7). It can be seen that the mixing area in the hemispherical pool is larger (in height) than that of the slice-type semicircular cavities. This fact, of course, agrees well with the essence of the ν-phenomenon. A major feature of the hemispherical geometry is that the cooled curved pool wall converges at its bottom, enabling the possibility for interaction between descending boundary flows at the lowest spherical segment. Since cross areas decrease towards the pool bottom, heights of stably stratified layers affected by penetration flows are larger at the lowermost part of the pool. Consequently, relative enhancements of the heat transfer rate must be the largest at the pool bottom. This can be seen in fig.8 which presents local Nu numbers $Nu_{dn} = f(\phi)$ for two selected fluid Pr numbers. Also, the ratios $\frac{Nu_{dn}(Pr=0.6)}{Nu_{dn}(Pr=7)}$ of the hemispherical pool are larger than those of the semicircular slices for angles $\phi < 30°$. In this sense, the present 3D computational modeling indicates that the heat

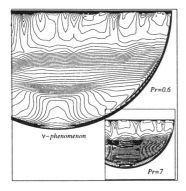

Table 1: *Integral heat transfer characteristics: results of 2D-grid analysis*, $Ra=10^{12}$, $Pr=0.6$.

Grid	# blocks	# nodes	Nu_{up}	Nu_{dn}
G-4/3975	4	3975	116	96
G-4/7200	4	7200	112	97
G-4/21000	4	21000	112	97
G-12/2485	12	2485	120	96
G-12/3988	12	3988	116	98

Figure 7: *Temperature fields in the hemisperical pool with all isothermally cooled walls*, $Ra = 10^{12}$, $Pr = 0.6$ and 7.

Table 2: *Some 3D results for semicircular and hemispherical pools*, $Ra=10^{12}$.

Cases	Pr	Nu_{up}	Nu_{dn}
Semicircular slice	0.6	142-148	95-92
Semicircular slice	7	164	94
Hemispherical pool	0.6	135	104
Hemispherical pool	7.	153	110
2D semicircular cavity	0.6	112	97
2D semicircular cavity	7	136	100

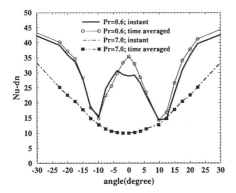

Figure 8: *Local Nusselt numbers on the lowermost part of the hemispherical pool surface,* $Ra = 10^{12}$.

Table 3: *Geometry and heating conditions for side-wall vs. internal heating analysis.*

Cases	Heating	Δz, m	H, m	Ra
SC-I-0.1	Internal	0.1	0.4	10^{12}
SC-S-0.1	Side wall	0.1	0.4	10^{12}
SC-I-0.4	Internal	0.4	0.4	10^{12}
SC-S-0.4	Side wall	0.4	0.4	10^{12}

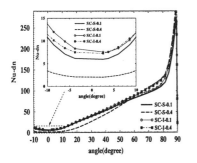

Table 4: *Integral heat transfer characteristics.*

Case	Nu_{up}	Nu_{dn}	$\Delta T_{max}, K$	$\Delta T_{av}, K$
SC-I-0.1	160	97	19.1	14.8
SC-I-0.4	146	97	23.6	15.2
SC-S-0.1	248	89	21.4	11.7
SC-S-0.4	248	91	35.	11.6

Figure 9: *Slicewise distribution of the depthwisely averaged Nusselt number on the curved surface of the semicircular cavity.*

207

transfer effect of fluid Prandtl number is even more significant in the hemispherical cavity than in the semicircular ones.

3.3 Side-wall heating vs. internal heat generation: comparative analysis

Table 3 presents the geometry and heating conditions for a comparative analysis of internal and side-wall heating. The calculated integral heat transfer characteristics in the internally heated and side-wall heated melt pools are presented in Table 4. It can be seen that Nu numbers on the downward surface, Nu_{dn}, in the internally heated and side-wall heated melt pools are quite similar. Nu numbers on the upper surface Nu_{up} in the side-wall heated melt pools are as much as ~ 1.5 times larger than those of the internally heated pools. Correspondingly, the volume-averaged driving temperature differences ΔT_{av} in the internally heated pools are larger than those of the side-wall heated ones.

Enhanced heat transfer to the upper isothermal surfaces in the side-wall heating cases is related to their flowfields. In such cases, flows ascend along the heated side walls and transfer the heat to the upper isothermal surfaces. Then the cooled fluids descend down in both the center upper pool portion and along the cooled curved walls. In the SC-S-0.1 case, the temperature field in the upper-half pool portion is very uniform. In addition, the flowfields in the examined cases of the side-wall heating (SC-S-0.1; SC-S-0.4) are transient and of complex structure. This is in contrast to flowfields calculated previously for $Ra = 10^{10}$ (see ref.[2]), where the flow patterns were of two-loop structure. In this sense, the flow structure in the side-wall-heated melt pools become similar to those of the internally heated pools, in which the mixing region in the upper pool portion is caused by the ascending plumes and falling down blobs. Fig.9 also indicates nearly identical local distributions of the heat transfer characteristics at the bottom part ($\phi < 45°$) of the curved surfaces in cases SC-I-0.1 & SC-S-0.1. Even though the general pictures are almost the same in these two cases, one can note, from calculated temperature fields, differences in the temperature gradients in stably stratified layers in the internally heated melt pool and side-wall heated melt pool.

The above agreement of heat transfer in cases SC-I-0.1 and SC-S-0.1, cannot be viewed as a general law applicable to all side-wall and internal heating cases with equivalent power input and slice thickness. From fig.9 one can see that the slicewise and depthwise distributions of the heat transfer characteristics on the downward curved surface of the 0.4m thick internally heated pool coincide with that of 0.1m thick one. Temperature and velocity fields obtained for the semicircular cavity with $\Delta z = 0.4$m are quite similar to those of the semicircular cavity with $\Delta z = 0.1$m, indicating no significant effects of the side walls and of the slice thickness in internally heated natural convection flows. *This feature is the major difference in the thermal hydraulics of internally heated and side-wall heated melt pools.* When increasing the slice thickness, the heat transfer rate to the bottom part of the curved surface decreases significantly in side-wall heating cases. Even though the averaged Nu numbers (see Table 4) are quite similar for the side-wall heated melt pools with $\Delta z = 0.1$m and $\Delta z = 0.4$m, the differences of the local Nusselt numbers $Nu_{dn} = f(\phi)$ might reach 200-250% for $\phi < 30°$; see fig.9. Although the volume averaged temperatures are nearly the same (11.6K for $\Delta z = 0.4m$ and 11.7K for $\Delta z = 0.1m$), the maximum temperature difference in the 0.4m-thick pool is much higher. From the isotherm pattern of the 0.4m thick side-wall heated semicircular cavity, one can see that the temperature gradients in the vertical direction ($\frac{dT}{dy}$) in the stably stratified layers are much smaller than that of the internally heated or 0.1m thick side-wall heated pools. Moreover, calculational results indicate no velocities in the middle part in the bottom half of the 0.4m thick side-wall heated pool. This observation is important because heat conduction is the major heat transfer mechanism in the stable stratification region. Thus, the larger the slice thickness (Δz), the smaller the thermal conduction to the middle section of the melt pools. Therefore, heat transfer from the heated walls to the middle section is limited. Furthermore, in the side-wall heating case there is remarkable peaking of the heat transfer in the heated wall regions. The uniformly distributed internal energy sources allow for better conditions for heat removal to the bottom wall, whereas heat supplied from the side walls is directed mainly upwards through the ascending boundary flows. That is why the physical dissimilarity becomes significant for the side-wall heated pools with a larger slice thickness (Δz).

4 Summary

Computational modeling has been carried out in order to explore the geometrical and 3D effects on very complex flows and heat transfer characteristics in naturally convecting volumetrically and side-wall heated melt pools. Rather high Ra numbers ($\propto 10^{12}$) were selected for the numerical analysis. Under such

Ra number conditions, uncertainties in turbulence modeling did not override other uncertainties of interest. On the other hand, it is believed that the flow situations modelled are quite representative of the physics of fluids in an internally heated melt pool in a wide range of Ra numbers. Nevertheless, extrapolation of the summary given below to Ra numbers higher than 10^{12} must be made with caution.

1. 3D computations of naturally convecting flows are capable of providing higher heat transfer rates to the cooled upper pool surface than 2D simulations. Therefore, 3D computational modeling is recommended for the analysis of a limited number of existing experiments.

2. The 3D modeling results confirm the presence of fluid Pr number related convection phenomena and the significance of such effects on heat transfer characteristics on the curved surface of major interest for the reactor safety application motivating this CFD analysis. The most significant effect of fluid Pr number on heat transfer rate distribution in the lowermost part of the curved wall is observed in the hemispherical pool. This is related to the simultaneous actualization of penetration phenomena of descending boundary flows and influence of convergent geometry of the spherically bottomed cavity.

3. The numerical analyses performed, have provided a more solid basis for obtaining insight into the physics of side-wall heated natural convection flows and heat transfer. They have also resolved the question whether side-wall heating is a good method of simulating of volume heat generation.

4. The calculational results indicate significant 3D effects in the side-wall heated melt pools when varying slice thicknesses. Minor effects of the slice thickness variations are found for the internal heating case.

5. Heat transfer data in experimental facilities using side-wall heating are irrelevant to heat transfer characteristics and their local distributions in decay-heated core melt pools of reactor safety applications.

6. If the purpose of related experiments is not to obtain safety relevant natural convection heat transfer data but to investigate physico-chemical behavior of core melt and melt-vessel interaction, one could use side-wall heating to simulate internal energy sources, since the flow and temperature fields in the side-wall heated melt pools involve similar flow field structure. Furthermore, it is possible to select such slice thicknesses that the heat transfer characteristics at the downward surface are similar in the side-wall and internal heating cases.

Nomenclature

Arabic letters

h	Total enthalpy, (J/kg)	R	Radius of semicircular cavity, (m)
H	Height of semicircular cavity, (m)	Ra	Rayleigh number, $Ra = \frac{q_v H^5}{\alpha \lambda \nu} g\beta$, (-)
Nu	Nusselt number, $Nu = \frac{q_t H}{\lambda(T_w - T_{ave})}$, (-)	T	Temperature, (K)
Pr	Prandtl number, $Pr = \nu/\alpha$, (-)	U	Velocity, $\mathbf{U} = (U, V, W)$, (m/s)
q_v	Volumetric heat generation rate, (W/m^3)	x, y, z	Co-ordinates, (m)

Greek letters

α	Thermal diffusivity, (m^2/s)	μ, ν	Dynamic, kinematic viscosity, (Pa·s, m^2/s)
β	Coefficient of thermal expansion, (1/K)	ρ	Density, (kg/m^3)
λ	Heat conductivity, (W/m·K)		

References

[1] Dinh,T.N. and Nourgaliev,R.R., 1995, On Turbulence Modeling in Large Volumetrically Heated Liquid Pools, *Intern. J. Nuclear Engineering and Design* (in press).

[2] Dinh,T.N., Nourgaliev,R.R., and Sehgal,B.R., 1995, On Heat Transfer Characteristics of Real and Simulant Melt Pool Experiments, *Proceeding of the Seventh International Topical Meeting on Nuclear Reactor Thermal Hydraulics NURETH-7*, Saratoga Springs, New York, September 10-15, 1995.

[3] Launder,B.E. and Sharma,B.I., 1974, Application of the Energy-Dissipation Model of Turbulence to the Calculation of Flow Near a Spinning Disc, *Letters in Heat Mass Transfer*, Vol.1, pp.131-138.

[4] CFDS-FLOW3D, 1994, Release 3.3, User Manual, June 1994.

[5] Nourgaliev,R.R., Dinh,T.N., 1995, An Investigation of Turbulent Transport Characteristics in a Volumetrically Heated Unstably Stratified Fluid Layer, *Research Report NPS-959ND, Royal Institute of Technology, Stockholm*, 40p.

Numerical Simulation and Analysis of Manipulated Transitional Flow over a Fence

Alexander Orellano and Hans Wengle
Institut für Strömungsmechanik und Aerodynamik
Universität der Bundeswehr München
D-85577 Neubiberg, Germany

1 Abstract

A direct numerical simulation of flow over a fence was performed for a Reynolds-number of 250 (based on fence height and maximum inflow velocity). The free shear layer created at the sharp edge of the fence undergoes transition to turbulence before reattaching on the bottom plate behind the flow obstacle.

This flow has been manipulated with ten stationary or oscillating wall jets located on a crosswind line before the separation in front of the fence. A 50 % reduction in the mean recirculating length (behind the fence) could be observed for the case with oscillating vertical wall jets (periodical blowing/suction). The disturbance frequency was selected to be close to the frequency of the Kelvin-Helmholtz instability mode.

Eigenmode-decompositions of the flow fields showed that in the most successful manipulation with oscillating jets, a pair of two-dimensional spatial modes, together with the corresponding temporal modes (oscillating at the forcing frequency), are responsible for an accelerated increase of the vorticity thickness in the free shear layer and for the accelerated transition.

2 Introduction

The understanding of flow separation and reattachment is important for the design of engineering applications, e.g. in turbomachinery and aeronautics. It has been shown in recent times that there are possibilities of controlling the laminar-turbulent transition process or the structures of fully turbulent flows to achieve a desired behaviour of the flow, such as acceleration/delay of transition, drag reduction, minimization/maximization of mixing or heat transfer, the reduction of separation, or the reduction of the size of recirculation zones. An improved understanding of the basic dynamics in the flows is required to carry out sucessfully the controlled manipulations to achieve such goals.

The pioneering experiments by Liepmann et al. [5] have probably been the beginning of the new era of management and control of laminar/turbulent flows. There are recent reviews on that subject by Ho and Huerre [3], Bushnell and Mc Ginley [1] and Fiedler and Fernholz [2]. Related to the topics of this paper, there are results from experimental work available, e.g. from Johnston and Nishi [4].

The objectives of this study are (a) to investigate numerically the effects of stationary and periodically oscillating wall jets on the size of the recirculation zone in front and behind a fence and (b) to analyze the computed flow fields with an eigenmode decomposition (Karhunen-Loève decomposition, proper orthogonal decomposition).

3 Numerical Simulation

The numerical scheme: The governing equations describing the flow quantities are derived from the integral conservation equations for mass and momentum. The resulting equations are solved numerically on a staggered and non-uniform cartesian grid, using second-order finite-differencing in time and space (explicit leap frog for time discretization, central differencing for convection terms and time-lagged diffusion terms). The problem of pressure-velocity coupling is solved iteratively. The geometry of the flow obstacle is approximated by simply blocking out the corresponding grid cells within a cartesian grid. As inflow condition we applied a uniform velocity profile. At the outflow section, the normal gradients have been set to zero. At the lateral sides of the computational domain periodic boundary conditions have been used. At the bottom plate we used no-slip boundary conditions, and at the upper side slip conditions have been applied.

The simulation: The geometry of the computational domain and of the fence is evident from figure 1. The origin of the coordinate system is located at the front side of the fence. Measured in units of the reference length L_{ref} (height of the fence) the dimensions of the computational domain are $(X, Y, Z) = (50, 10, 5)$ with $(N_x, N_y, N_z) = (164, 104, 52)$ grid points.

Here we present results of the flow over a fence manipulated by ten jets located on a crosswind line at $X = -2.5$. The jets were realized as boundary conditions at the wall. The first type of manipulation (case B3) introduces a certain lateral length scale by using stationary jets. In addition these jets (skewed and pitched under 45°) create counter-rotating longitudinal vortices. The second type of manipulation (case C2) introduces (in addition to the length scale) a time scale by using vertical jets with a time-dependent amplitude (periodical blowing/suction with Strouhal number $= L_{ref} \cdot f/U_{inflow} = 0.15$). As reference case serves a simulation without any manipulation (case A1).

Recirculation length: Figure 1 shows an isosurface of the streamwise velocity component for a sample of the instantaneous flow (left side) and for the mean flow (right side). The reference case A1 exhibits a recirculation length of 1.5 fence heights in front of the obstacle and 23 reference lengths after the obstacle.

The flow perturbed by stationary jets (case B3) shows a locally reduced recirculation length in front of the obstacle. The longitudinal vortices created immediately behind the jets are convected downstream and loose first some of their strength. Then, in the recirculation zone in front of the obstacle, the vorticity in streamwise and vertical direction increases in strength due to the strong curvature of the streamlines. In the region of detachment of the flow at the edge of the fence the longitudinal vorticity decreases again while a strong increase of lateral vorticity is created by the developing shear layer. The ($u = 0.0$)-isosurface of case B3 in the region after the fence shows very little change compared to the reference case A1. The distribution of the mean velocity component $<u>$ at the wall (fig. 2) shows a similar behaviour for the cases A1 and B3. The similar streamwise distribution of the vorticity thickness (fig. 3) for case A1 and B3 indicates that the geometry of the shear layer could not be altered by the introduction of stationary longitudinal vortices.

A totally different behaviour of the flow after the obstacle could be obtained by manipulating the flow with jets oscillating at the Kelvin-Helmholtz frequency. A reduction of the recirculation length of about 50% could be reached. The exponential growth of the perturbation introduced by forcing the flow with oscillating jets leads to a stronger roll-up process inducing an enhancement of the vorticity thickness (fig. 3).

Turbulent kinetic energy: Isolines of the turbulent kinetic energy (averaged in the lateral direction) for the three cases are shown in fig. 4. The maximum of the turbulent kinetic energy is located in the region of reattachment. This suggests that the transition process is also triggered by the fluid splashing onto the bottom plate.

The comparison of case A1 (no manipulation) with case B3 shows that the distribution of the turbulent kinetic energy and its intensity have not been affected by the longitudinal vortices induced by stationary jets in front of the obstacle. However, the introduction of oscillating jets accelerates the transition process (case C2) and enhances the maximum of the turbulent kinetic energy.

4   Eigenmode-decomposition

The analyses of turbulent flow by a Karhunen-Loève-decomposition (Lumley [6]) is based on the idea that a turbulent flow field can be expanded by a superposition of spatially fixed coherent structures $\phi^n(x,y,z)$ weighted by time-dependent coefficients $a^n(t)$: $u(x,y,z,t) = \sum_n a^n(t) \cdot \phi^n(x,y,z)$. The spatial modes ϕ^n can be deduced by calculating the eigenvectors of the two-point correlation tensor. Then, the temporal coefficients $a^n(t)$ can be evaluated using the spatial modes (Manhart and Wengle [7]).

We used 1000 samples of the velocity field with a time-intervall of $\Delta t =$

$L_{ref}/U_{in} = 0.15$ to construct the correlation tensors. In the following the eigenmode decomposition of the case A1 and C2 will be presented.

Spatial behaviour of the coherent structures: The four fluctuating velocity modes containing most of the energy are shown in fig. 5. The reference case A1 (left) is compared to the flow manipulated with vertical oscillating jets (case C2). The fluctuating energy of case A1 is produced mainly by 3-dimensional structures with relatively long periods. The modes 2 and 3 exhibit for both cases a phase shift describing a coherent structure convectively transported downstream. The modes 4 and 5 of the manipulated flow exhibit a totally different behaviour compared to the equivalent modes of case A1. They contain most of the fluctuating energy responsible for the roll-up process of the shear layer.

Temporal behaviour of the coherent structures: Fig. 6 shows the temporal coefficients for the reference case A1 and the manipulated case C2. It is evident that the perturbed flow includes coherent structures with well defined frequencies compared to the reference case that exhibit coherent structures (modes) with a broadband frequency. The spectra of the temporal coefficients are similar for the mode pairs (2,3) and (4,5) for the manipulated flow. The peak-frequency of the mode-pair (4,5) is equal to the forcing frequency indicating the strong dependence of this coherent structure on the oscillating jets in front of the obstacle.

References

[1] D.M. Bushnell and C. McGinley. Turbulence control in wall flows. *Ann. Rev. Fluid Mech.*, 21:1–20, 1989.

[2] H.E. Fiedler and H.H. Fernholz. On management and control of turbulent shear flows. *Prog. Aerospace Sci.*, 27:305–387, 1990.

[3] C.M. Ho and P. Huerre. Perturbed free shear layers. *Ann. Rev. Fluid Mech.*, 16:365–424, 1984.

[4] J.P. Johnston and M. Nishi. Vortex generator jet - means for flow separation control. *AIAA Journal*, 28(6):989–994, June 1990.

[5] H.W. Liepmann and D.M. Nosenchuck. Active control of laminar-turbulent transition. *J. Fluid Mech.*, 118:201–204, 1982.

[6] J.L. Lumley. The structure of inhomogeneous turbulent flows. In A.M. Yaglom and V.I. Tatarski, editors, *Atmospheric Turbulence and Radio Wave Propagation*, pages 166–178, Moscow, 1967. Nauka.

[7] M. Manhart and H. Wengle. A spatiotemporal decomposition of a fully inhomogeneous turbulent flow field. *Theoret. Comput. Fluid Dynamics*, 5:223–242, 1993.

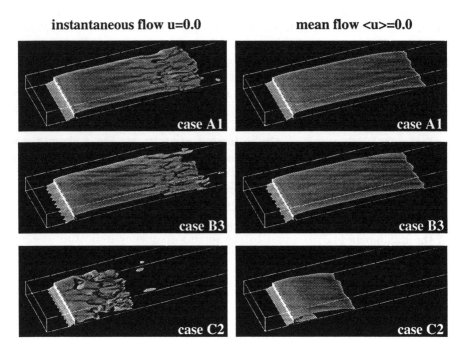

Figure 1 Isosurfaces $u = 0.0$ and $<u> = 0.0$

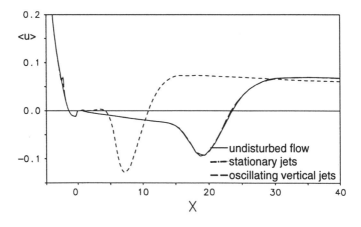

Figure 2 $<u>$-velocity distribution at the wall ($Z = 0.015$)

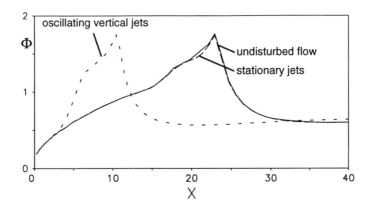

Figure 3 Vorticity thickness of the shear layer

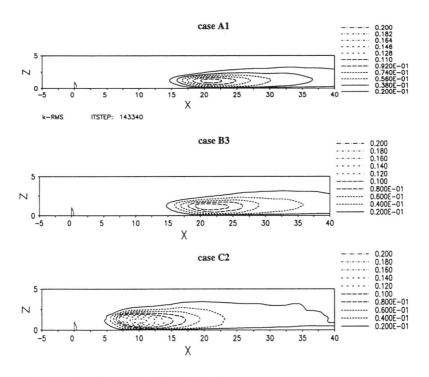

Figure 4 Isolines of the turbulent kinetic energy

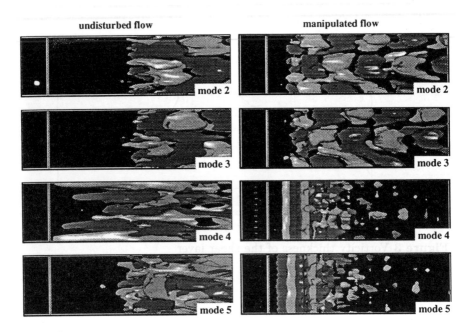

Figure 5 Spatial Karhunen-Loève modes: (u)-velocity

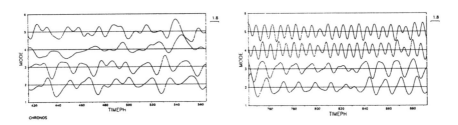

Figure 6 Temporal Karhunen-Loève modes

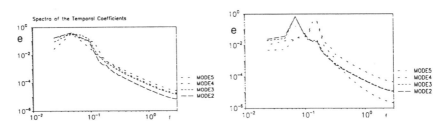

Figure 7 Fourier analyses of the temporal modes

Calculation of 3D Turbulent Boundary-Layer Flows in an S-Shaped Channel

R.D. Parker and J. Bruns
Institut de Machines Hydrauliques et de Mécanique des Fluides, Ecole Polytechnique Fédérale de Lausanne, CH-1015 Lausanne, Switzerland

1 Summary

Calculations are made of the 3D boundary-layer flow over the test wall in an S-shaped channel in order to test the performance of turbulence models under conditions of anisotropic turbulence. The external inviscid flow driving the boundary-layer is obtained by integrating the Euler equations along streamlines on the basis of the measured pressure distribution. Comparison of the measured $-\overline{uv}$ profiles with those calculated using isotropic eddy viscosity models reveals clearly those parts of the flow where anisotropic effects are likely to be important. The data should therefore provide a clear test for anisotropic models. Details are given of how turbulence models based on Rotta's anisotropy relations are implemented into the non-orthogonal, surface oriented tensor coordinate system used in the boundary-layer code, by means of a local projection technique.

2 Introduction

In complex three-dimensional flows the principle axis of the Reynolds stress tensor is not, in general, aligned with that of the strain tensor, rendering inapplicable the use of turbulence models based on the Boussinesq approximation. Even so such models remain the main method for treating turbulence in industrial flow calculations.

A large variety of models based on the Reynolds averaged equations have been developed to take account of this turbulent anisotropy. These range from relatively simple quasi-3D extensions of popular eddy viscosity models [1, 2, 3, 4, 5], which in some cases, e.g.: see Lindberg, pp 133-138 of [1] and [5], include an additional transport equation for modelling the direction of the stresses, to the large multi-equation models which solve an equation for each stress component. A major task is validating such models against experimental flow test-cases in order to prove their ability to reproduce the effects of anisotropy and other features of the turbulence.

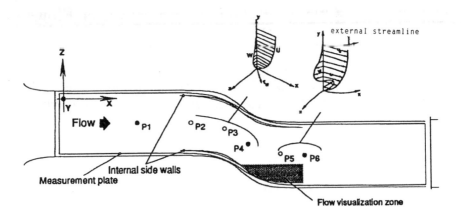

Figure 1 Diagram of S-shaped duct showing measurement positions and pictures of the typical three-dimensional mean velocity and "cross-over" profiles.

An experiment conceived for this purpose is the EPFL S-shaped channel or S-duct [6], figure 1. The S-shaped geometry of the channel induces a pressure gradient driven 3D boundary layer flow over the test wall. The boundary layer becomes highly sheared and exhibits the interesting feature of cross-over profiles downstream from the inflection point. This shearing inevitably also causes the turbulence to become anisotropic, as is evidenced by differences, of around 10 degrees at the point P3 and up to 30 degrees off the centreline, measured in the angles:

$$\beta_\tau = tan^{-1}\left(\frac{\overline{vw}}{\overline{uv}}\right) \text{ and } \beta_g = tan^{-1}\left(\frac{\partial W/\partial y}{\partial U/\partial y}\right). \tag{2.1}$$

Detailed measurements [6] have been made of the velocity flow field and turbulence correlations in the S-duct, yielding a valuable database of results which are compared here with results of calculations made with a 3D boundary-layer code called SOBOL.

The code [7] uses a fast space-marching algorithm to solve the 3D boundary-layer equations on a Surface Oriented Locally Monoclinic Coordinate (SOLMC) system define by the metric:

$$g_{ij} = \begin{pmatrix} g_{11} & g_{12} & 0 \\ g_{12} & g_{22} & 0 \\ 0 & 0 & 1 \end{pmatrix} \tag{2.2}$$

Generalized coordinates are used over the body surface, with the third coordinate aligned along normals to the surface. This allows almost any surface coordinate geometry to be used. For the S-duct the streamlines of the external flow solution are used to generate the surface grid. Coordinate stretching in the surface normal direction is used to accomodate the down-stream growth of the boundary layer.

Previous measurements and calculations have been made on an earlier configuration of the S-duct having a smaller spanwise pressure gradient. These are compiled in the proceedings of the 1st ERCOFTAC workshop in reference [1]. The models tested showed varying degrees of agreement but all models left plenty of room for improvement. More importantly, when agreement was found, it was often not clear whether this was due to improved modelling of the anisotropy or some other aspect of the calculation.

In the current configuration the curvature of the side walls has been increased until the flow just separates over a small region. This was done to maximize the three-dimensionality of the boundary layer. By comparing measurements with calculations using isotropic turbulence models the parts of the flow significantly affected by the anisotropy are clearly revealed and can therefore be used as tests for anisotropic models. In the previous study the effects of anisotropy were not so easily distinguished, perhaps because of the weaker three-dimensionality of the flow but also because of a greater scatter in the data. The greater accuracy of the present hot wire probe measurements is a result of improved statistics made possible by recent advances in the data aquisition technology.

3 Outer Boundary Condition

The boundary condition prescribed on the outer surface of the boundary layer is deduced from a set of external flow streamlines calculated from a fit to static pressure measurements taken over the test wall. By assuming an ideal, isentropic gas the density, pressure and internal energy of the external flow were obtained from measurements of the $C_p = \frac{P-P_{ref}}{1/2\rho_{ref}V_{ref}^2}$ using:

$$\begin{cases} \frac{\rho}{\rho_{ref}} = 1 + \frac{1}{2}M_{ref}^2 C_p + O(\delta^4) \\ \frac{p}{p_{ref}} = 1 + \frac{1}{2}\gamma M_{ref}^2 C_p + O(\delta^4) \\ \frac{e}{e_{ref}} = 1 + \frac{1}{2}(\gamma-1)M_{ref}^2 C_p + O(\delta^4). \end{cases} \quad (3.3)$$

The magnitude of the external inviscid flow velocity along a streamline is obtained from the C_p using $|\mathbf{V}_e|/V_{ref} = \sqrt{1.-C_p}$. The streamlines of the external inviscid flow are then integrated downstream from measured starting profiles using:

$$\frac{d\beta}{dx} = \frac{\frac{\partial C_p}{\partial x}\tan\beta - \frac{\partial C_p}{\partial z}}{2(1-C_p)} \quad (3.4)$$

where β is the angle between a streamline and the x-axis. The C_p and its gradients were obtained from a fit to the pressure measurements using a bi-cubic spline. Details of the fitting process and sensitivity studies are given in ref. [8].

4 Tensor Application of Anisotropic Turbulence Model

The implementation of turbulence models is complicated by the fact they are often formulated for 2D flows and usually only in terms of Cartesian coordinates. Various attempts have been made to reformulate the more successful models such as Johnson and King [9] in a generalized 3D form, usually by replacing the convection and production terms in the model equation by their generalized 3D counterparts; e.g. [4]. The modelling assumptions, however, make this difficult to do for all terms. In particular the expressions derived by Rotta [2] and used in the model of Savill, Gatski and Lindberg (SGL) (pp 158-165 of [1]) for taking into account the anisotropy of the turbulence are expressed in terms of a local Cartesian coordinate system and are not easily generalised to tensor form.

An alternative approach is to simply apply the model in its original form in terms of the local Cartesian coordinate system, by aligning it at each point with the direction of the external inviscid flow, and then projecting out the necessary components from the tensor system in which the variables are defined. The effect of the turbulence model can then be fed back to the mean flow equations by projecting the local Cartesian components of the Reynolds stress back to their corresponding tensor form.

The local Cartesian component, U, of the mean flow vector, $\mathbf{V}$, in the direction of the external inviscid flow, $\mathbf{V}_e$, can be obtained from the contravariant components of the mean flow using:

$$U = \frac{\mathbf{V} \cdot \mathbf{V}_e}{|\mathbf{V}_e|} = \frac{a_{\alpha\beta}\, \mu_\gamma^\alpha(x^3)\, V^\gamma\, \mu_\eta^\beta(\delta)\, V_e^\eta}{\sqrt{g_{\eta\zeta}\, V_e^\eta\, V_e^\zeta}} \qquad (4.5)$$

where $g_{\alpha\beta}$ is the metric tensor and $a_{\alpha\beta}$ is its value on the surface. Here the "shifter" tensor μ_β^α is used to transform the vectors, defined at different points on a normal, in terms of the corresponding base vectors at the surface. This is necessary to ensure the validity of the dot product since in general the base vectors will vary around the grid as a function of the local curvature. The monoclinic (or shell) property of the metric allows the shifter to be defined simply in terms of the Kronecker delta tensor δ_β^α and the curvature tensor $b_\beta^\alpha x^3$ by:

$$\mu_\beta^\alpha = \delta_\beta^\alpha - b_\beta^\alpha x^3 \qquad (4.6)$$

where x^3 is the shift along the normal coordinate. In the same manner the local Cartesian component W of the mean flow orthogonal to U, is taken in the direction of a cross product between the external inviscid flow vector and the surface normal base vector as follows:

$$W = \frac{\mathbf{V} \cdot \mathbf{g}_3 \times \mathbf{V}_e}{|\mathbf{g}_3 \times \mathbf{V}_e|} = \frac{\sqrt{a}\, e_{3\alpha\gamma}\, \mu_\beta^\alpha(\delta)\, V_e^\beta\, \mu_\eta^\gamma(x^3)\, V^\eta}{\sqrt{g\, e^{3\alpha\beta}\, e^{3\gamma\eta}\, g_{\beta\eta}\, g_{\alpha\zeta}\, V_e^\zeta\, g_{\gamma\xi}\, V_e^\xi}} \qquad (4.7)$$

where e_{ijk} is the permutation tensor and a and g are the determinants of the metric and surface metric tensors respectively.

Once U and W have been evaluated at all the points along a normal the values of $\partial U/\partial y$, $\partial W/\partial y$, and the other auxiliary variables can be calculated and used to compute the scalar eddy viscosity, ν_t, with, for example, the SGL (pp 158-165 of [1]) turbulence model. The local Cartesian components of the Reynolds stresses: $-\overline{uv}$ and $-\overline{wv}$ can then be obtained using:

$$\begin{bmatrix} -\overline{uv} \\ -\overline{wv} \end{bmatrix} = \begin{bmatrix} \nu_{11} & \nu_{12} \\ \nu_{21} & \nu_{22} \end{bmatrix} \cdot \begin{bmatrix} \partial U/\partial y \\ \partial W/\partial y \end{bmatrix} \tag{4.8}$$

where

$$\nu_{11} = \nu_t \left(\frac{U^2 + TW^2}{q^2} \right) \tag{4.9}$$

$$\nu_{22} = \nu_t \left(\frac{W^2 + TU^2}{q^2} \right) \tag{4.10}$$

$$\nu_{12} = \nu_{21} = \nu_t (1 - T) \frac{UW}{q^2} \tag{4.11}$$

are Rotta's anisotropy relations [2] based on modelling the pressure-strain term of the Reynolds stress equations. An alternative set of anisotropy relations was derived by Moreau [3].

The calculated Reynolds stresses must be converted into contravariant tensor components $-\overline{u^\alpha u^3}$ so that they can be included explicitly in the next iteration of the mean flow equations. By assuming the vector relationship implied by equation 4.8 we are effectively treating the $-\overline{uv}$ and $-\overline{wv}$ correlations as components of a 'stress vector' which is related to a 'strain vector' via a contraction with a viscosity tensor of rank 2. This is a simplification of the general case, in which the the stress and strain are 2nd rank tensors related by a double contraction with a 4th rank 'viscosity' tensor. The simpler form follows from the use of the boundary layer approximation. The relationship between the local Cartesian and the contravariant tensor components of the Reynolds stress can therefore be obtained by the same projection method used to extract U and W; that is:

$$\begin{aligned} -\overline{uv} &= |\mathbf{V}_e|^{-1} a_{\alpha\beta} \mu_\delta^\beta V_e^\delta \mu_\gamma^\alpha (-\overline{u^\gamma u^3}) &= A_{1\gamma} (-\overline{u^\gamma u^3}) \\ -\overline{wv} &= |\mathbf{g}_3 \times \mathbf{V}_e|^{-1} \sqrt{a}\, e_{3\alpha\gamma} \mu_\beta^\alpha V_e^\beta \mu_\delta^\gamma (-\overline{u^\delta u^3}) &= A_{2\delta} (-\overline{u^\delta u^3}). \end{aligned} \tag{4.12}$$

These can then be inverted to determine the contravariant components $-\overline{u^1 u^3}$ and $-\overline{u^2 u^3}$ using:

$$\begin{bmatrix} -\overline{u^1 u^3} \\ -\overline{u^2 u^3} \end{bmatrix} = \frac{\rho}{\det(A_{\alpha\beta})} \begin{bmatrix} A_{22} & -A_{12} \\ -A_{21} & A_{11} \end{bmatrix} \cdot \begin{bmatrix} -\overline{uv} \\ -\overline{wv} \end{bmatrix}. \tag{4.13}$$

This approach is particularly suited to the boundary layer technique where the use of the space marching algorithm on a surface oriented generalized coordinate system [7] ensures that the turbulence model can be solved locally at each point, rather than as part of a global system of coupled equations, where such local changes of coordinate system are likely to be a problem.

5 Results and Discussion

Results are presented here for the 3D but isotropic versions of the models by: Johnson-King (JK) [9], Cebeci-Smith (CS) [10] and Baldwin-Lomax (BL) [11] at the points marked in figure 1.

We first observe that, except for the point P6, the JK model predicts quite well the meanflow profiles U and W and therefore indicates that the anisotropy is relatively weak, at least along the centre line. The improvement over the CS and BL models is not, however due to the inclusion of history effects since running the JK model without the non-equilibrium correction produces only minor differences. It is probably therefore a consequence of the feature in the JK model of integrating the eddy viscosity along the path of maximum Reynolds stress.

The absense of side inflow boundary conditions reduces the domain of validity of the calculation. This may explain the poorer agreement for P6 since this point lies just outside the domain [8]. It is intended to increase the size of the domain of validity by employing measured side wall boundary conditions.

The much poorer agreement for the turbulent stresses, $-\overline{uv}$ and $-\overline{wv}$, is consistent with the lack of anisotropic corrections in the models tested. In particular a local deviation of the measurements from an otherwise reasonable agreement in the $-\overline{uv}$ profiles is observed to move outwards from the wall as it goes downstream. The distinctness of this feature plus the fact that it is not at all captured by the isotropic models should make it a point to look for when testing anisotropic turbulence models.

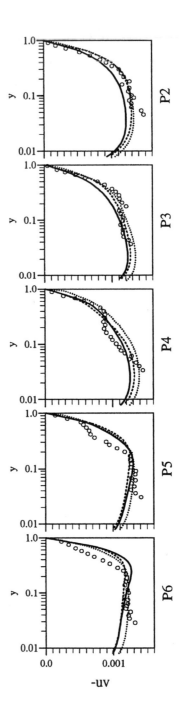

222

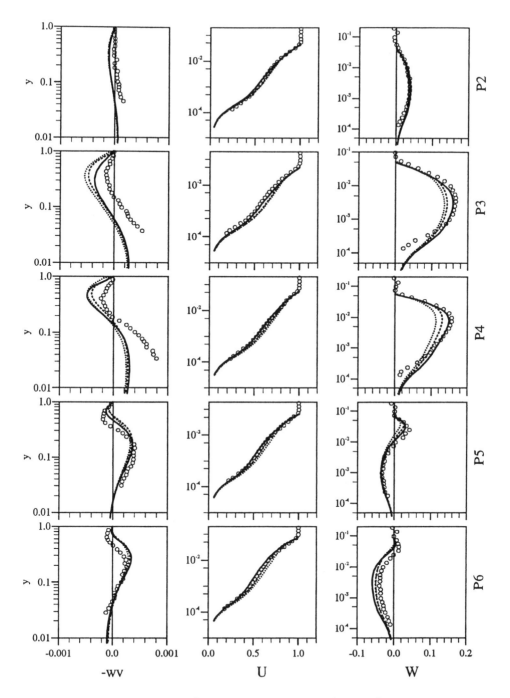

Figure 2 Profiles of $-\overline{uv}/U_e^2$ (on previous page) and $-\overline{wv}/U_e^2$ vs. y/δ and U and W vs. y for calculations with the JK (—), CS (- - -) and BL (....) turbulence models and measurements at the points: P2, P3,...,P6

The next phase of this study will not only involve the testing of such models using the projection technique described but also a comparison with points away from the centre-line where measurements (see Ryhming et al. pp 197, [1]) indicate that the anisotropy is stronger.

Acknowledgements

The authors wish to thank Prof. I.L. Ryhming for his guidance and stimulating discussions. Useful discussions were also had with Prof. P.A. Monkewitz, Dr. T.V. Truong and Prof. H.H. Fernholz.

References

[1] Ed. by: PIRRONEAU, O., RODI, W., RYHMING, I.L., SAVILL, A.M. and TRUONG, T.V., Proceedings of 1st. ERCOFTAC Workshop on *Numerical Simulation of Unsteady Flows and Transition to Turbulence*, Lausanne, Cambridge University Press, (1992).

[2] ROTTA, J.C., *A family of turbulence models for three-dimensional boundary layers*, Turbulent Shear Flows 1, pp. 267-278, Ed. by Durst,F; Launder, B.E.; Schmidt, F.W. and Whitelaw, J.H., Springer-Verlag, (1979).

[3] MOREAU, V., *A study of anisotropic effects in three-dimensional, incompressible turbulent boundary layers driven by pressure gradients*, Ph.D Thesis No.755, EPFL, Lausanne, Switzerland, (1988).

[4] ABID, R., *Extention of the Johnson and King turbulence model to 3-D flows*, AIAA Paper 88-0223 (1988).

[5] BETTELINI, M.S.G., FANNELOP, T.K., *Modifications to the Johnson-King model for three-dimensional boundary layers*, Z. Flugwiss. Weltraumforsch., 16, pp. 325-330, (1992).

[6] BRUNS, J., TRUONG, T.V., *Flows in an "S"-shaped duct*, ICEFM-94, pp. 679, Proceedings of the 2nd. International Conference on Experimental Fluid Mechanics, 4-8 July, (1994), Torino, Italy.

[7] MONNOYER DE GALLAND, F., *Calculation of three-dimensional atteached viscous flow on general configurations using second-order boundary-layer theory*, Z. Flugwiss. Weltraumforsch. 14, 95-108, (1990).

[8] PARKER, R.D., *Boundary-layer calculations of the S-duct, part-1*, Internal report no. T-94-29, IMHEF-DGM, EPFL, CH-1015, Lausanne, Switzerland, (1994).

[9] JOHNSON, D.A., KING, L.S., *A mathematically simple turbulence closure model for attached and separated turbulent boundary layers*, AIAA Journal, vol.23, no.11, (1985), 1684-1692.

[10] CEBECI, T., SMITH, A.M.O., *Analysis of turbulent boundary layers*, Academic Press Inc., London, (1974).

[11] BALDWIN, B., LOMAX, H., *Thin layer approximation and algebraic model for separated turbulent flows*, AIAA Journal, 78-257, (1978).

EXPERIMENTAL ACTIVITIES WITHIN COST-F1, SUBGROUP A-7,
ON LAMINAR-TURBULENT TRANSITION: A SUMMARY REPORT

F. Pittaluga
IMSE, Institute of Thermal Machines & Energy Systems, Univ. of Genova
via Montallegro 1, 16145 Genova, Italy

SUMMARY

In the broad field of the laminar-turbulent transition phenomena, so widely relevant to the advancement of scientific knowledge and technological application, COST F1 Subgroup A7 has been carrying out advanced research activity, in connection with both the development of sophisticated theoretical prediction methods, and the setting up of experimental equipment, suitable to yield most reliable and detailed test-cases, aimed at validating purposes. The paper deals with a first-tentative effort to set-up and present in an organized, though still partial, way, the experimental facilities available to COST Action F1's partners, for conducting "reference" researches on transition. In this connection, but just for informative purposes, a few experimental achievements are also reported, most recently obtained on some of the listed facilities.

INTRODUCTION

Transition phenomena are quite difficult to be physically interpreted and theoretically predicted, whilst, at the same time, the various influences, traceable to them, acting in the broad fields of aero- and thermo-fluid-dynamics can be important and, many times, even critical. For these reasons, a specific "transition" research group, named Subgroup A7, was constituted among the participants to COST Action F1: this group, under the coordination of Dr.A.M. Savill of the University of Cambridge, UK, and of the writer of this paper as deputy coordinator, has recently and formally merged with the ERCOFTAC Transition SIG, and has been carrying out advanced theoretical and experimental research activity, as attested, e.g., in refs.[1, 2]. Specifically, the fields of interest are those of development, evaluation and improvement of theoretical prediction methods for laminar-turbulent transition (and re-transition) phenomena, under a wide variety of flow conditions and degrees of free-stream disturbances, in connection with general thermo fluid-dynamics and turbomachinery applications. To this end, crucial has turned out the availability of detailed, complete and reliable experimental test-cases to be utilized as references for assessing validity, and cross comparing results, of the different computational activities, usually performed by each group with "blind" modality.

In this context, and in order to give an overall, though still not exhaustive, informative picture of the operative facilities, equipment, and instrumentation available within COST Action F1 for transition studies, the first part of the paper will be dealing with a synthetic overview of the information on this subject which, in some more detailed and organized form, is on-line accessible on a specific data-bank (see ref.[3]). For each facility, on-going, or near-term planned, research activities are also indicated, with particular specification of those addressed toward production of reference test-cases for purposes of theoretical predictions' validation. Then, in a second part of the paper, most recent (i.e. latest months) experimental outcome from a few of the above facilities is briefly reported and discussed, aimed at confirming the actual "vitality", in addition to validity, of the experimental research in the field.

A final objective of the paper is to stimulate all Institutions and Laboratories with experimental potential in the field of transition to contribute a continuous extension and updating to the data-bank, in order to provide a complete and organized reference tool quite relevant to both research advancement and technology-oriented utilizations.

SYNTHESIS OF THE TRANSITION FACILITIES DATA-BANK

The data-bank [3] is organized into 5 Sections, named A, B, C, D, E, corresponding respectively to Water Tunnels, Low Speed Wind Tunnels, High Speed Wind Tunnels, Hypersonic Facilities, and Blade Cascade Tunnels. At the end of this paper, the list is given (reference numbers preceded by letter R) of the institutions or laboratories where facilities are located, together with the persons in charge.

In Section A, Water Tunnels, 10 facilities are listed, here briefly reported.

Tunnel A.1 (see ref.[R-1]) has test section dimensions 0.1 x 0.05 m (with 3D capability), max velocity 0.2 m/s, turbulence level 0.1% (for V < 0.05 m/s). It features a concave test section with curvature radius R=0.1 m ; boundary layers suction; constant head pressure; open and closed circuit circulation; controlled dye injectors. Diagnostic instrumentation includes Laser Induced Fluorescence (LIF), 1-component LDA; digital image processing. Researches: experiments in Goertler instability; non-linear evolution of perturbations; hystogramme of perturbations in weak turbulence. Test-case data could include: boundary layer profiles with high curvature and strong non-parallel flow; downstream evolution of the Goertler vortex; and influence of the side-wall circulation in the test section

Tunnel A.2 (ref.[R-1]) has test section dimensions 0.05 x 0.05 m (3D capability), max velocity 0.1 m/s. It is provided with a constant head pressure; open and closed circuit circulation; controlled dye injectors ; bluff bodies of different cross section shapes. The instrumentation is the same as in tunnel A.1. Researches: experiments in wakes, and global mode of instability for the vortex shedding. Possible test-case data could be produced on the spatial downstream evolution of the fluctuating velocities of shedding.

Facility A.3 [R-1] has test section dimensions 0.03 x 0.1 m (2D capability), max velocity 0.1 m/s. It is provided with a constant head pressure; open and closed circuit circulation; controlled dye injectors ; variable length from 0.05 to 0.60 m. Instrumentation as in tunnel A.1. Researches in confined plane jets: self sustained oscillations, influence of the geometry, mechanisms of feedback. Test-case data on space-time evolution of fluctuating velocities in oscillating jets.

Test rig A.4 [R-2] is a Rotating Water Tunnel with test section dimensions 0.3 x 0.2 m (0.5 m mean radius; 0.55 m outer wall; 60 deg bend). Velocity ranges from 0 to 5 m/s. Features: closed return; 6:1 contraction; mounted on a rotating table (both directions, up to 20 rpm); fully automated; 2 degree of freedom traversing gear for imaging; transparent side walls and ceiling (plexiglas); 6 stations for 2-degree of freedom traversing gear for probes; inlet boundary layer suction and jet injection or perturbation control. Diagnostics: hot-film anemometry, dye, hydrogen bubble visualization, pressure probes, PIV. Researches: growth of Goertler vortices without rotation , preliminary studies of stabilizing or destabilizing effects of rotation. Test-case data: effect of rotation on Goertler vortex-induced transition, i.e. primary and secondary instabilities.

Tunnel A.5 [R-3] has test section's cross section of 0.04 x 0.04 m and a length of 40 m (2D capability). Max (unit) Reynolds number is 12400. Open or closed loop; all-transparent;

continuous; electromagnetic flowmeter; several curved ducts appropriate for research on Dean flow and chaotic advection in twisted duct flow; special pump for non-Newtonian fluids. Instrumentation: 2 component LDA; micro-thermocouples; laser induced fluorescence (LIF); CCD camera and video imaging; digital image processing. Researches: Dean flow; twisted duct flow; chaotic advection. Data on curved duct and boundary layer flows.

Tunnel A.6 [R-4] has test section dimensions 0.02 x 0.02 m (2D capability) and a velocity range from 0.05 to 1 m/s. Turbulence level 1 %. It is a closed circuit facility with 25:1 contraction; operates in continuous mode, with suction provisions on models. Diagnostics: 2-components LDA; dye injection. Researches: experimental analysis of boundary layers; wakes of cylinders and of wings; flow patterns induced by riblets.

Rig A.7 (see ref.[R-4]) is a Water Cavity with an inside-rotating disk. The cavity is 1 x 1 x 1 m (cubic) and the disk diameter is 0.5 m. Disk rotation is from 0.1 to 10 round/s. Instrumentation includes hot-film anemometry. Researches on flow characterization near the surface of the disk. Test-case data could be set-up on detailed analysis of cross-flow transition

Water tunnel A.8 [R-4] is a closed-circuit rig, with 10:1 contraction ratio and test section dimensions of 0.02 x 0.128 m (2D and 3D capability). Velocity can range from 0.01 to 1.0 m/s, and turbulence level is about 1%. Hot-film anemometry and electro-chemical visualization are available. Researches deal with the behavior of coupled (interacting) wakes downstream of grids made up of cylindrical bodies

Facilities A.9 and A.10 (see ref.[R-3] for both) are two Electro-Hydrodynamic Water Tunnels with test section dimensions respectively 0.04 x 0.2 m (L = 0.4 m) and 0.04 x 0.4 m (L = 1.0 m). Capability is 2D. Max unit Reynolds numbers are 1440 and 720. Turbulence levels are negligible. Main features, for both, are: continuous operation in open or closed loop; all-transparent; electro-magnetic flowmeters: the two larger walls are electrodes so that an a-c electric current can pass through the flow; special pumps for non-Newtonian fluids. Instrumentation is the same as for tunnel A.5. Researches: direct resistance heating (DRH) of ionic Newtonian or non-Newtonian very viscous fluids, and electroconvection.

In Section B, Low Speed Wind Tunnels, of the data-bank, 8 facilities are listed.

Wind Tunnel B.1 (see ref.[R-5]) has test section dimensions 0.5 x 0.9 m, 2D capability, and a velocity range from 5 to 50 m/s. Turbulence level is less than 0.3 %. Main features: closed-circuit, controlled velocity fluctuations, controlled turbulent length scales. Instrumentation: pitot and pitot-static probes, preston probes, floating-element skin-friction probes, hot-wire anemometry, semi-conductor temperature and pressure transducers. Researches being carried on: effects of free-stream turbulence on turbulent boundary-layers; turbulence in sudden expansions; turbulent mixing-layers with zero mean stresses; hot-wire data from very near a wall and in non-isothermal flows. Test-case data available on transitional boundary layers in turbulent flows with controlled velocity and length scales [2].

Wind Tunnel B.2 [R-6] has test section dimensions 1.5 x 0.25 m (2D capability), and a max velocity of 30 m/s. Reynolds number is 10 million (max). Turbulence level:0.04-0.08 %. Test section walls are flexible to control pressure distribution. Diagnostic instrumentation: pitot probes, preston tubes, multi-component hot-wire anemometry, 3-component LDA. Researches: transitional and turbulent forced-convection boundary-layers with variable pressure gradients; airfoil design. Possible test-case data on the transitional boundary layer development in adverse pressure gradients with low free-stream turbulence (less than 1%).

Tunnel B.3 [R-2] has test section dimensions 0.75 x 0.12 x 3.4 m , and a velocity range between 0 and 25 m/s. Turbulence level is 0.5 %. Main features: closed return; 9:1 contraction; fully automated 5-degree of freedom traversing gear; transparent (plexiglas) test section with adjustable ceiling. Diagnostics: hot-wire anemometry; smoke-wire visualization; preston tubes; miniature floating-element skin-friction balance; PIV. Researches: generation of hairpin-type vortices into sub-critical laminar boundary layer by slot injection or vortex generators; investigation of near-wall vortex dynamics. Test-case data on vortex induced transition, development of secondary instability and development of synthetic hairpin vortices.

Facility B.4 (see ref.[R-2]) has test section dimensions of 0.80 x 0.25 x 2.5 m and a velocity range from 0 to 50 m/s. Turbulence level is 0.2 %. Features: closed return; 9:1 contraction; fully automated 3-degree of freedom traversing gear; transparent side walls (in plexiglas) and ceiling (optical glass); tiltable floor and ceiling; test plate with trailing-edge flap. Diagnostics: hot-wire anemometry; smoke-wire visualization; preston tubes; 2-component LDA. Researches: riblets in favourable and adverse pressure gradient flows; generation of hairpin-type vortex array into subcritical laminar boundary layer by slot injection or vortex generators; investigation of near-wall vortex arrays. Test-case data could be set-up on vortex induced transition; development of secondary instability; passive and active control of near-wall vortex arrays.

Facility B.5 [R-2] is a Rotating Low-Speed Wind Tunnel with test section dimensions same as water-tunnel A.4. Velocity range: from 0 to 0.5 m/s. Features are exactly the same as in facility A.4, except that here return is open. Diagnostics: hot-wire anemometry; smoke; pressure probes; PIV. Researches and possible test cases (here performed in air) are the same as in facility A.4.

Tunnel B.6 [R-3] has test section dimensions 1.2 x 1.1 m (L = 1.7 m). Capability is 2D and 3D. Max unit-Reynolds number: 2 million. Turbulence level: 0.03 - 8.0 %. Main features: open loop; transparent test section; continuous; a specially designed and constructed large scale model for research on instability and transition on concave-convex-flat boundary layers; vertical movement (periodic or continuous) of the model; variation of sweep angle, attack angle, and position of after-body; completely similar model with extra heat transfer provisions, to be studied in above said flow configurations. Diagnostics: 2 component LDA; multicomponent hot-wire anemometry; pitot tubes; preston tubes; infra-red thermography; micro thermo-couple instrumentation; laser-sheet visualization; aerosol generator for visualization; CCD camera and video imaging; digital image processing. Researches: instability and transition in concave and/or convex 2D & 3D boundary layers; effects of longitudinal large vortical structures on heat transfer; film cooling; wakes and jets in cross flow. Detailed test-case data can be yielded on transition due to curvature effects, sweep and heat transfer.

Facility B.7 (ref.[R-3]) is a wind tunnel with cross section dimensions 0.3 x 0.3 m , length L = 1.5 m , and 2D capability. Max unit-Re is 4 million and turbulence level 0.15 %. Features: open circuit; transparent test section; continuous operation. Instrumentation: 2 component LDA; multicomponent hot-wire anemometry; pitot tubes; preston tubes; balance for lift and drag measurements; micro-thermocouple instrumentation; laser sheet visualization; aerosol generator; CCD camera and video imaging; digital image processing.

Tunnel B.8 [R-7] is a very-low-turbulence facility, with test section dimensions 0.8 x 1.2 x 7.0 m (2D capability). Velocity range: 0 - 68 m/s , with mean velocity non-uniformity less than 0.1%. Noise spl (f> 60 Hz) is 69 db for V=35 m/s ; and 83 db at 60 m/s. Turbulence level: less than 0.02 % (stream- and cross-wise), for velocities up to 25 m/s. Temperature variation: less than 0.3 C (in time). Features: various screens for turbulence generation; flat plate specially designed for transition studies in boundary layers at near zero pressure gradient; 5-degree of freedom traversing gear (computer-controlled). Instrumentation: multiple-channel hot-wire anemometry; hot-wire probes manufacturing (diam. down to 0.6 microns); data acquisition up to 1 MHz.

Tunnel C.1 (ref.[R-5]) is a high-speed facility with test section dimensions 0.1 x 0.1 x 2.2 m (2D & 3D capability). Mach number range is from 0 to 0.82 (min velocity is 0.1 m/s, with error less than 1 %). Turbulence level: 0.1 %. Main features: intermittent blow-down type; testing time up to 5 minutes (at M = 0.82); suction on bottom and side walls; rotatable flap in the sonic throat to produce controlled oscillations. Instrumentation as in tunnel B.1.

Researches: unsteady turbulent flow parameters assessment; probe calibration; high speed flow investigations in the turbulent regime; informative investigation on turbulence evolution and on boundary layer transition in flows with periodic velocity fluctuations. Possible test-case: data on transitional boundary layer development on the walls of a closed channel with various characterizations of the incoming flow.

Hypersonic tunnel D.1 (see ref.[R-8]) has an axysimmetric test section with a diameter of 0.210 m. Mach number ranges: 6.9 to 7.14 (6 600 000 <Re/m< 33 000 000, 650 K <Tt< 800 K). At Ma = 8.15 these values are: 16 000 000 <Re/m< 34 000 000, Tt = 800 K. Main features: test duration from 1 to 20 s; possible 20 tests/day (2 s each). Diagnostics: pressure and temperature probes; schlieren; laser; fourier analysis; densitometry; hot-wire anemometry (in progress); infrared thermography. Researches: typical of hypersonic flows for re-entry problems. Test-case data on flow parameters around a double ellipsoid.

Facility E.1 (ref.[R-9]) is a low-speed blade-cascade tunnel with test section dimensions 0.5 x 0.3 m (2D & 3D capability). Max velocity is 80 m/s and cascade inlet turbulence level is 0.5 %. Main features: continuous blow down; large-scale blade cascade; fully automated 3-axis probe traversing equipment; local and overall cascade performance analysis. Diagnostics: 3-channel hot-wire and glue-on hot-film sensors; fast-sampling (up to 1 MHz) data acquisition, with hold capability; analysis techniques for 1-, 2-, 3-wire probes with data processing for fluctuations analysis in turbulent flows (time & frequency domains); 2-component LDA anemometer with optic fiber probes and BSA processors. Researches: investigation of the fluctuating characteristics of wake flows downstream of turbine blades; detailed analysis of boundary layer transition on convex and concave turbine blade surfaces. Test-case data on transitional boundary layer development on a complete turbine blade.

Tunnel E.2 [R-9] is a transonic blade-cascade facility with test section dimensions of 0.05 x 0.280 m (2D capability). Mach number range is from 0.1 to 1.7 , whilst the turbulence level is controllable between 0.05 % and 6 % . Features: blade number up to 16; flexible walls; continuous operation (suction mode); max blade deflection 150 deg. Diagnostics: pressure probes; traversing gears; schlieren; closed-circuit tv; strain-gage pressure transducers; heated thin-film multiple-sensor surface probes with spectral analysis of the signals; 2-component LDA. Researches: turbine blade cascade performance analysis; profile optimized redesign; boundary layer fluctuations analysis under varying free-stream turbulence levels; transition/ separation/relaminarization detection. Test-case data on transitional boundary layer development on the suction side of a root-section turbine blade [4].

MOST RECENT EXPERIMENTAL OUTCOME

In order to attest the active, and productive, continuous advancement of experimental research in the fields here discussed, three latest achievements will be now very briefly reported. They have been selected purposely to support the following consideration: whilst, in transition research, most recent theoretical advancements have begun to take advantage, in addition to "modeling" criteria, also of detailed "simulation" approaches (i.e. LES), a corresponding counterpart is becoming available also on the experimental side, namely the digital analyses to be performed on to hot-wire and surface thin-film fluctuating signals. In this connection, the experimental activity needs to be mentioned which is being carried out on above-listed facility B.1 (refs.[2, R-5]), in which investigations have dealt with the receptivity of a laminar boundary layer to local intensities and length scales of outer stream disturbances. Lately, a complete characterization of tunnel test section in terms of the distributions of these latter parameters has been achieved, in correspondence of two different grid-type turbulence generators. Figures 1 and 2 give a synthetic picture of this outcome, which has made it possible to yield, correspondingly, detailed test-case data for zero-pressure gradient boundary layer development through the (by-pass) transitional regime.

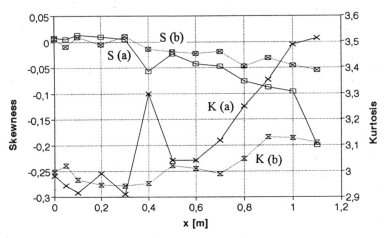

Fig. 1   Outer-stream turbulence parameter distributions in tunnel B.1 for two different inlet grids (sizes a, b). S(a) and K(a) : skewness and kurtosis for grid a ; S(b) and K(b) : skewness and kurtosis for grid b.

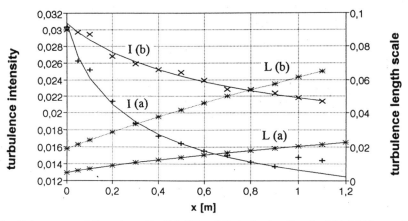

Fig. 2   Outer-stream turbulence intensities I(a), I(b) and dissipative length scales L(a), L(b) in tunnel B.1

A second quick look is here made at the experimental data coming from blade-cascade facility E.2 (refs.[4, R-9]), where the characterizations, in terms of spectral content, of the inlet boundary layers and of the blade-surface heat transfer fluctuations, for different tunnel turbulence levels, are providing deeper, often unexpected, interpretations of transitional phenomena. To give an example (details are discussed in [4]), Fig.3a presents, for a situation of high free-stream turbulence, the power density spectrum of the surface fluctuations on the forward suction side of a turbine blade, right ahead of a "transitional" separation bubble, pulsating, as clearly visible in the figure, with a frequency of about 900 Hz. Since downstream the flow is detached, one would expect a rapid prosecution of "transition" into a fully turbulent condition. Surprisingly, no such thing is taking place: the downstream sensor (inside the separated flow) yields the spectral trace in Fig.3b, with fluctuation levels lower by more than a decade, no memory of the 900 Hz frequency and a fluctuating-signals trace of a laminar-like character. This phenomenon, attributable to a "fluctuation-suppressive" capacity of the separated layers, extends downstream (as discussed in [4]), up to the reattachment point, wherefrom a true "transition" process appears to take place.

Finally, a very recent, still unpublished, outcome is here presented, in connection with blade-cascade facility E.1 [R-9]. Here, transition situation refers again to suction surfaces of turbine blades but, unlike the previous example, the flow is here incompressible and there is no separation, rather, the process is a natural-transition one. Figures 4a and 4b show, respectively, fluctuations traces taken shortly ahead, and (in our interpretation) exactly after, the "transition point". What is here happening can be interpreted as follows. Fig.4a reflects the process of transition in the phase governed by the progressive growth and lateral merging of turbulent spots, which appear as active "bursts" superposed to a laminar-like background-trace. Here, the intermittency factor could be easily evaluated from the traces. Fig.4b shows, on the contrary, "trailing-flow" regions lagging beneath a more turbulent-like background-trace, symptom that the turbulent spots have begun to merge (i.e. transition has just occurred), but these latter still leave behind themselves a kind of temporarily "suppressed", laminar-like, regions, whose existence has been discussed e.g. in [5]. To be stressed here, is the important fact that there is an analytic feature of the fluctations traces quite responsive to their positive (fig.4a), and negative (fig.4b), asymmetries, i.e. their third-order central-moments about the mean ("skewness"). Fig.5 shows the skewness distribution of the sensor traces all the way from the leading edge (LE) to the trailing edge (TE) of the suction surface of the instrumented blade: quite clearly, the transition location is evidenced by the zero-crossing point, from positive to negative values, of the said distribution.

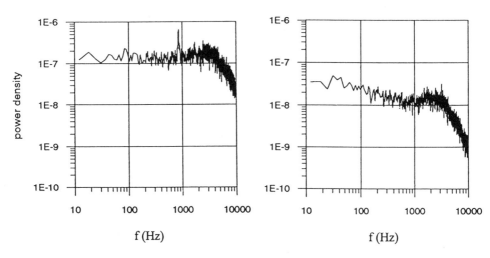

Fig. 3a Power density content of fluctuations on a turbine blade suction side (tunnel E.2) just upstream of a separation bubble (high free-stream turbulence)

Fig. 3b Corresponding power density content measured a few mm downstream, inside the said separation bubble

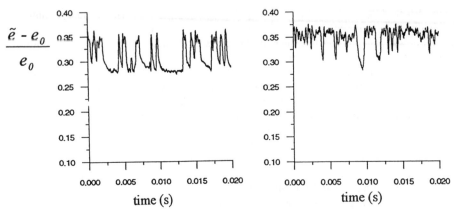

Fig. 4a Fluctuating traces on a turbine blade suction side (tunnel E.1) just upstream of transition "point".

Fig. 4b Corresponding traces as measured exactly after transition point.

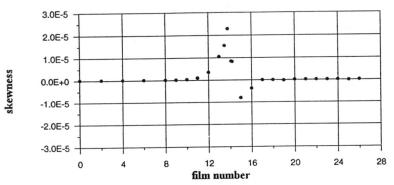

Fig. 5 Complete blade suction-side distribution of surface-fluctuations' skewness (tunnel E.1). Transition point localized at the zero-crossing point (from positive to negative values) of the distribution.

CONCLUSIONS

The data-bank on experimental facilities here presented will be progressively extended and continuously updated. Where possible, and particularly in connection with data taking for reference test-case production, information will be asked, and added, about equipment calibration procedures and experimentation protocols, in order to help assess the uncertainty range and the repeatability level of measured variables. An important target to be fully attained in future, but already with significant experiences positively performed within COST-ERCOFTAC Transition SIG, would be a "fully interactive" experimental-theoretical approach, by which not just experimental data are used for validating theoretical models, but the outcome of these latter could re-address the experimentation, for deeper integrated analyses.

Implied hope is that, through a detailed and systematic information about its consistency, the whealth of advanced-technology equipment and operational competences here presented could be organically utilized and addressed toward multi-national concerted objectives of wide breadth, reflecting its full potential: that of a real European asset for both research and application.

BIBLIOGRAPHIC REFERENCES

[1] SAVILL, A.M.:"A Summary Report on the COST ERCOFTAC Transition SIG Project Evaluating Turbulence Models for Predicting Transition", ERCOFTAC Bulletin, 24 (1995), pp. 57-61.

[2] SAVILL, A.M., PITTALUGA, F.:"COST-ERCOFTAC Transition SIG , COST Action F1 Workshop- Institute of Thermomechanics, Prague, April 13, 1995", ERCOFTAC Bulletin, 25 (1995), pp. 36-37.

[3] ERCOFTAC World Wide Web - URL Access: http://imhefwww.epfl.ch/lmf/ERCOFTAC.

[4] CAMPORA, U., PITTALUGA, F., UBALDI, M., ZUNINO, P., MARETTO, L.:"Turbine Blade Cascade Transition Zone Response to Tunnel Turbulence Level", EUROMECH Colloquium 330 on Laminar-Turbulent Transition, Inst. of Thermomechanics, Prague (April 1995), Extended Abstract Book, pp. 11-12.

[5] SCHUBAUER, G.B., KLEBANOFF, P.S.:"Contributions on the Mechanics of Boundary Layer Transition", NACA, Report 1289 (1956).

REFERENCE INSTITUTIONS

[R-1] ESPCI , Ecole Superieure de Physique et de Chimie Industrielles, URA 857, CNRS, Paris, France. (Contact: Dr. Wesfreid).

[R-2] Institut de Machines Hydrauliques et de Mecanique des Fluides (IMHEF), EPFL Lausanne, Switzerland (Contact: Dr. T.V. Truong; Mr. A. Pexieder; Mr. B. Tanguay).

[R-3] Thermofluids and Complex Flows Research Group, LTI, URA , CNRS 869, ISITEM, University of Nantes, Nantes, France (Contact: prof H. Peerhossaini).

[R-4] Institut de Mecanique Statistique de la Turbulence, Universite' d'Aix-Marseille II, UM 33, NRS, Marseille , France (Contact: Dr. Cl. Beguier; Dr. M.P.Chauve; Dr. Fulachier; Dr. P. LeGal).

[R-5] Institute of Thermomechanics, Academy of Sciences, Prague, Czech Republic (Contact: RNDr. P. Jonas).

[R-6] Dept. of Aerospace Engineering, Delft Technical University, Netherlands (Contact: Dr. R.A.W.M. Henkes).

[R-7] Royal Institute of Technology, Department of Mechanics, Stockholm, Sweden (Contact: prof. H.Alfredsson).

[R-8] CEAT-LEA, CNRS, Poitiers, France (Contact: Dr.T. Alziary; Dr. D. Aymer).

[R-9] Institute of Thermal Machines and Energy Systems, IMSE, Univ. of Genova, Genova, Italy (Contact: prof. F. Pittaluga; prof. P. Zunino) .

THE MODELLING OF THE WAKE OF A TORUS BY THE GINZBURG-LANDAU EQUATION

M. Provansal and T. Leweke
Institut de Recherche des Phénomènes Hors d'Equilibre/Laboratoire de Recherche en Combustion, U.M.R. 138 C.N.R.S. Centre de Saint-Jérôme - Service 252, F-13397 Marseille Cedex 20, France

SUMMARY

The wake of a torus is studied in the Reynolds number range [50, 300]. In the periodic regime different modes of vortex shedding, parallel closed rings or oblique helical lines of vortices, have been observed. We present their limits of stability as functions of the control parameter. At higher Reynolds number a chaotic regime of vortex shedding is characterized by a discontinuity in the velocity-frequency relation. Both these regimes are well described by the Ginzburg-Landau equation which allows to interprete many phenomena as consequences of the Eckaus or Benjamin-Feir instabilities.

INTRODUCTION

In recent years the wake behind a circular cylinder has been extensively studied in the context of the transition to turbulence in open flow systems. The first step in this process is the Bénard-von Kármán vortex street, which is periodic and laminar in the Reynolds number range $50<Re<180$ ($Re=Ud/v$, with U: free stream velocity, d: cylinder diameter, v: kinematic viscosity). A complete analytical description of the transition stationnary-instationnary flow is not available. Important features of this flow regime have been described by a phenomenological Ginzburg-Landau model using a one-dimensional complex amplitude equation [1]-[4]. At higher Re the wake undergoes a transition to a less ordered state characterized by a loss of periodicity, a broadening of the velocity spectra, and discontinuities in the frequency-velocity relation. This transition range, which extends from $Re=180$ to $Re=350$ approximately, has not been clearly understood so far ; the origin of the instability of the ordered periodic vortex shedding is *not* caused by the onset of turbulence in the free shear layers[5]. We present experimental results concerning the case of a bluff ring of circular cross section. The ring geometry is chosen because of the absence of body ends, which are known to have a strong influence on the whole wake in the periodic regime (see e.g. [6]-[8]). A comparison with circular cylinder wakes is given.

EXPERIMENTAL RESULTS

Experiments were carried out in a low speed wind tunnel, having a 25 x 25 x 100 cm^3 test section. Here, we present results for a ring having a mean outer diameter D=56.9 mm and a thickness d=3.03 mm (aspect ratio $\pi D/d=59$). Vibrations were monitored by a sensitive Laser-photodiode set-up and it turned out that they did not interfere with the flow phenomena in the wake. A straight circular cylinder used for comparison has a diameter d=2.01 mm, a length L= 157 mm (aspect ratio L/D=78), and was fitted with angled endplates to ensure parallel vortex shedding in the periodic regime (see [6]). For rings the Reynolds number is based on the ring thickness d.

In the periodic regime the power spectrum of the near wake velocity fluctuations exhibits a single sharp peak with its harmonics (Fig. 1, $Re=158$). This frequency varies continuously with *Re* in the case of a ring [3]. Oblique vortex shedding induced by the flow at the cylinder ends [6]-[8] is frequently observed in cylinder wakes. The torus geometry is more suitable for the study of intrinsic phenomena in bluff body wakes than finite cylinder wakes, which are strongly influenced by end effects. Various modes of vortex shedding coexist in a large Reynolds number range[2,3]. An increase of the Reynolds number from subcritical values select the parallel mode where the vortex lines are closed rings parallel to the torus. Experiments have shown that for higher Reynolds number a small perturbation, such a transverse jet, gives the way to new helical modes. Quantitative measurements of the phase between two probes lead to a precise description of these various modes [9]. The limits of stability of these modes are recalled on Figure 2.

At a Reynolds number Re_t around 180, the wake enters the transition regime. Above Re_t the velocity fluctuations are rather irregular and chaotic. Their spectra show a large band around a given center frequency f_o (Fig. 1, $Re=183$). With reference to Williamson ([10], see below) we call this kind of vortex shedding mode A. This first transition is hysteretic: when decreasing Re, the wake restabilizes only below Re_t. At higher Reynolds numbers a second peak appears in the velocity spectra at a higher frequency (Fig. 1, $Re=227$). It is visible in the range $200<Re<300$ and its intensity never exceeds the one at f_o. Finally, above $Re \approx 300$ (Fig. 1, $Re=353$), the vortex shedding becomes "more periodic" again. However, the background noise level continues to increase with Re. We call this mode B.

When plotting the frequencies of the different peaks in the spectra in the form of the dimensionless Strouhal number $S=f_{(o)}d/U$ (Fig. 3.a), several observations can be made. The transition from periodic vortex shedding to mode A is characterized by a discontinuity in the *S-Re* relationship. During the passage from mode A to mode B, this relationship is *continuous* but a discontinuity in its slope is visible at $Re \approx 260$. Third transition mode (A^*) is very weak: its frequencies align well with the periodic curve.

We have analyzed the same data for the circular cylinder case. The scenario is similar to the one described above. However, the important difference is that here the transition from mode A to mode B is characterized by a *second* discontinuity in the frequency law. This is in good agreement with the measurements of Williamson, who first examined in detail this transition range for circular cylinders [10]. These features of the transition regime are the same for all rings investigated, i.e. for aspect ratios up to 100.

RESULTS FROM THE GINZBURG-LANDAU MODEL

These phenomena have been analyzed in the context of the phenomenological Ginzburg-Landau (GL) model mentioned above. With the appropriate boundary conditions (zero amplitude of oscillation at the ends of straight cylinders) this model accurately describes most 3D aspects in bluff body wakes: oblique shedding, chevrons shapes of vortices, end cells with different frequencies [1]-[4]. The discussion below concerns the ring wake, for which the following GL model equation is used with periodic conditions:

$$\frac{\partial A}{\partial t} = \sigma(1+ic_0)A + \mu(1+ic_1)\frac{\partial^2 A}{\partial (D\varphi/2)^2} - l(1+ic_2)|A|^2 A, \quad (1)$$

where $\sigma = k(\nu/d^2)(Re-Re_c)$, φ is the spanwise coordinate (azimuthal angle), and k, μ, $l>0$, c_0, c_1, and c_2 are model parameters (see Refs. [1], [2], [4], [9], [11], [12] for details). With an appropriate rescaling Eq. (1) reduces to:

$$\frac{\partial A}{\partial t} = (1+ic_0)A + (1+ic_1)\frac{\partial^2 A}{\partial z^2} - (1+ic_2)|A|^2 A. \quad (2)$$

Here we call z the dimensionless spanwise coordinate. The dynamical features of the model are mainly determined by the values of c_1 and c_2 ; c_0 represents a simple frequency shift. The influence of the finite spanwise length L of the domain is negligible for values of L large enough.

When varying the parameters with Re, the GL model gives an accurate description of the entire periodic regime and a great part of the transition range, up to $Re \approx 300$. From the experimental determination of the GL parameters in the periodic regime, we propose the following simplified dependence of c_0, c_1, and c_2 on Re:

$$c_0 = \frac{2\pi}{k} \cdot \frac{Ro}{Re-Re_c} + c_2, \qquad (3)$$

where $Ro = f \cdot d^2/v = S \cdot Re$ is the dimensionless frequency of parallel vortex shedding. Experimentally one finds an approximately linear dependence: $Ro = -5.125 + 0.215 \cdot Re$ for the present ring, which is very close to the circular cylinder law. For k we take the experimentally determined value for circular cylinders: $k=0.2$ [11], [13], [14]. Eq. (3) is valid in the periodic regime, but we conserve it for the transition range, too.

$$c_1 = c_2 + 2.7, \qquad (4)$$

which is compatible with experimental results [1], [4].

$$c_2 = \begin{cases} -3 & \text{for} \quad Re \leq 100 \\ -4.1 + 0.011 \cdot Re & \text{for} \quad Re > 100 \end{cases} \qquad (5)$$

For $Re<100$, c_2 is approximately constant. Eqs. (4) and (5) say that for $Re>100$ the wake follows the trajectory in the (c_1, c_2)-plane shown in Fig. 4. It is known that for $1+c_1 c_2 < 0$ (domain bounded by the line marked BF) the uniform periodic oscillation becomes unstable by resonant excitation of sidebands with wave numbers and frequencies very close to the initial ones [9]. The part of the parameter space depicted in Fig. 4 has been investigated by Shraiman et al. [16] using numerical simulations. They showed that the BF-unstable domain is essentially divided into two regions (A and B) corresponding to two different dynamical behaviours of the solutions to Eq. (2). In region A they are characterized by space-time phase dislocations and a small correlation length ("defect chaos" in [16]), in region B they exhibit no such dislocations and the spanwise correlation is much higher ("phase turbulence"). We identify these regimes to the modes A and B in the transition range of bluff body wakes.

We have carried out numerical simulations of Eq. (2), with parameters given by Eqs. (3)-(5) and periodic boundary conditions in a domain of length $L=100$. For comparison, the same quantities as measured in the experiments were calculated (see [1] and [4] for the relations between the model parameters and physical variables); the results are shown in Figs. 5 and 6, with the same representations as in Figs. 1 and 3. Despite the simplified assumptions made about the parameters, the correspondence between the two sets of results is remarkable.

For each mode the spectra have essentially the same shape in both cases (Fig. 5). The transition between periodic and defect regime is discontinuous in frequency *and*, as pointed out by Shraiman et al. [16], hysteretic, just as in the experiments (compare Figs. 3.a and 6). The passage from mode A to B is continuous in frequency, although the change in the slope is less pronounced.

CONCLUSION

A simplified model such the Ginzburg-Landau equation allows to describe many mechanisms occuring in vortex shedding instability. At low Reynolds number oblique or helical modes are destabilized by the Eckaus instabilty of plane waves. For Re higher than Re_t a chaotic regime appears similar to the defect chaos regime. This study shows the importance of spanwise effects which have been emphasized by recent 3D numerical simulations [17],[18], [19].

This work has benefited from the financial support of action Incitative en Mécanique des Fluides DPST8 Ministère de l'Enseignement Supérieur et de la Recherche.

REFERENCES

1. P. Albarède and P. A. Monkewitz, Phys. Fluids A **4**, 744 (1992).
2. T. Leweke, M. Provansal, and L. Boyer, C. R. Acad. Sci. Paris **316**, Série II, 287 (1993).
3. T. Leweke, M. Provansal, and L. Boyer, Phys. Rev. Lett. **71**, 3469 (1993).
4. P. Albarède and M. Provansal, J. Fluid Mech vol **291**, 191-222. (1995).
5. M. S. Bloor, J. Fluid Mech. **19**, 290 (1964).
6. C. H. K. Williamson, J. Fluid Mech. **206**, 579 (1989).
7. M. König, H. Eisenlohr, and H. Eckelmann, Phys. Fluids A **2**, 1607 (1990).
8. M. Hammache and M. Gharib, J. Fluid Mech. **232**, 567 (1991).
9. T.Leweke and M. Provansal, J. Fluid Mech vol **288**, 265-310. (1995).
10. C. H. K. Williamson, Phys. Fluids **31**, 3165 (1988).
11. M. Provansal, C. Mathis, and L. Boyer, J. Fluid Mech. **182**, 1 (1987).
12. L. D. Landau, C. R. Acad. Sci. URSS **44**, 311 (1944).
13. K. R. Sreenivasan, P. J. Strykowski, and D. J. Ohlinger, in *Forum on Unsteady Flow Separation*, edited by K. N. Ghia (ASME, New York, 1986), FED-Vol. 52, pp. 1-13.
14. M. Schumm, E. Berger, and P. A. Monkewitz, submitted to J. Fluid Mech. (1994).
15. T. Leweke and M. Provansal,Eur. Phys. Lett. **27**, 655(1994).
16. B. I. Shraiman, A. Pumir, W. van Saarlos, P. C. Hohenberg, H. Chaté and M. Holen, Physica (Amsterdam) **57D**, 241 (1992).
17. D.Barckley and R.D.Henderson Three-dimensional Floquet stability analysis of the wake of a circular cylinder. Submitted to J. Fluid Mech. (1995).
18. H.Zhang,U.Fey, B.R.Noack, M.König and H.Eckelmann Phys. Fluids. **7**, 779 (1995).
19. C.Dauchy, J.Dusek and P.Frauniè 12ème Congrés de Mécanique, Strasbourg (1995).

FIGURES

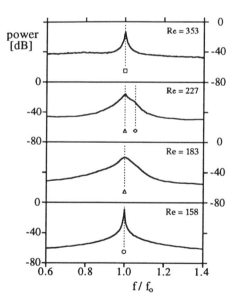

Fig. 1. Spectra of the longitudinal velocity fluctuations in the wake of a ring of aspect ratio $\pi D/d$=59 (long time average). 1: periodic mode, s: mode A, u: mode A*, n: mode B.

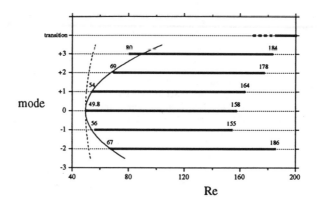

Fig. 2. Stability domains of the periodic vortex shedding modes for the ring with aspect ratio 59.

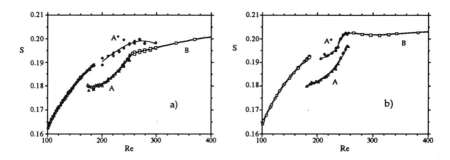

Fig. 3. Strouhal number as function of the Reynolds number in the transition range. a) ring, aspect ratio $\pi D/d=59$, b) circular cylinder, aspect ratio $L/d=78$. The bold line indicates the peaks in the spectra containing the highest energy. Symbols as in Fig. 1.

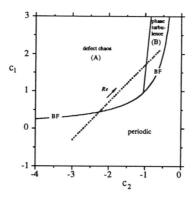

Fig. 4. Parameter space for the Ginzburg-Landau model (following Shraiman et al. [16]) and trajectory (.....) for the transition regime according to Eqs. (4) and (5).

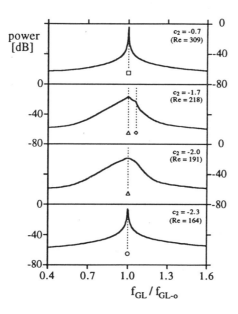

Fig. 5. Time spectra of the GL-amplitude $A(z, t)$ obtained from numerical simulations of Eq. (2) with c_0 and c_1 given by Eqs. (3) and (4), $k=0.2$, $L=100$, and c_2 as indicated. The Reynolds number was obtained from Eq. (5). Symbols as in Fig. 1.

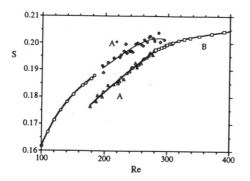

Fig. 6. S-Re relationship resulting from the GL model. For a given c_2, the Reynolds number was calculated from Eq. (5), and the Strouhal number from the relation $S = Ro/Re = (k/2p) \cdot f_{GL} \cdot (ReRe_c)/Re$. Symbols as in Fig. 1.

Diagonalization of the Reynolds-averaged Navier–Stokes equations with the Reynolds-stress Turbulence Model

Patrik Rautaheimo and Timo Siikonen
Laboratory of Applied Thermodynamics
Antti Hellsten
Laboratory of Aerodynamics
Helsinki University of Technology, Espoo, Finland

1 Introduction

Because of the need for a general turbulence model that would work for arbitrary cases, the Reynolds-stress model (RSM) has gained popularity in turbulence modeling. Most RSM applications have been for incompressible flow. In that case Reynolds-stress forces act only in the momentum equations. Still, the coupling with the flow equations is a non-trivial task, see e.g. [1]. In the case of a compressible flow, Reynolds-stresses, as well as pressure, act not only in the momentum equations but also in the energy equation. This makes the implementation more complicated. Because the coupling between the Navier-Stokes equations and the Reynolds-stress equations is non-isotropic, time integration and upwinding are more complex than with algebraic or two-equation models.

In this study, Shima's low-Reynolds number RSM [2] is coupled with a compressible flow solver [3] based on Roe's method [4]. A new non-isotropic coupling method is introduced. The implicit solution method is modified in order to take account of the Reynolds stresses. The convergence is improved by using a multigrid acceleration. The model is applied for a channel flow and for a flow over an airfoil.

2 Governing Equations

The Reynolds-stress equations can be written in the following Cartesian tensor form

$$\rho \frac{D\widetilde{u_i'' u_j''}}{Dt} = P_{ij} + \Phi_{ij} - T_{ij} - \epsilon_{ij} - d_{ij} \quad (2.1)$$

where u_i'' is the Favre-averaged fluctuation velocity component in i-direction, tilde denotes Favre-averaging, and P_{ij}, Φ_{ij}, T_{ij}, ϵ_{ij} and d_{ij} are the production, the velocity pressure-gradient correlation, the turbulent transport, the dissipation rate and the diffusion terms, respectively. The turbulent transport, the velocity pressure-gradient and the dissipation rate must be modeled, whereas the production term is exact.

In this study Shima's model [2] is applied. Some modifications were needed in the velocity pressure-gradient term. The dissipation transport equation was taken from Chien's $k - \epsilon$ model [5] because it was found to be stable and well behaved. Shima's original equation was applied for the airfoil test case.

3 Numerical Method

Diagonalization of the flow equations

In the evaluation of the inviscid fluxes Roe's method [4] is applied. A rotation operator is used for the velocity components and also for the Reynolds-stresses. The flux is calculated as

$$\hat{F} = T^{-1} F(TU). \tag{3.2}$$

Here T is a rotation operator that transforms the dependent variables to a local coordinate system normal to the cell surface. In this way, only the Cartesian form F of the flux is needed. Velocity components are rotated from global to local coordinate system as

$$\begin{pmatrix} \hat{u} \\ \hat{v} \\ \hat{w} \end{pmatrix} = T \begin{pmatrix} u \\ v \\ w \end{pmatrix} \tag{3.3}$$

where the hats refer to the local Cartesian coordinate system. Matrix T is the rotation matrix that is determined by the normal and the tangent vectors of the cell face.

Correspondingly, the Reynolds-stresses are rotated from the global to the local coordinate system by the following formula

$$\hat{I} = T I T^T \tag{3.4}$$

where the components of the Matrix I are the Reynolds-stresses. Transformation from the local back to the global coordinate system is given by

$$I = T^T \hat{I} T. \tag{3.5}$$

The Cartesian form of the flux is

$$F(U^l, U^r) = \frac{1}{2} \left[F(U^l) + F(U^r) \right] - \frac{1}{2} \sum_{k=1}^{K} r^{(k)} |\lambda^{(k)}| \alpha^{(k)} \tag{3.6}$$

where U^l and U^r are the solution vectors evaluated on the left and right sides of the cell surface, $r^{(k)}$ is the right eigenvector of the Jacobian matrix $A = \partial F/\partial U$, the corresponding eigenvalue is $\lambda^{(k)}$, and $\alpha^{(k)}$ is the corresponding characteristic variable obtained from $R^{-1} \delta U$, where $\delta U = U^r - U^l$. A MUSCL-type approach has been adopted for the evaluation of U^l and U^r. In the evaluation of U^l and U^r, primary flow variables $V = (\bar{\rho}, \tilde{u}, \tilde{v}, \tilde{w}, \tilde{e})$, and conservative turbulent variables ($\overline{\rho u_i'' u_j''}$, $\rho \epsilon$) are utilized. Jacopian matrix A can be split in following ways

$$A = R \Lambda R^{-1} = M L \Lambda L^{-1} M^{-1} \tag{3.7}$$

where R and R^{-1} are the right and left eigenvector matrixes written by conservative variables, L and L^{-1} play the same role, with respect to the primitive variables, as the matrixes R and R^{-1}, Λ is the eigenvalue matrix, and M and M^{-1} are the transformation matrixes between the conservative and the primitive variables.

A coupling between the Navier–Stokes and the Reynolds-stress equations is introduced, since the Reynolds-stresses may be connected with the pressure [6]. In the i-momentum equation, the resulting effective pressure can be defined as

$$p_i^* = \bar{p} + \widetilde{\bar{\rho} u_i'' u_i''}. \tag{3.8}$$

In order to utilize Roe's method, the Jacobian of the flux vectors must be diagonalized. This requires that the Jacobian matrix of the flux vector has a complete set of eigenvectors. Unfortunately, linearly independent eigenvectors cannot be found if the non-isotropic pressure field of Eq.(3.8) is applied.

Since the non-isotropic pressure field is difficult to handle, the turbulent pressure is usually approximated by the mean of three components

$$p^* = \bar{p} + \frac{2}{3} \bar{\rho} k. \tag{3.9}$$

Using this, the flux vector can be divided into the isotropic and non-isotropic parts. The Jacobian of the isotropic part can be diagonalized. The effect of the non-isotropic part on a solution is small, and, consequently, it can be evaluated using central differences. In this approach, there is no need to rotate Reynolds-stresses in the local cell face coordinates. This diagonalization is similar to one of the $k - \epsilon$ model and can be found in [3].

The second method of diagonalization utilizes the production term P_{ij}. The production term is exact in RSM and it can be included into the vector F. This is not a conservative form of the vector F, but RSM is never in a strong conservative form because of the source terms.

The eigenvalues, i.e. the characteristic speeds of the combined matrix are

$$\begin{aligned} \lambda^{(i)} &= u, u + c, u - \sqrt{\widetilde{u'' u''}}, u - \sqrt{\widetilde{u'' u''}}, u + c, \\ & u, u + \sqrt{\widetilde{u'' u''}}, u + \sqrt{\widetilde{u'' u''}}, u, u, u, u \end{aligned} \tag{3.10}$$

where c is the speed of sound. For an arbitrary equation of state, the speed of sound is

$$c^2 = p_e\, p/\rho^2 + p_\rho + 3\widetilde{u'' u''}. \tag{3.11}$$

Using the primitive variables, the characteristic variables are

$$\delta W = L^{-1} \delta V = R^{-1} \delta U \tag{3.12}$$

where $\delta V = V^r - V^l$. The left eigenvector matrix L^{-1} and right eigenvector matrix R are relatively complicated and can be found in [7].

In matrix L^{-1} and R there are terms that have to be limited to avoid unnatural behaviour between turbulent and unturbulent regions. For example, the term $\widetilde{u''v''}/\sqrt{\widetilde{u''u''}}$ or $\delta(\widetilde{u''v''})/\sqrt{\widetilde{u''u''}}$ must not get very large values when $\widetilde{u''u''}$ goes to zero. It can be shown that

$$\sqrt{\widetilde{u''u''}}\sqrt{\widetilde{v''v''}} \geq |\widetilde{u''v''}|. \tag{3.13}$$

Using this $\widetilde{u''v''}/\sqrt{\widetilde{u''u''}}$ can be limited as

$$\frac{|\widetilde{u''v''}|}{\sqrt{\widetilde{u''u''}}} \leq \sqrt{\widetilde{v''v''}}. \tag{3.14}$$

Time Integration Method

In the implicit stage the approximate factorization is done assuming isotropic Reynolds-stresses. The algorithm consists of a backward and forward sweep in every coordinate direction. The sweeps are based on a first-order upwind differencing. In addition, the linearization of the source term is factored out of the spatial sweeps.

The matrix inversion resulting from the source-term linearization is performed before the spatial sweeps. The matrix D is approximated by using the following pseudo-linearization

$$D = \frac{\partial Q}{\partial U} = -\frac{Q}{|\Delta U_{\max}|}. \tag{3.15}$$

In this way, the maximum change of U caused by Q is limited to $|\Delta U_{\max}|$. The value of $|\Delta U_{\max}|$ is evaluated as in [3].

A multigrid method is used to accelerate the convergence. In order to stabilize the multigrid cycle with Reynolds stresses, several devices have been developed. The implementation for the multigrid cycling is described in [3].

4 Test Calculations

Channel Flow

The model was checked by calculating a fully developed flow in a plane channel. Several methods were compared. The numbering of the test cases can be seen in Table 1. The second-order upwind scheme was used. The results were compared with those of Kim et al. [8] DNS data and the Reynolds-stress budgets were compared with Mansour et al. [9] data. The DNS data is at $Re_m = \rho u_m \delta/\mu \approx 2800$, where u_m, δ and μ are the mean velocity, the channel half-width and molecular viscosity, respectively. Because a compressible flow solver is used, the Mach number was set to 0.2. This introduced a 1% change in density across the channel. Calculations were performed with five multigrid

Table 1 Description of the test cases.

Case 1	Isotropic flux difference splitting with diffusion of Daly et al. [10]
Case 2	Non-isotropic flux difference splitting with diffusion of Daly et al.
Case 3	Non-isotropic flux difference splitting with the scalar diffusion [11]
Case 4	Same as case 2 with a single grid

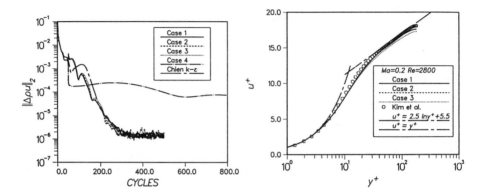

Figure 1 Convergence of the L_2-norm of the x-momentum residual and mean velocity profiles in wall coordinates.

Table 2 Mean flow variables.

	DNS	Case 1	Case 2	Case 3
δ^*/δ	0.141	0.135	0.133	0.135
θ/δ	0.087	0.081	0.080	0.082
$H = \delta^*/\theta$	1.62	1.66	1.67	1.64
$c_f = \tau_w/(\frac{1}{2}\rho u_m^2)$	8.18×10^{-3}	8.45×10^{-3}	8.43×10^{-3}	8.72×10^{-3}
$Re_\tau = \rho u_\tau \delta/\mu$	180	181	181	184
$Re_m = \rho u_m = \delta/\mu$	2800	2817	2819	2819
$Re_c = \rho u_c \delta/\mu$	3250	3256	3252	3259
u_m/u_τ	15.63	15.55	15.62	15.36

levels at $CFL = 10$. The first 50 iteration cycles were performed with a $k - \epsilon$ model and the Reynolds-stresses were not coupled with flow equations. A converged solution was obtained after 350 iteration cycles.

The convergence history of the L_2-norm of the x-momentum is shown in Fig. 1. In this case also the difference in the convergence rate between the isotropic, non-isotropic methods and $k - \epsilon$ model is marginal, whereas the effect of the multigrid acceleration is significant. However, two times higher CFL numbers could be used with the $k - \epsilon$ model.

Mean flow variables are presented in Table 2. As can be seen in Table 2, the difference between isotropic and the non-isotropic flux difference splitting was small. The velocity

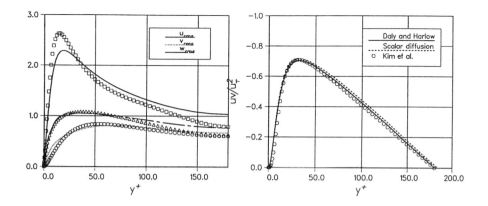

Figure 2 Comparision of the calculated Reynolds-stresses and the DNS-data in the plane channel.

profiles are also compared in Fig. 1. Velocity profiles in the viscous sublayer agree well with the DNS data and the universal profile. On the outer layer the velocity profiles are not completely satisfactory. The Reynolds-stresses can be seen in Fig.2, where $u_{rms} = \sqrt{\widetilde{u''u''}/u_\tau^2}$. Turbulent intensities agree well with the DNS data, except that u_{rms} peak level is low and the near wall values v_{rms} are not satisfactory. Shear-stress in Fig. 2 agrees with DNS.

Onera A-airfoil

For this case the experimental data was provided by Capbern et al. [12] and Gleyzes [13]. The Reynolds number and the Mach number are 2.1×10^6 and 0.15, respectively. The angle of attack is 13.3 degrees. The RSM is compared with the two-equation model of Chien and the Cebeci-Smith algebraic turbulence model. The dissipation equation is the same as in Shima's original article [2]. More detail can be found in [14]. In this case only the isotropic flux splitting method was used. Skin friction coefficient and pressure coefficient distributions can be found in Fig. 3.

5 Summary

The Reynolds-averaged Navier–Stokes equations with a low-Reynolds number RSM have been solved using an implicit method with a multigrid acceleration for convergence. In the evaluation of fluxes Roe's method is applied and the turbulence equations are coupled with the inviscid part of the flow equations. A new non-isotropic coupling of the Navier–Stokes and the Reynolds-stress equations was introduced.

This paper has focused attention on the problem of coupling the Reynolds-stresses with Navier–Stokes equation in a compressible case. The main conclusion is that when

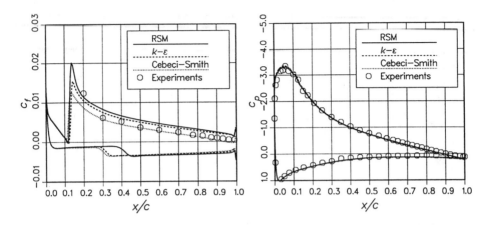

Figure 3 Skin friction coefficient and pressure coefficient distributions around the Onera A-airfoil.

Roe's flux-difference splitting is applied, the Reynolds stress equations can be coupled with Navier-Stokes equations if the effect of the production term is included in the Jacobian matrix. The resulting eigenvectors and the characteristic variables have a fairly complex form. In the present examples the new coupling method had only a minor effect on the results and on the convergence rate. However, with a computational approach this form can be rearranged and simplified. As a result the computing time is increased by only 13% in comparison with the non-isotropic coupling. The iteration sweep in RSM was roughly 2.5 times slower than in Chien's $k - \epsilon$ model.

References

[1] F.S. Lien and M.A. Leschziner. A general non-orthogonal collocated finite volume algorithm for turbulent flow at all speeds incorporating second-moment turbulence-transport closure, part 1: Computational implementation. *Computer Methods in Applied Mechanics and Engineering*, pages 123–148, 1994.

[2] N. Shima. A Reynolds-stress model for near-wall and low-Reynolds-number regions. *Journal of Fluids Engineering*, 110:38–44, 1988.

[3] T. Siikonen. An application of Roe's flux difference splitting for the $k - \epsilon$ turbulence model. Report, Series A A-15, Helsinki University of Technology, Laboratory of Aerodynamics, 1994. ISBN 951-22-2059-8.

[4] P.L. Roe. Approximate Riemann solvers, parameter vectors, and difference schemes. *Journal of Computational Physics*, 43:357–372, 1981.

[5] K. Chien. Predictions of channel and boundary-layer flows with a low-Reynods-number turbulence model. *AIAA Journal*, 20(1):33–38, Jan 1982.

[6] D. Vandromme. Turbulence modeling for turbulent flows and implementation in Navier-Stokes solvers. In *Introduction to the Modeling of Turbulence*. von Karman Institute for Fluid Dynamics Lecture Series 1991-02, 1991.

[7] P.P. Rautaheimo and T. Siikonen. Numerical methods for coupling the Reynolds-averaged Navier–Stokes equations with the Reynolds-stress turbulence model. Report 81, Helsinki University of Technology, Laboratory of Applied Thermodymics, 1995. ISBN 951-22-2748-7.

[8] J. Kim, P. Moin, and R. Moser. Turbulence statistics in fully developed channel flow at low Reynolds number. *Journal of Fluid Mechanics*, 177:133–166, 1987.

[9] N.N. Mansour, J. Kim, and P. Moin. Reynolds-stress and dissipation-rate budgets in a turbulent channel flow. *Journal of Fluid Mechanics*, 194:15–44, 1988.

[10] B.J. Daly and F.H. Harlow. Transport equations of turbulence. *Physics of Fluids*, 13:2634–2649, 1970.

[11] L. Davidson. Prediction of the flow around an airfoil using a Reynolds stress transport model. *Journal of Fluids Engineering*, 117:50–57, March 1995.

[12] C. Capbern and C. Bonnet. Operation decrochage. Rapport final de synthese. Technical Report 443.535./89, Report Aerospatiale, Toulouse, 1989.

[13] C. Gleyzes. Operation decrochage. resultats de la deuxieme campagne d'essais a f2- (mesures de presion et velocimetrie laser). CERT 57/5004.22- OA 72/2259 AYD, ONERA, 1989.

[14] A. Hellsten. Reynold-stress model in computational aerodynamics (in Finnish). Master's thesis, Helsinki University of Technology, May 1995.

3 D NUMERICAL SIMULATION OF A NATURAL DRAUGHT COOLING TOWER FLOW

E.Razafindrakoto, J.L.Grange
Electricité de France, Direction des Etudes et Recherches,
6 quai Watier, 78401 Chatou Cedex, France
L.Fabre, H.Delabrière
Electricité de France, Direction de l'Equipement,
12-14, Avenue Dutriévoz, 69628 Villeurbanne Cedex, France.
S.Delaroff
SIMULOG, 1 rue James Joule, 78286 Guyancourt, France

ABSTRACT

Although in many circumstances a simple model is adequate for the estimation of the performance of a cooling tower, some problems require a three dimensional model. The present paper proposes a coupling algorithm with a fully 3D air flow model using the finite element method and a 1D model of air-water heat and mass transfer exchange. Numerical simulations of a natural draught counter-flow cooling tower on a 1300 MW power plant are tested and presented on a 2D axisymmetric case and on a fully 3D case.

INTRODUCTION

In order to obtain a better estimation of air-water heat and mass transfert in a cooling tower, a 3D model for a natural draught counter-flow cooling tower is being developped at EDF, coupling on each time iteration an available industrial finite element code (N3S) [1] for the computation of air flow in the tower and in the area surrounding the tower, with a 1D vertical model (module provided by the code TEFERI [3]) for air-water heat and mass transfer exchange computation in the packing zone.
On each time step, an explicit coupling algorithm is used as follows :
 - for air flow, N3S solves momentum averaged Navier-Stokes equations (closed with a $k - \varepsilon$ turbulence model) and transport-diffusion equations for temperature and humidity. The obstrued zones such as packing, cooled water recuperators and piles (see Figure 1) are represented by head loss source terms. In the exchange zone, N3S gives air values entries to 1D module which calculates in return additional heat and humidity source terms depending on hot water temperature, flow rate and packing characteristics.
 - the 1D module solves water-air exchange equations (wet air enthalpy, humidity and water temperature) on each vertical packing zone, with Poppe's formulation or Merkel approximation [2].
In this paper we present in part one the air flow equations and the 1D water-air exchange equations. Part two describes the numerical scheme and the coupling algorithm. In last part, application results are given on an industrial tower on a 1300 MW power plant in 2D axisymmetric case and 3D case using Merkel approximation.

I. GOVERNING EQUATIONS

I.1 AIR FLOW EQUATIONS

The governing equations for air flow are averaged Navier-Stokes equations with a standard $k - \varepsilon$ turbulence model. They are described as follows.

For pressure p and velocity **u** :

$$\begin{cases} \text{div}(\rho \mathbf{u}) = 0 \\ \rho \left(\dfrac{\partial \mathbf{u}}{\partial t} + \mathbf{u}.\text{grad}\mathbf{u}\right) + \text{div}(\mu_e \text{grad}\mathbf{u}) + \text{grad}p = - (\rho - \rho_0)\mathbf{g} + \dfrac{1}{2}\rho|\mathbf{u}|\,\mathbf{K}.\mathbf{u} \end{cases} \quad (1)$$

where ρ, ρ_0, $\mathbf{g}$, $\mathbf{K}$ denote density, reference density, gravity and head loss coefficients tensor and $\mu_e = \mu + \mu_t$ with μ and μ_t the laminar and turbulent viscosities,

For temperature T and air humidity X :

$$\rho C p \left(\dfrac{\partial T}{\partial t} + \mathbf{u}.\text{grad}T\right) - \text{div}(\lambda_e \text{grad}T) = f_T \quad (2)$$

$$\rho \left(\dfrac{\partial X}{\partial t} + \mathbf{u}.\text{grad}X\right) - \text{div}(\sigma_e \text{grad}X) = f_X \quad (3)$$

where Cp is the thermal air capacity, $\lambda_e = \lambda + \lambda_t$ (resp. $\sigma_e = \sigma + \sigma_t$) with λ and λ_t (resp. σ and σ_t) the laminar and turbulent diffusivity coefficients for T (resp for X). Variable density ρ is given as function of T and X.

The system is closed with

$$\mu_t = \rho C_\mu \dfrac{k^2}{\varepsilon} \;;\; \lambda_t = \dfrac{\mu_t}{Prt} \;;\; \sigma_t = \dfrac{\mu_t}{\sigma_x} \quad (4)$$

where turbulent kinetic energy k and dissipation ε derived from standard k - ε transport-diffusion equations (Prt , σ_x are the turbulent Prandtl number and its equivalent for X).

I.2 MODELLING THE EXCHANGE ZONE

In the exchange zone, the cooling of water in contact with air is both ensured by convection and evaporation. We assume that air and water flows are one dimensional and depend only of the vertical z axis. Given an elementary volume dV with an exchange area dS=adV (a is the exchange surface by unit volume), the following notations are adopted :
- h , he, hv, the mass enthalpies respectively of wet air (referred to the mass of dry air), water and steam and Cp, Cpe, Cpv their mass thermal capacities,
- hv0 the latent heat of vaporization of water, Ta and Te the air and water temperatures, hse the mass enthalpy of satured wet air at Te,
- Xse, Xsa the saturating steam content of wet air at air temperature Te or Ta,
- Fa, Fe the mass fluxes of air and water in the exchange fill,
- α the heat transfert coefficient by convection and βx the evaporation coefficient.

For unsatured air, the mass and heat transferts are determined by the following equations :

dFe = βx (Xse-X).a.dz	(Dalton's evaporation law)	(5a)
dFe = Fa.dX	(mass balance)	(6)
Fa.dh = hv.dFe + α (Te-Ta)	(evaporation and convection effects)	(7)
Fa.dh = he.dFe + Fe.dhe	(energy balance)	(8)

where Cpx = Cp + XCpv is the mass thermal capacity of the wet air,
h = Cpx.Ta + X.hv0 , he = Cpe.Te and hv = Cpv.Te + hv0.

This leads to the following system (Poppe's formulation) :

$$\begin{cases} \dfrac{dh}{dz} = \dfrac{\beta x.a}{Fa}(hse - h + K) \\ \dfrac{dX}{dz} = \dfrac{\beta x.a}{Fa}(Xse - X) \end{cases} \text{and} \quad \begin{cases} \dfrac{dTe}{dz} = \dfrac{Fa}{Fe}\left[\dfrac{1}{Cpe}\dfrac{dh}{dz} - Te\dfrac{dX}{dz}\right] \\ \dfrac{dFe}{dz} = Fa\dfrac{dX}{dz} \end{cases} \quad (9)$$

where K is defined from the Lewis number $Le = \dfrac{a}{\beta x\, Cpx}$ by :

$$K = (Le - 1) [(hse - h) - hv.(Xse - X)] . \quad (10a)$$

For satured air, (5), Cpx, h and K must be replaced by :

$$dFe = \beta x (Xse-Xsa).a.dz \tag{5b}$$
$$Cpx = Cp + Xsa.Cpv + (X-Xsa).Cpe$$
$$h = (Cp+Xsa.Cpv)Ta + X.hv0 + (X-Xsa).CpeTa$$
$$K = (Le - 1) [(hse - h) - hv.(Xse - Xsa)] + Le\, Te\, Cpe\, (X - Xsa) \tag{10b}$$

Merkel approximation assumes that :
- the air is satured in the packing, so dFe is neglected in energy balance (8),
- the Lewis number is equal to 1.

Therefore, system (9) is reduced to :

$$\begin{cases} \dfrac{dh}{dz} = \dfrac{\beta x.a}{Fa} (hse - h) \\ \dfrac{dTe}{dz} = \dfrac{Fa}{Fe} \dfrac{1}{Cpe} \dfrac{dh}{dz} \end{cases} \tag{11}$$

Systems (9) or (11) are solved by a Runge-Kutta algorithm of 4-th order with given mass fluxes Fa, Fe and an exchange law with the correlations form as :

$$\frac{\beta x.a}{Fe} = \lambda \left(\frac{Fa}{Fe}\right)^n \tag{12}$$

where coefficient λ and exponent n are characterics of the exchange filling and obtained from measurements.

2. TIME DISCRETISATION AND COUPLING

Time discretisation is based on fractional step method [1] : convection, diffusion and pressure-continuity are separated from each other and solved in that order. This splitting is shown to be first order accurate. Knowing the values of variables C^n, $\mathbf{u}^n$, p^n at time t^n, we have to calculate in an advection step convected quantities $\widetilde{C}$, $\widetilde{\mathbf{u}}$ and in a diffusion step C^{n+1}, $\mathbf{u}^{n+1}$, p^{n+1} their values at time t^{n+1} as follow :

- convection step solved by a characteristics method :

$$\begin{cases} \dfrac{\partial \widetilde{C}}{\partial t} + \mathbf{u}^n.\text{grad}\widetilde{C} = 0 \\ \widetilde{C}(x, t^{n+1}) = C(x,t^n) \end{cases} \tag{13}$$

- diffusion step :

$$\alpha^n \frac{C^{n+1} - \widetilde{C}}{\Delta t} - \text{div}(v^n \text{grad} C^{n+1}) = f \tag{14}$$

- pressure-continuity step :

$$\begin{cases} \rho \dfrac{\mathbf{u}^{n+1} - \widetilde{\mathbf{u}}}{\Delta t} + \text{grad}\, p^{n+1} = f \\ \text{div}(\rho \mathbf{u}^{n+1}) = 0. \end{cases} \tag{15}$$

The finite element method is used for diffusion and pressure-continuity steps with P1-ISOP2 triangles or tetrahedra. Assuming a logarithmic wall velocity profile, wall functions are used on the wall boundaries to compute the friction shear stress at each time step. For the boundaries with open flow, zero normal derivative is specified on all variables.

Let Ta^n be the air temperature after the exchange with the hot water (e.g., return values exchange module). At time n+1, the problem is to enforce T^{n+1} being as close as possible to Ta^n. For this, two methods can be prescribed :

a) thermal exchange effects are taken into account by forcing T^n values on exchange zone to Ta^n before the advection step. It will then be convected and diffused at the next time step. This approach can be interpreted in term of a corrector temperature source f_T equals to $\rho Cp \dfrac{\widetilde{Ta^n} - \widetilde{T^n}}{\Delta t}$ on the exchange domain and zero outside.

b) temperature equation (2) can be solved under the constraint of minimising the difference $Ta^n - T^{n+1}$. Then, classical linear constraint algorithm can be applied.

The same approaches are used to take into account the exchange effects on air humidity.
In this paper our results are obtained with the first approach because of its easy implementation, but the second approach will eventually be tested also.

3. APPLICATIONS ON A NATURAL DRAUGHT COOLING TOWER

Numerical simulations are done on a tower of a 1300 MW EDF's power plant in a simplified 2D axisymmetrical simplified geometry (lintel and hot water inlet tube are not represented) and in a fully 3D geometry with uniform hot water temperature profile and Merkel approximation.
Directional head loss terms are added in the vertical direction to represent packing zone and cooled water recuperator zone (for 3D case), and in a radial direction for piles (for 3D case). The head loss tensor coefficients in equations (1) are obtained from experimentally derived drop correlations and are expressed as a parabolic correlation as follows.
$K_{ij} = C^0_{ij} + C^1_{ij} |u| + C^2_{ij} |u|^2$
The constants C_{ij} are obtained from pack tests.
Exchange coefficient λ and exponent n in (12) are taken as constant on all the exchange zone.
Additionally, Boussinesq approximation is used for buoyancy terms.

3.1 CALCULATIONS

The full computational domain of the calculations is approximately 12 times the heigth of the tower in the vertical direction and 27 times its base in the horizontal direction with a P1-IsoP2 finite element space discretization (see Figures 2 and 3). Table 1 gives the number of elements (NELT), of pressure nodes (N1) and of velocity or of scalar nodes (N2).
Ratio $r = h_{max} / h_{min}$ of the coarse mesh size (near the exterior boundary domain) and of the fine mesh size (around the packing or around the lintel zone) are approximately 800 for the two cases. Finite element modelling allows such a distorted mesh, but in return a small time step value had to be used to respect CFL condition stability in the fine mesh zone.
Results are obtained after running on 12000 time step iterations for the 2D case and on 6200 iterations for the 3D case and required 0.1s time CPU (resp. 0.4s) per 1000 nodes per time step and 0.26 Megawords (resp. 0.65 Megawords) per 1000 nodes on Cray C98 for the 2D (resp. 3D) case

Table 1 : Configuration of calculations

CASE	N1	N2	NELT	Δt	CFLmax
2D AXI	1949	7545	3647 triangles	0.08s	1.2
3D	12852	96836	67876 tetrahedra	0.08s	0.75

3.2 NUMERICAL RESULTS

Major phenomena are well predicted by simulations for air flow (see Figures 2, 4 and 5). Particularly, in the packing and in the recuperator zones, the effects of directional charge loss appear clearly. Velocity field is constrained to be in the vertical direction and the discontinuity of the static pressure through these zones is well picked up. It also can be seen in figure 7, that heat transfer is inversely proportional to air velocity module (low effects near wall).
Comparison with experimental data concern only averaged cooled water temperature, which is the interesting cooling tower perfomance parameter.
The 2D case shows an averaged cooled water temperature 0.8 °C above the measurement value whereas the 3D computation 2.6 °C above. Moreover, for the 3D case, the solid beam (the hot water inlet tube) above the packing middle zone decreases heat exchange around it, figure 6. Additionally, predicted total air flow through the tower is not high enough. Indeed, the numerical predicted ratio of total air flow through the tower on total water flow through the pack is 64% in 2D case and 59% in 3D which is too small in comparison with experimental values generally estimated to be more than 70%. The draught of the air flow could be improved by using variable density model instead of Boussinesq approximation.
Although precisions are not as good as expected, these first results shows clearly the feasibility of the approach to predict major phenomena in cooling tower by multidimensional numerical simulations.
In any case, more comparisons are needed with experimental profiles data both on air and water for a better appreciation of results.

4. CONCLUSION

Coupling 3D air flow code with 1D mass and heat transfer air-water model has been implemented and tested in a numerical 2D axisymmetric and a fully 3D simulation of a natural draught counter-flow cooling tower. Major air flow phenomena are well predicted by simulations, but the estimation of averaged cooled water temperature results and the total air flow through the tower are not sufficient and must still be improved. Future improvements could be :
- improving the coupling algorithm (as described in paragraph 2.a for example),
- using variable density model and space-time varying exchange coefficients.
Additionally, more experimental comparisons are certainly needed to validate the approach.

REFERENCES

[1] Pot G., Bonnin O., Moulin V., Thomas B.
"Improvements of finite element algorithms implemented in CFD code N3S for industrial applications"
Proceedings of the 8th International Conference on Numerical Methods in Laminar and Turbulent Flow, vol.8, pp.631-642, 1993, Swansea, UK.
[2] "Les réfrigérants atmosphériques industriels".
"Collection EDF DER", 1991, Eyrolles, Paris.
[3] Bourillot C.
"TEFERI, numerical model for calculating the performance of an evaporative cooling tower".
"Electric Power Research Institute, Special Report CS-3212 SR", 1983.

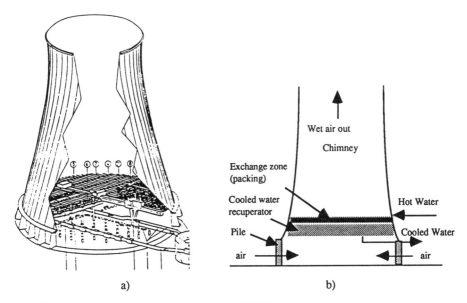

Figure 1 : a - a counter-flow cooling tower (SCAM)
b - a schematic natural -draught counter flow cooling tower

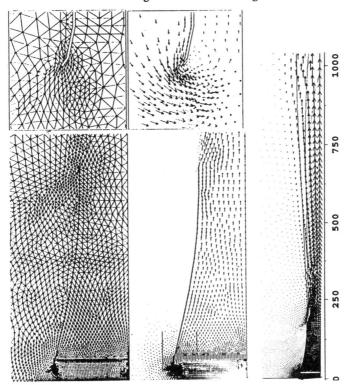

Figure 2 : mesh and velocity field, 2D axisymmetric case

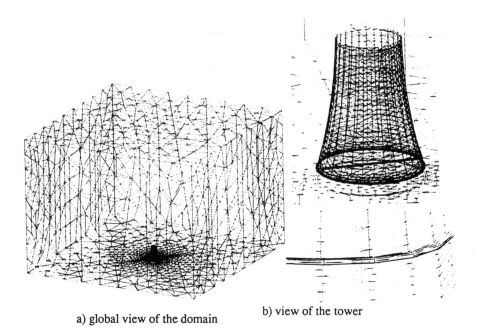

a) global view of the domain

b) view of the tower

Figure 3 : mesh views of the 3D case,

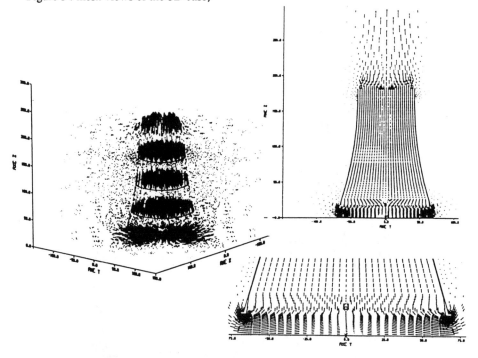

Figure 4 : velocity fields, 3D case
- some sections 3D views
- at the section y=0

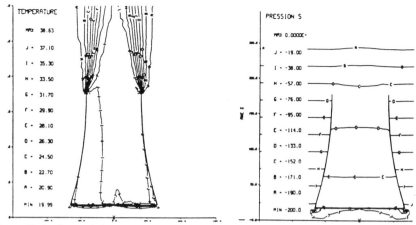

Figure 5 : isolines of air temperature and static pressure at the section y=0 , 3D case

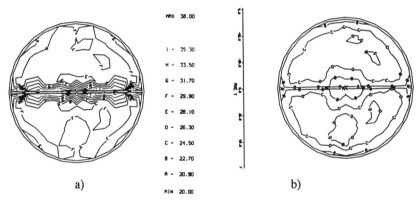

Figure 6 : isolines of the cooled water temperature a) and the air temperature b) below the packing, 3D case.

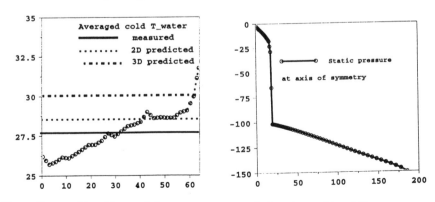

Figure 7 : - profile of the cooled water temperature below the packing, 2D case. Comparison of the averaged cooled water temperatures
- profile of the static pressure along axis of symmetry, 2D case.

An Evolutionary Far-Field Boundary Treatment for the Euler Equations

S. Rham   C. Albone
Low Speed and Basic Aerodynamics Department. Defence Research Agency. Farnborough. United Kingdom.

Summary

A new, evolutionary treatment of far-field boundary conditions is devised and implemented in the flow algorithm of the FAME suite of codes. Continual, local calculation of flow variables on the far-field boundary, based on physical concepts, is used to construct new conditions that can be used for *non-remote* boundaries. The consequent use of a smaller computing region results in significant improvements in the speed of convergence of flow calculations and so reduces computational costs, with little loss in solution accuracy.

Introduction

The aerodynamic community has spent many years developing numerical schemes for solving the Euler equations, since these provide a reasonable approximation to the flow about the aircraft throughout the speed range. These equations represent the infinite Reynolds number limit of the equations that govern fluid motion, and describe wave propogation through an inviscid medium.

The requirements of such solution methods, which are speed and accuracy, are mutually conflicting; for example prohibitive run times are often associated with the fine meshes required for accuracy. This report details an investigation into one aspect of the solution of the Euler equations which aims to speed up the solution procedure whilst not degrading the solution accuracy (beyond some given tolerance).

The unbounded nature of external aerodynamics means that any disturbances can propogate through the fluid an 'infinite' distance from the aircraft. Numerical methods require the definition of some finite boundary to the computational domain. This 'artificial' boundary requires the prescription of some flow variables ; the boundary conditions. These must be both consistent with the mathematical model (e.g. the correct number of prescribed boundary conditions) and consistent with the true physical detail of the flow (e.g. Mach number, angle of attack etc.). The normal treatment of these boundary conditions uses a one-dimensional wave decomposition approach[1] and requires the value of the flow variables to be those of the undisturbed stream. For this reason the artificial boundary must be located far away from the body (hence the name 'far-field' boundary). Results shown later highlight the errors in the solution that arise when this boundary, using this treatment of the boundary conditions, is too close to the body.

We will use a 'decay-rate' theory to specify one of the incoming waves, and the fact that there is zero vorticity in an undisturbed stream, to impose conditions for the other two necessary pieces of information at inflow. The aim of this is to bring the boundary in much closer to the body, gaining an increase in speed with no loss in solution accuracy.

The Test Case

This work aims at investigating the effect, on some aspect of solution quality, of changing the location of the far field and the conditions imposed on it. To focus clearly on this, the flow-field is kept as simple as possible. Thus, only subsonic flow past a sphere is considered, with the undisturbed stream Mach number as 0.4.

The Feature-Associated Mesh Embedding (FAME)[2] method is chosen as the vehicle to test any ideas.

The FAME system uses a number of overlying/embedded meshes associated with geometric or physical features, and ensures an overall mesh of high quality for any complexity of configuration.

For a sphere, the FAME mesh generation system produces three regions of mesh (Figure 1): type-1 curvilinear meshes aligned with the body and far-field boundary (black), and the type-0, cartesian background mesh.

The extent of both type-1 meshes into the field, and hence the region covered by the type-0 mesh is easily controlled in the FAME mesh generation procedure.

We must ensure that, as changes are made to the position of the far-field boundary, differences in the flow solution arise solely as a consequence of these changes. Accordingly, the body-fitted type-1 mesh and the type-0 mesh remain fixed as the far-field boundary is moved. Indeed this movement is effected by simply adding or subtracting cells on the far-field type-1 mesh, whilst maintaining a constant radial expansion ratio.

The flow algorithm employs time-marching to obtain the steady state solution to the (homentropic) four-equation Euler model. It uses [2] a first-order, non-conservative, upwind, finite-difference scheme ; second-order accuracy is gained via a forcing, defect-correction term. A multigrid acceleration scheme is used on the background type-0 mesh.

New Far-Field Treatment

It was stated in the introduction that the Euler equations describe wave propagation. We must prescribe all pieces of incoming information (Riemann invariants) via boundary conditions. All outgoing Riemann invariants are extrapolated from the flow solution.

We begin by considering the way in which the incoming Riemann invariant, R, behaves near the far-field boundary. Here

$$R = q_n - A$$

where q_n is the outward normal velocity component and A is a function of the speed of sound, c, and the ratio of specific heats, γ, defined as $A = \frac{2c}{\gamma-1}$.

We assume that, in the region of the far-field boundary, R decays monotonically to its undisturbed stream values (Figure 2).

$$R = R_\infty + \frac{\kappa}{r^\alpha} . \tag{1}$$

R_∞ represents its value in the undisturbed stream, r is some measure of the distance between the body and the far-field boundary, κ is some constant of proportionality and α is the (unkown) decay rate index. Existing treatments simply assume that r is very large so that R on the far-field boundary (denoted R_0) is set equal to R_∞. In this new treatment we wish to place the 'far-field' close to the body; hence R_0 must be determined for small values of r. If we assume that α is known then there is a linear relationship between R and $\frac{1}{r^\alpha}$. If we know the value of R_∞ and, from the flow calculation, the value of R at the first mesh point inside the flow field, R_1, then R_0 (on the far-field boundary) can be determined by using equation (1).

Given α, plotting R against $\frac{1}{r^\alpha}$ gives a straight line (Figure 3). For the three points, $r = \infty$, $r = r_0$ (the far-field boundary) and $r = r_1$ (the first point into field) to lie on a straight line, we require

$$\frac{R_0 - R_\infty}{\frac{1}{r_0^\alpha} - 0} = \frac{R_1 - R_0}{\frac{1}{r_1^\alpha} - \frac{1}{r_0^\alpha}} \; .$$

After some straightforward algebra this gives

$$R_0 = (1 - (1 - \frac{\Delta r}{r_0})^\alpha) . R_\infty + (1 - \frac{\Delta r}{r_0})^\alpha . R_1 \qquad (2)$$

where $\Delta r = r_0 - r_1$ is the length of the outermost mesh interval. The standard choice of setting undisturbed stream conditions on the far-field boundary is equivalent to an infinite value of α.

So far, it is proposed that the value R_0 on the boundary be imposed as a combination of R_∞ and R_1, using some assumed value of α in equations (1) and (2). But, α is unknown and may vary throughout the field. We will use information from further inside the field to determine α locally, as described below.

Let R_2 be the value of R two mesh intervals into the field, at $r = r_2$. Then, if the assumed decay rate index, α, of R near the far-field was consistent with local flow conditions, the value R_2 would lie on the straight line through R_∞, R_0 and R_1 when plotted against $\frac{1}{r^\alpha}$ as shown in Figure 3. It is, however, unlikely that specifying a constant value of α everywhere on the far-field sphere is consistent, and, as is the case, the point is more likely to lie on the dashed line. A mechanism is needed that iteratively drives the calculation of α to a value that takes account of information at $r = r_2$.

Let a decay rate index denoted α^n, where the superscript refers to the time level, be imposed in equation (2) for some given number of time steps, such that the situation shown in figure 3 holds.

We construct a range of decay rate indices, β_i, either side of α^n

$$\beta_i = \alpha^n \pm i . \Delta \alpha \qquad i = 0, \ldots, i_{max}$$

where $\Delta\alpha$ is some increment in the decay rate index and i is an integer; the maximum value of i, i_{max}, controls the range of decay rate index covered. Typically $\Delta\alpha$ is 0.01 and i_{max} is 10.

We then consider a series of graphs of R vs $\frac{1}{r^{\beta_i}}$, figure 4, and fit a least-squares, straight line, $\overline{R}$, through the four points in the usual way for linear regression analysis.

Then for each value of β_i, $i = 0, \cdots, i_{max}$, we compute the RMS error E_i.

The value of i at which E_i is a minimum, i_{min}, is evaluated and then at the next time step the value of α used is given by

$$\alpha^{n+1} = \beta_{i_{min}} \qquad (3)$$

that is some 'optimum' value of α based on information at four points.

To evaluate the other two incoming Riemann invariants at inflow, the two tangential velocity components, q_{t_η} and q_{t_ζ}, we use the knowledege that, in an undisturbed stream, there is zero vorticity. (We will consider here how one of the tangential velocity components is treated; the treatment extends directly to the other component.)

Consider part of the mesh near the far-field boundary, as in figure 5. Here points 1, 2 and 3 lie on the far-field boundary, $r = r_0$, and points 4, 5 and 6 lie one cell into the computational domain, $r = r_1$, with computed values for the normal velocity component, q_n, and tangential velocity component, q_{t_η}, in this case. The normal velocity components on the far-field boundary have been treated according to the decay rate theory as described above. We wish to use the fact that in an undisturbed stream, near inflow, the circulation in cells between the far-field boundary and the penultimate grid surface must be zero.

We construct a secondary cell, shown by the dashed lines in figure 5 joining the midpoints of cell edges, and set the circulation, Γ, given by

$$\Gamma = \oint q.ds \qquad (4)$$

in this secondary cell to zero.

Here, for example, q_l is taken as the average of four normal velocity components at points 1, 2 5 and 6, q_u is taken as the tangential velocity component, q_{t_η}, at point 2. The integration is performed around the perimeter of the secondary cell, with the appropriate values of ds given by the corresponding side lengths of the secondary cell.

The way the process is carried out is as follows. A fixed value of α is used on the Riemann invariant R for an arbitrary number of time steps (in this case 600); the zero-circulation is imposed from the outset of the calculation. Then the calculation of the error term, E_i, over the range of decay rate indices, β_i, commences and the decay-rate index that yields the minimum error is used at the next time step. This is done for every point on the far-field, totally independently of any other value of α calculated around the boundary and the iterative process repeated to give quite impressive results, shown next.

Results

Three calculations will be considered; two with unmodified FAME, one in which the far-field boundary is far away, r=47, and the other with the boundary very much closer, r=2, and the third in which the new far-field boundary treatment, described above, is applied, at r=2. (Here r is the distance between the body centre and the far field, measured in sphere radii.)

In each case the calculation was stopped when some specified convergence criterion was satisfied. In all cases this was that, over a period of 100 time steps, the range of both the front and rear surface-pressure coefficients did not vary by more than 0.00001.

Two parameters are used to measure the effect of bringing in the far field and then modifying the boundary treatment; the front and rear pressure coefficients on the body. The results for the three calculations are shown in Figures 6 and 7, showing front and rear surface pressure coefficient respectively plotted against the number of cycles in the iterative scheme that solves the flowfield equations.

The case with the standard FAME method (r=47 FAME), is the definitive solution†; both values of pressure coefficients are still oscillating after 2500 time steps, in fact the solution did not statisfy the convergence criterion.

The next case, r=2 FAME (red), shows the errors introduced when the boundary is brought closer and undisturbed stream conditions imposed. As expected, bringing the far-field boundary much closer with no change to the treatment of far-field variables greatly increases the convergence rate; the rate of convergence is related directly to the time taken for the damping of waves travelling between, and reflecting from, the solid surface and far-field boundaries. Clearly as the distance between the far-field boundary and the body is reduced, the time taken for a signal to travel between the boundaries decreases.

The dark blue line shows the convergence history for the new scheme in which α is calculated. A fixed value of $\alpha = 2$ was used for the first 600 cycles. The convergence criteria were the same as the other four cases and were satisfied in 1800 cycles. The dramatic change after 600 time steps in front and rear pressure coefficient to reduce the error to less than 0.06 at the front and almost nothing at the rear is particularly encouraging for what is such a relatively simple method.

Finally we consider an example to illustrate the potential gain of this new method in a practical design environment. In this case both the standard boundary treatment and the new method were used with the boundary close in. The total number of cycles in each case was 800; in the new method α was fixed at 2 for the first 400 cycles only.

Here the information is summarised in tabular form.

Algorithm	Radii	Cycles	CPU secs.	Error at Front	Error at Rear
FAME	r=47	2500	3742	0.00	0.00
FAME	r=2	800	1053	0.304	0.217
NEWFAME	r=2	800	1128	0.047	0.011

Here the earlier notation is used for the type of conditions imposed on the far-field boundary and the distance this is located from the sphere centre. The CPU times given are for the Cray 2 supercomputer. The errors are absolute differences in pressure coefficient, based on the default solution with the boundary far away and run for 2500 cycles. This table shows clearly the bonuses of the new method : a cost reduction by two thirds with little degradation in solution quality. In particular the computational overhead associated with this new treatment is practically negligible.

Conclusions

The boundary conditions normally used require the far-field boundary to be located far from the object body. This leads to poor convergence properties. Moving the boundary closer to the body and not modifying the conditions improves convergence but degrades solution quality. The understanding of wave propogation has been used to devise a scheme which computes the rate at which a specified flow quantity decays in the region of a far-field boundary. A phsically correct treatment of circulation is also introduced . This has led to an unusual treatment of the flow variables on that boundary, that actually evolves

---

[0]†Note here that this is not an assessment of the absolute accuracy of the FAME method, but one of relative accuracy when modifications are made.

with the flow solution and has enabled the boundary to be located much closer to the object body, improving convergence rates but with minimal loss of accuracy.

This new approach to satisfy the far-field boundary conditions is independent of the numerical scheme used, and should work equally well for methods based on central differencing, as well as the upwind technique used in FAME.

References

[1] Pulliam, T.H.: "Characteristic boundary conditions for the Euler equations." Proc. Symposium on Numerical Boundary Conditions Procedures. NASA CP2201, 165-182 (1981).

[2] Blaylock, T.A., Onslow, S.H., Albone, C.M.: "Mesh generation and flow solution for complex configurations using the FAME system." Procedings of the Royal Aero. Soc. European Forum on Recent Developments and Applications in Aeronautical CFD pp 8.1-8.14 (1993).

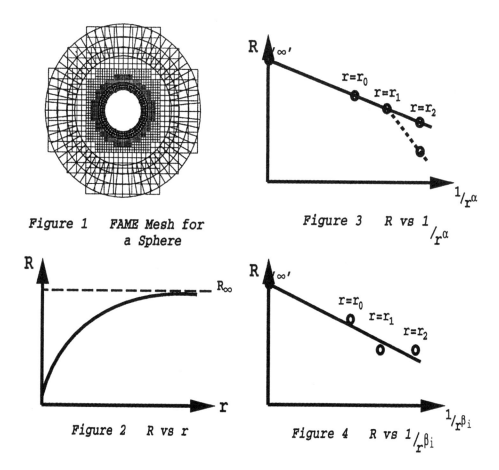

Figure 1   FAME Mesh for a Sphere

Figure 2   R vs r

Figure 3   R vs $1/r^\alpha$

Figure 4   R vs $1/r^{\beta_i}$

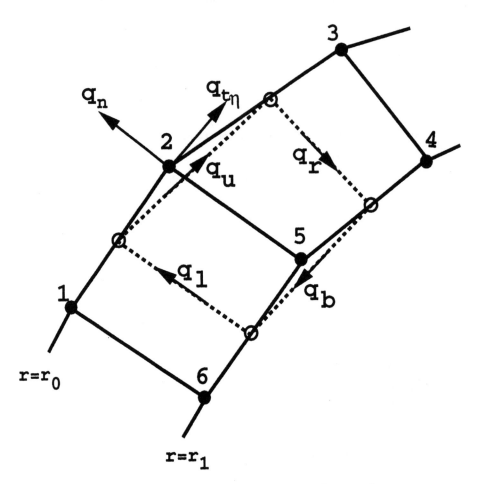

Figure 5 Zero circulation in boundary cell

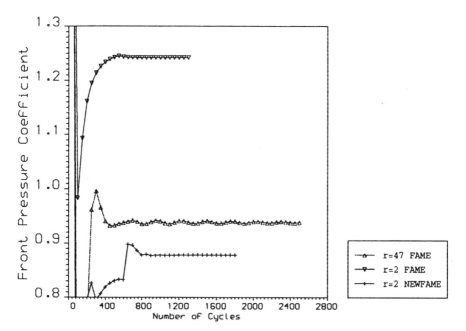

Figure 6 Convergence of front pressure coefficient

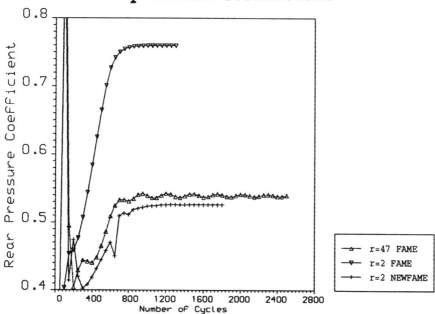

Figure 7 Convergence of rear pressure coefficient

3-D Calculations Showing the Effects of Stratification on the Evolution of Trailing Vortices

by

Robert E. Robins and Donald P. Delisi
Northwest Research Associates, Inc.
PO Box 3027
Bellevue, WA 98009-3027 USA

Summary

We have computed numerical solutions to the 3-D Navier-Stokes equations for a stratified, incompressible fluid. Our results show that stratification significantly modifies the phenomena that occur during the vertical migration and linking instability of trailing vortices. In particular we show that stratification accelerates the development of the linking instability and suppresses the vertical migration of the trailing vortices.

Introduction

The linking instability of the vortices that extend behind the trailing edge of aircraft wings was first noted by Scorer [1]. This instability was later analyzed by Crow [2], who carried out a linear perturbation analysis based on the kinematic relation between velocity and vorticity in an unstratified incompressible fluid. Crow found: (i) the wavelength of the maximum perturbation growth (which we denote as λ^*) is $8.6 b_0$, where b_0 is the initial distance between the parallel vortices; (ii) the perturbation amplitude grows by a factor of e ($\cong 2.72$) in a time equivalent to $1.21 T_0$, where T_0 is the time required for the vortices to descend a distance equal to b_0 (i.e., $T_0 = b_0/V_0$, where V_0 is the initial descent speed, given by $\Gamma_0/2\pi b_0$, Γ_0 being the initial magnitude of the circulation about each of the vortices); (iii) the angle to the horizontal of the plane in which the initial maximum growth occurs is 48°; and (iv) λ^* increases as the size of the vortex cores decreases.

In recent work, Robins and Delisi [3] present 3-D calculations which agree with and extend the results of Crow. For their results, λ^* is at first near the value of $8.6 b_0$ predicted by Crow, and then decreases to a value between $5 b_0$ and $8 b_0$ as the vortices link and rings are formed. It is also found that the instability initially grows in a plane at an angle near 48° to the horizontal, even when the initial perturbation is in a horizontal plane, and that this angle increases as the instability grows. This work included calculations starting from both single and multiple wavelength perturbations. (What we mean by single and multiple wavelength cases is discussed below under Numerical Approach.) The single wavelength case permits greater numerical resolution, but the multiple wavelength case provides for greater realism.

In this paper, we consider the effects of stratification on the development of the linking instability, and we restrict our attention to multiple wavelength cases. We show results for three cases: no stratification, Froude number (Fr) = 2, and Fr = 8, where Fr = V_0/Nb_0, N being the intrinsic frequency of the stratified fluid. N is defined as $-(g/\rho_0)d\rho/dz$, where $\rho(z)$ is the ambient density profile of the fluid and ρ_0 is a constant representative value of $\rho(z)$. We assume that $\rho(z)$ is linear so that N is a constant. An important time scale is the intrinsic period, $2\pi/N$. For the vortices considered below, this period is $12.69T_0$ and $50.78T_0$ for Fr = 2 and 8, respectively.

Numerical Approach

To solve the equations of motion, we use highly accurate finite-difference approximations for the spatial discretizations, and ensure incompressibility by using a projection method to evolve the solution forward in time. A second-order Adams-Bashforth method is used for the advection phase, and a fast Poisson solver is used for the projection phase of the time-stepping. Horizontal differences are computed using fast Fourier transforms, and a sixth order compact scheme is used for vertical derivatives. Horizontal boundary conditions are periodic, and "no surface" conditions (second derivatives equal zero) are imposed at the top and bottom. The two-thirds rule is used to avoid aliasing errors, and Fourier and compact low-pass filters are used in the horizontal and vertical directions, respectively, to control the build-up of small scale energy. Both the compact derivative and the compact filter schemes are based on the work of Lele [4]. The computational domain may be doubled in width or translated vertically when migrating vortices approach the cross-axial or vertical boundaries, respectively.

The initial conditions for the trailing vortices are specified so that the axes of the clockwise and counter-clockwise vortices deviate slightly from being parallel. For some of the calculations reported in [3], we used initial vortices having axes defined by single wavelength sinusoids and a Gaussian cross-axis distribution of vorticity. The deviation from being parallel was specified by defining each vortex's locus of maximum vorticity to be sinusoidal, with a peak-to-peak amplitude, δ, which is two orders of magnitude smaller than b_0. The vortices were perturbed in opposite directions in a common horizontal plane so that the distance between the vortex centers varied from $b_0 - \delta$ to $b_0 + \delta$. The Gaussian distribution was defined by a core radius r_0, which is the radius at which the vorticity falls off to 1/e of its peak value.

In an effort to model greater realism, we also did calculations reported in [3] that were initialized by vortices formed from several component vortices, each having a different perturbation wavelength. It is this approach that we adopt in the current study. More specifically, all cases described below are initialized by trailing vortices formed from thirteen component vortices ranging in wavenumber from 4 to 16. The computational length in the axial direction is chosen to be $45.5b_0$, so the wavelengths of the component vortices are (in units of b_0) 11.38, 9.10, 7.58, 6.50, 5.69, 5.06, 4.55, 4.14, 3.79, 3.50, 3.25, 3.03, and 2.84. Each component is a nearly parallel pair of counter-rotating vortices, with axes perturbed from parallel as described above for the single wavelength cases. We choose the total peak-to-peak amplitude of each sinusoid to be $0.02b_0$, and set its Gaussian core radius to be $0.16b_0$. The energies of the various component vortex pairs are chosen to scale as $m^{-5/3}$, where m is wavenumber, and the relative phases are randomly selected. The energies are normalized so that the circulation for the composite trailing vortex is the same as for the single wavelength cases described in [3]. Figure 1 shows a plot of the relative energies for the component vortices.

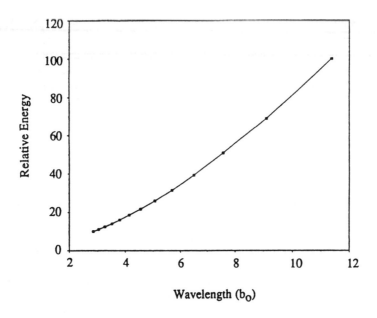

Fig. 1 Energy spectrum of the component vortices used to initialize the trailing vortices.

For the unstratified case, the number of grid points in the axial, cross-axial, and vertical directions is 257 x 65 x 91, and the corresponding extents (in terms of b_0) are 45.5, 5.333, and 7.5. For the Fr = 2 case, the grid point selection is 257 x 129 x 145 with corresponding extents 45.5, 8, and 12. And for the Fr = 8 case, the grid point selection is 257 x 129 x 121, and the corresponding extents are 45.5, 5.333, and 10. The cross-axial domain extent for this case is doubled at the computational time of $6T_0$. For all cases, the Reynolds number is 2930. The time step was $T_0/449$ and $T_0/494$ for the unstratified case and the stratified cases, respectively.

All calculations were performed on a Cray C-90 at the US Army Corps of Engineers Waterways Experiment Station. Cray library routines for multiple FFTs and multiple solution linear solvers were used whenever possible.

Results

Figure 2 shows a comparison of calculations from the unstratified and Fr = 2 cases. Visualized are constant surfaces of vorticity magnitude equal to 20 per cent of the vorticity magnitude maximum at the indicated times. The view is at an angle from above. Differences in the two cases are evident at $t = 4T_0$ when the Fr = 2 vortices can be seen to move closer together than in the unstratified case. At t = 6T₀ (about one-half the intrinsic period for the Fr = 2 case), the differences are accentuated as linking is well under way for the Fr = 2 case, but has just begun for the unstratified case. Also it is clear that the axial scale of the instability growth is smaller for the Fr = 2 case. By 8 T₀, linking is finally underway for the unstratified case; for the Fr = 2 case, the vorticity associated with the vortex pair is now of the same magnitude as the vorticity resulting from the detrainment of light fluid, and further linking and ring formation has been suppressed.

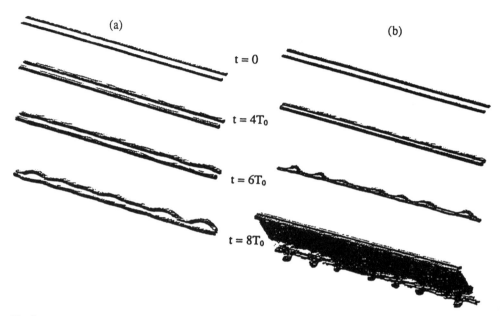

Fig. 2 Vorticity magnitude surfaces for the evolving trailing vortices. (a) shows the unstratified case and (b) is for Fr = 2.

Figures 3, 4, and 5 depict the density and axial vorticity fields at t = 8T₀ for the Fr = 2 case. Figure 3 shows cross-axial vertical slices at an axial position where rings had started to form, and Fig. 4 shows similar slices at an axial position where linking had begun. The contrast between the cross-axial widening of the perturbation in Fig. 3 and its suppression in Fig. 4 is quite striking. Figure 5 shows the density field in a vertical plane midway between the two vortices at t = 7T₀. Downward pointing arrows indicate where rings had started to form. On either side of the arrows there are density gradients pointing towards the arrows which are responsible for the widening seen in Fig. 3. Thus, stratification effects first have forced the vortices closer together, accelerating the linking process, and then have set up axial density gradients which have decreased the axial scale of the perturbation and transformed the formation of rings into cross-axial widening of the perturbation within narrow axial regions. In addition, the rising of light fluid in regions where linking had occurred suppresses the vertical migration of the vortex system.

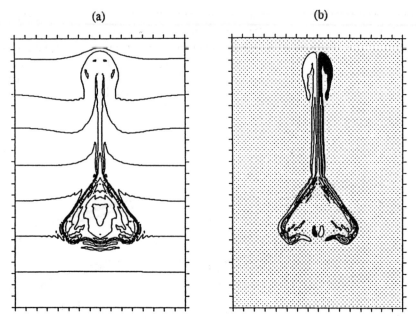

Fig 3 (a) density and (b) vorticity contours (for the Fr = 2 case, at t = 8 T_0) in a vertical slice perpendicular to the axes of the vortices at an axial position where rings had started to form (see Fig. 2b, t = 6T_0). Each tic mark represents 0.5 b_0.

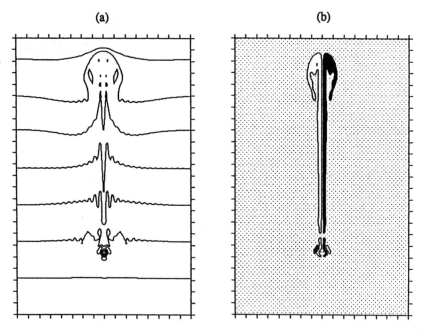

Fig 4 (a) density and (b) vorticity contours (for the Fr = 2 case, at t = 8T_0) in a vertical slice perpendicular to the axes of the vortices at an axial position where the vortices have linked (see Fig. 2b, t = 6T_0). Each tic mark represents 0.5 b_0.

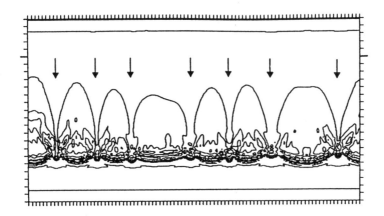

Fig. 5 Density contours (for the Fr = 2 case, at t = 7T_0) in a vertical plane midway between the trailing vortices. Arrows indicate the axial positions where rings had started to form. Each tic mark represents 0.5 b_0. The vortices' starting height is indicated by elongated tic marks.

Figure 6 shows a comparison of calculations from the unstratified and Fr = 8 cases. The quantity visualized and the viewing direction are the same as in Fig. 2. For the times shown, the Fr = 8 case is quite similar to the unstratified case. The main distinction is that linking for the Fr = 8 case is proceeding more rapidly than for the unstratified case. Since the final time shown, 12T_0, is only about one-quarter of the intrinsic period for the Fr = 8 case, we can be sure that stratification effects will increase in significance for later times.

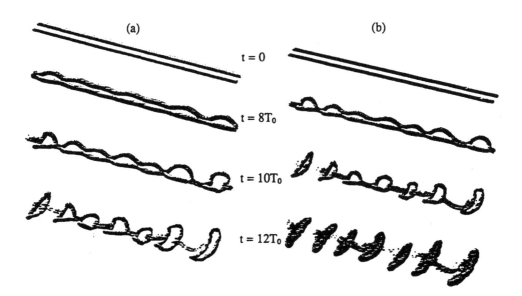

Fig. 6 Vorticity magnitude surfaces for the evolving trailing vortices. (a) shows the unstratified case and (b) is for Fr = 8.

In summary, our results show that stratification significantly modifies the phenomena that occur during the vertical migration and linking instability of trailing vortices. In particular, the results indicate that stratification accelerates the development of the linking instability and suppresses the vertical migration of the trailing vortices.

References

[1] Scorer, R.S., "Condensation Trails," *Weather*, Vol. X, No. 9, 1955, p. 281.

[2] Crow, S.C., "Stability Theory for a Pair of Trailing Vortices," *AIAA Journal*, Vol. 8, No. 12, 1970, pp. 2172-2179.

[3] Robins, R.E., and D.P. Delisi, "Nonlinear Development of Linking Instability for a Pair of Trailing Vortices," submitted to *AIAA Journal*.

[4] Lele, S., "Compact Finite Difference Schemes with Spectral-like Resolution," *Journal of Computational Physics*, Vol. 103, 1992, pp. 16-42.

Large Eddy Simulation of Turbulent Flow past a Backward Facing Step with a new Mixed Scale SGS Model

P. SAGAUT [†], B. TROFF [†], T. H. LÊ [†], TA PHUOC LOC [‡]
[†]: O.N.E.R.A., BP 72, 92322 Châtillon Cedex, France
[‡]: O.N.E.R.A. and L.I.M.S.I.(CNRS), BP 133, 91403 Orsay Cedex, France

Abstract

In this paper, a computational fluid dynamic prediction method is proposed, based on the resolution of the full unsteady incompressible Navier-Stokes equations. An original numerical method is elaborated, corresponding to the three-dimensionnal cartesian version of the PEGASE code for DNS and LES of incompressible flows. A new improved subgrid-scale model, the Mixed Scale Model, is proposed. This model is based on both the largest resolved scale gradients and the smallest resolved scale kinetic energy. The problem of turbulent flow past a backward facing step is used in the present study to evaluate the potentiality of LES for the prediction and the analysis of separated flows. Numerical results obtained by LES at a Reynolds number equal to 11 200 are compared with experimental data of Eaton & Johnston [3].

1 Introduction

Fluid flow phenomena such as separation, recirculation, reattachment and relaxation towards a fully developed equilibrium flow appear in many engineering applications related to aeronautics industries.

Such phenomena involve some highly unsteady behavior of the parameters of interest for engineering studies, like lift, drag or heat transfer. The knowledge of the extrema of these parameters in space or time, and of their characterisctic spatial and temporal frequencies do necessitate the resolution of the unsteady Navier-Stokes equations.

Because the Reynolds number of industrial flows are very often far out of range of the Direct Numerical Simulation (DNS) technique, the Large Eddy Simulation (LES) has to be employed.

At present time, the overwhelming majority of the LES studies has dealt with strongly homogeneous flows, and the determination of a numerical method and a subgrid scale model both well suited for the computation of complex inhomogeneous flows still remains an open question.

A first step consists in studying geometrically simple flows involving complex physical mechanisms. The flow over a backward facing step belongs to that category, and is a good example of academic flow involving separation.

2 Governing Equations

The filtered Navier-Stokes equations are used for the LES of an incompressible fluid. Considering the conservative form for the convection term, equations for the grid filter level quantities $\overline{u}$ and Π can be written under the following form:

$$\frac{\partial \overline{u}}{\partial t} + \nabla \cdot (\overline{u} \otimes \overline{u}) = -\nabla \Pi + \frac{1}{Re}\nabla^2 \overline{u} - \nabla \cdot m \qquad (2.1)$$
$$\nabla \cdot \overline{u} = 0$$

where Π is the modified pressure defined by :

$$\Pi = \overline{p} + \frac{1}{3}\tau_{kk}I, \qquad \tau = \overline{\overline{u} \otimes \overline{u}} - \overline{u} \otimes \overline{u} \qquad (2.2)$$

and m is a model for the anisotropic part of the subgrid scale stress tensor τ.

By taking the divergence of equation (2.1), the Modified Pressure Poisson Equation (MPPE) is obtained:

$$-\nabla^2 \Pi = \frac{\partial(\nabla \cdot \overline{u})}{\partial t} + \nabla \cdot \nabla \cdot (\overline{u} \otimes \overline{u}) + \nabla \cdot \nabla \cdot m \; . \qquad (2.3)$$

3 Subgrid Scale Models

In order to check the influence of SGS model on results obtained, Smagorinsky model and the proposed model have been considered in the present study.

3.1 The Smagorinsky Model

The subgrid scale tensor is achieved in standard way by the Smagorinsky model:

$$m_{ij} = -2\nu_{SM}\overline{S}_{ij}, \qquad \overline{S}_{ij} = \frac{1}{2}\left(\frac{\partial \overline{u}_i}{\partial x_j} + \frac{\partial \overline{u}_j}{\partial x_i}\right) \qquad (3.4)$$

$$\nu_{SM} = c_s \overline{\Delta}^2 \left|\overline{S}\right|, \qquad \left|\overline{S}\right| = \sqrt{2\overline{S}_{ij}\overline{S}_{ij}} \qquad (3.5)$$

where $\overline{\Delta}$ represents the characteristic lengthscale of the grid filter and is evaluated at each mesh node by the quantity $(\Delta x \Delta y \Delta z)^{\frac{1}{3}}$.

The main well-known drawbacks of this model come from the fact that it depends exclusively on large scales and its eddy viscosity does not vanish towards a wall and for laminar flow.

3.2 A new Subgrid Scale Model: the Mixed Scale Model

In order to alleviate the difficulties mentioned above, we proposed the Mixed Scale Model (MSM) [8] based on both large and small scales. Furthermore this model will vanish when the flow is fully resolved on the computational mesh and at the solid wall. The subgrid-scale viscosity of this model is defined as follows:

$$\nu_{SM}(\alpha) = C_M \left(|\mathcal{F}(\overline{u})| \right)^\alpha q^{\left(\frac{1-\alpha}{2}\right)} \overline{\Delta}^{(1+\alpha)} \tag{3.6}$$

with

$$\mathcal{F}(\overline{u}) = \overline{S} \text{ or } \nabla \times \overline{u}$$

where $0 \leq \alpha \leq 1$ and where q is the fluctuating kinetic energy evaluated by the following formula:

$$q = \frac{1}{2} \overline{(u_i')} \, \overline{(u_i')} \, . \tag{3.7}$$

The filtered characteristic subgrid-scale velocity $\overline{(u_i')}$ is calculated using Bardina's *similarity hypothesis* [1]:

$$\overline{(u_i')} \approx (\overline{u}_i)' . \tag{3.8}$$

The fluctuating resolved scale $(\overline{u}_i)'$ is extracted from the resolved velocity field by employing a *test filter* with a characteristic cutoff lengthscale equal to $2 \times \overline{\Delta}$:

$$(\overline{u}_i)' = \overline{u}_i - \tilde{\overline{u}}_i \, . \tag{3.9}$$

So the quantity q can be understood as the energy at the cut-off wave number of the grid filter or the subgrid-scale energy. The MSM can be interpreted as a generalization of the spectral Chollet-Lesieur SGS model [2] in the case of the grid filter energy interpretation, or of the Turbulent Kinetic Energy model presented in Bardina *et al.* [1] in the other case. The small-scale dependency of the model ensure that it will adapt itself to the local state of the flow, and then will disappear in fully resolved regions of the flow and near solid boundaries.

The following simulation was performed using $\alpha = 1/2$.

4 Numerical procedure and convection treatment

The ONERA PEGASE Navier-Stokes solver for DNS and LES of incompressible flows [9] has been extended to 3-D configurations [8]. The basic form of the code solves the Navier-Stokes equations on a non-uniform cartesian grid with a second order accurate hybrid Finite-Difference/Finite Element method. The Finite Element discretization is achieved using Q1 elements.

Time integration is performed with a first order Euler scheme. Incompressibility is enforced by employing a first-order time accurate projection method. Details of the method and validation benchmarks are available in [7].

For LES the convection terms are splitted into skew-symmetric form:

$$\frac{1}{2}(\nabla \cdot (\overline{u} \otimes \overline{u}) + \overline{u} \cdot \nabla \overline{u}). \qquad (4.10)$$

The resulting two parts of the non-linear term are treated in different ways: the conservative part is discretized with the centered second order-accurate scheme issued from the Finite Element approach, while the convective part with a third-order upwind scheme coming from a Finite Difference discretization. The centered part is discretized in a way similar to the Group Finite Element Method proposed by Fletcher [4].

The selected third-order upwind scheme is close to the one proposed by Kawamura and Kuwahara [5]:

$$\left(f\frac{\partial g}{\partial \xi}\right)_i + O(\Delta \xi^3) = f_i \frac{1}{12\Delta \xi}(-g_{i+2} + 8(g_{i+1} - g_{i-1}) + g_{i-2,j,k})$$
$$+ |f_{i,j,k}|\frac{\beta}{12\Delta \xi}(g_{i+2,j,k} - 4g_{i+1,j,k} + 6g_{i,j,k} - 4g_{i-1,j,k} + g_{i-2,j,k}).$$

The parameter β was taken equal to 1. It must be noted that the chosen value of the upwinding parameter β does ensure that the artificial dissipation doesn't mask the subgrid-scale model action [9].

The higher order non-linear terms arising from subgrid-scale modeling were discretized under conservative form with centered coupling second-order accurate 27-point stencil schemes. The modified pressure is calculated from the MPPE by a Jacobi-preconditioned Bi-CGSTAB solver.

5 Results

The calculations are performed on a $201 \times 31 \times 51$ non-uniform cartesian grid with mesh refinement near solid boundaries and a time step of $\Delta t = 0.005$ was used. The computational box is taken equal to $20H \times 4H \times 2.5H$, where H is the step height. The expansion ratio of the channel is $5/3$.

The case studied corresponds to Eaton and Johnston experiments [3]. The Reynolds number is taken equal to 11200.

Calculations were performed using both Smagorinsky model and the MSM. The two constants of the SGS models were set to 0.22 and 0.1, respectively for the Smagorinsky model and the Mixed Scale model. These values correspond to same rate of dissipation of kinetic energy in the homogeneous isotropic case. Computed time averaged recirculation zone length is 7.22 H with Mixed Scale model and 9.95 H with Smagorinky model. The experimental value is 6.97 H, leading to the conclusion that the new SGS model properly alleviate the Smagorinsky model over-dissipative character, and is able to capture the right growth shear rate of the Kelvin-Helmholtz instability in the separated shear layer. Mean flow profiles and longitudinal turbulence intensity are given on figures 1 and 2.

Good agreement is observed between numerical results and experimental data, particularly concerning the time-averaged velocity. Satisfactory qualitative comparisons is obtained for the longitudinal turbulence intensity, associated with an over-evaluation of the peak level. This can indicate that the three-dimensional momentum tranfer process linked to the 3D instability of the shear layer is under-estimated, probably due to a lack of spatial resolution in the spanwise direction.

Two characteristic temporal frequencies of the flow are evaluated from the comptational data. The first one is the low frequency associated to the flapping of the recirculation bubble. Corresponding computed Strouhal numbers for Smagorinsky model and MSM are respectively 0.043 and 0.025. The MSM result falls in the theoretical range of 0.014-0.028 given in [6]. The second frequency is the high frequency associated with the vortex shedding phenomenon. Computed Strouhal numbers for the Smagorinsky model and MSM are 0.083 and 0.066. The experimental value [3] is 0.074. So the two models appear to give satisfactory results concerning the high frequency, but the MSM seems to better predict the low frequency phenomenon than the Smagorinsky model does.

Spatial distribution of the intermittency factor γ on the lower wall is compared to the experimental results on figure 3. A good agreement is observed, demonstrating that the dynamics of the flow is correctly captured by the simulation.

6 Conclusion

The PEGASE code for DNS and LES of incompressible flows was used to perform the LES of the flow past a backward facing step. The selected case corresponds to the experimental configuration of Eaton and Johnston [3]. A new subgrid scale model, called the Mixed Scale Model (MSM), based on both the low and high frequencies parts of the resolved field, is proposed.

The results obtained do prove that the MSM greatly improve those obtained with the Smagorinsky model, especially concerning the time-averaged zone length.

That seems to indicate that the MSM is well suited for the simulation of transitional flows, with no excessive damping like in the Smagorinsky model case.

Good results were obtained for the mean flow and various temporal quantities (Strouhal numbers, intermitency factor), but discrepancies between computed results and experimental data are observed concerning turbulence intensity. These differences may be due to a lack of spatial resolution in the spanwise direction.

Future work will deal with the analysis of the dependency of the results to various parameters of the simulation, such as the size of the computational domain or the spatial resolution.

Bibliography

[1] Bardina, J., Ferziger, J.H., Reynolds, W.C., Improved turbulence models based on large eddy simulation of homogeneous, incompressible, turbulent flows. *Technical Report TF-19, Stanford University, Dept. of Mechanical Engineering* (1983).

[2] Chollet, J.P., Lesieur, M., Parametrization of small scales of three dimensional isotropic turbulence using spectral closures. *J. Atmos. Sci.*, 38 (1981), 2747-2757.

[3] Eaton, J.K., Johnston, J.P., Turbulent flow reattachment: an experimental study of flow and structure behind a backward-facing step. *Report MD-39, Stanford University, Thermosciences Division, Dept. of Mechanical Engineering* (1980).

[4] Fletcher, C.J.A., Computational techniques for fluid dynamics, Springer-Verlag, 1988.

[5] Kawamura, T., Kuwahara, K., Computation of high Reynolds number flow around a circular cylinder with surface roughness, *AIAA Paper, 84-0340*, 1984.

[6] Kiya, M., Sasaki, K., Structure of large-scale vortices and unsteady reverse flow in the reattaching zone of a turbulent separation bubble. *J. Fluid Mech.* 184 (1985), 463-491.

[7] Lê T.H., Troff B., Sagaut P., Dang-Tran K., Ta Phuoc Loc, PEGASE: a Navier-Stokes solver for direct numerical simulation of incompressible flows. *submitted to Int. J. Numer. Methods Fluids*.

[8] Sagaut, P., Simulations numériques d'écoulements décollés avec des modèles de sous-maille, *Thèse de Doctorat de l'Université Paris 6, 1995*.

[9] Sagaut P., Troff B., Lê T.H., Ta Phuoc Loc, Two-dimensional Simulations with Subgrid Scale Models for Separated Flow. *Direct and Large-Eddy simulation I*, Voke, Kleiser and Chollet editors, Kluwer Academics, 1994.

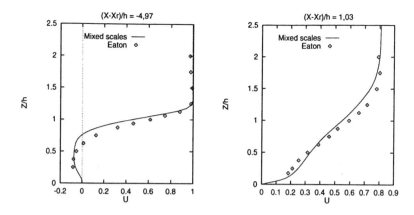

Figure 1 Time-averaged velocity profile

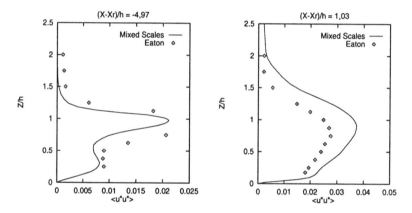

Figure 2 Longitudinal turbulence intensity

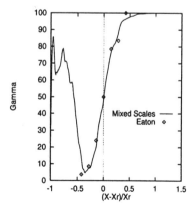

Figure 3 Intermittency factor on the lower wall

EULER AND NAVIER-STOKES FLOW ANALYSIS IN A FRANCIS RUNNER

M. Sallaberger
A. Sebestyen
Sulzer Hydro AG, Computaional Fluid Dynamics
Hardstrasse 319, CH-8023 Zürich, Switzerland

SUMMARY

Sulzer Hydro is intensively using both viscous and inviscid flow calculations for the design of hydraulic turbine components. Therefore, detailed knowledge on the limitations of the methods is becoming increasingly important. In this study, the flow in the runner of a Francis turbine is investigated by both a viscous and an inviscid three-dimensional method. The equations of viscous and inviscid, three dimensional flow are solved by a Finite Volume method. A detailed comparison of the flow fields at various points of operation is given. The results of the calculations indicate that the Euler approach gives a satisfactory prediction of the flow field even at off-design points. The Navier Stokes method enables the engineer to study complex three-dimensional flow phenomena in detail and is also able to predict losses in the flow field qualitatively.

INTRODUCTION

Growing capabilities and recent advances have enlarged the application of Computational Fluid Dynamics (CFD) in the field of hydraulic machinery. Inviscid and viscous flow simulations in parts of a turbine or even the entire machine are increasingly used in order to study the behavior of the flow. Hence, Computational fluid dynamics has become a powerful design tool for engineers. Based on the results of CFD, the cavitation behaviour is improved concurrently with an increase in efficiency and performance. Detailed knowledge of the power, the weakness and the limitations of the methods used is therefore of great importance.

At the best efficiency point, frictional and turbulence effects are small for the flow in water turbines. Due to the accelerated flow, boundary layers are thin along the channel, so there is practically no danger of severe flow problems. Flow separation at off-design is vorticity driven rather than an effect of small Reynolds numbers and it takes place generally as a consequence of flow incidence at the cascade inlet. These physical conditions render Euler-calculations of hydraulic turbines useful even at off-design points.

On the other hand, the demands of accuracy and reliability of flow predictions are increasing rapidly. More and more detailed knowledge of flow phenomena is needed and also the practical geometry limits, i.e. where flow separation may occur, is an important aspect from the designer's point of view. Flow separations are associated with diffusive phenomena, which result in dissipation of kinetic energy. Due to the zero-slip condition on solid surfaces, kinetic energy will be reduced towards solid boundaries and finally completely destroyed at the hub, shroud and blade surfaces. Wall boundary layers interact with each other and have effects on the core flow that typically gives rise to secondary flow.

The Euler code was developed at the Swiss Federal Institute of Technology (EPFL) in Lausanne in a cooperation between EPFL and Swiss industrial partners. The Euler code has increasingly been used in the industrial design work for the past decade. Today Sulzer Hydro is successfully applying the inviscid code as one of its tools in the daily design-routine for a large variety of hydraulic turbomachinery. Comparisons between CFD-designs and measurements have proven the power of the inviscid CFD-technology. The accuracy and reliability of CFD-based hydraulic designs are widely accepted from both manufactures and customers.

For the solution of the viscous flow, Sulzer Hydro uses the Tascflow3D Navier-Stokes code (ASC/Canada). Sulzer Hydro has used the viscous code intensively for some years and has not only enlarged its computational capabilities but also enriched its CFD-experiences in the computational design of various (rotating and stationary) parts of water turbines. From that point of view, the comparison of inviscid and viscid flow-computations is of great importance and furthermore the purpose of this paper.

A high-head Francis turbine with very low specific speed, designed more than ten years ago, is investigated. The steady flow field in the runner is calculated at four points of operation. The operating points are normalized with the parameters of the best efficiency point and defined in terms of a flow coefficient $\varphi^* = \varphi/\varphi_{opt}$ and a pressure coefficient $\psi^* = \psi/\psi_{opt}$. Three of the points are at the optimum energy-height of the turbine $\psi^* = 1.0$ at a flow coefficient of $\varphi^* = 0.9$, $\varphi^* = 1.0$ and $\varphi^* = 1.05$. The fourth point is set at the maximum of turbine energy-height at $\psi^* = 1.1$ with a value of the flow coefficient of $\varphi^* = 1.0$.

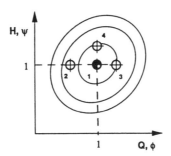

Fig. 1 Investigated points of operation

The grids are typical for what is used in the everyday design work. However, each code has its requirements and the grids are therefore not identical. For Euler, a mesh size of 47x21x13 was used, whereas for Navier Stokes a 99x29x21 mesh was applied.

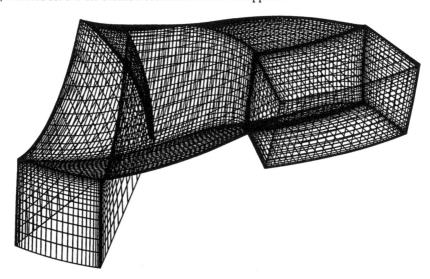

Fig 2 Computational mesh for Navier Stokes calculation

NUMERICAL METHOD

The physical model is represented by the equations describing the conservation of mass and momentum for incompressible flow. Although the steady state flow is of interest, the solution is based on the time-dependent equations of continuity (1) and motion (2).

$$\frac{\partial \rho}{\partial t} + \frac{\partial}{\partial x_j}(\rho u_j) = 0 \qquad (1)$$

$$\frac{\partial}{\partial t}(\rho u_i) + \frac{\partial}{\partial x_j}(\rho u_i u_j) = -\frac{\partial p}{\partial x_i} + \frac{\partial}{\partial x_j}(\tau_{ij} - \overline{\rho u'_i u'_j}) \qquad (2)$$

$$\tau_{ij} = \mu \left(\frac{\partial u_i}{\partial x_j} + \frac{\partial u_j}{\partial x_i} \right). \qquad (3)$$

The numerical method used in both the Euler and the Navier-Stokes code is of the Finite Volume type. The equations are solved in a cartesian coordinate system that rotates with the runner, i.e. in the relative frame of reference.

In the Euler method, the viscous terms in the system of equations above are omitted. For solution of the incompressible equations, artificial compressibility is added to the system. The density in the time dependent term of the continuity equation is replaced by an artificial density. With help of an artificial equation of state, the pressure is introduced into the continuty equation and thus a better numerically conditioned set of equations is obtained. In the Euler code, the equations are solved by an explicit scheme, thus the steady flow is calculated by the solution of the unsteady equations and marching on in time. Artificial viscosity is added to the system in order to ensure stability .

For the solution of the viscous flow the unsteady Navier-Stokes equations are simplified to the Reynolds-averaged Navier-Stokes equations (2) by time averaging of the fluctating quantities. The herefrom resulting additional unknown terms, the Reynolds stresses, are expressed by an eddy viscosity. Thus in the transport equations both a molecular viscosity and a turbulent viscosity appear. Hence, the system of four scalar equations governing the flow field contains five unknowns, the pressure p, three velocity components u_x, u_y, u_z and the turbulent viscosity. Therefore, an additional equation is needed in order to determine the eddy viscosity and to close the system. In this program, the standard k-ε model is used to calculate the eddy viscosity and the flow adjacent to solid walls is assumed to be governed by a logarithmic-law condition.

At the inflow boundary, the velocity profile is defined for both methods. At the outflow boundary, the static pressure is fixed when using Euler. While for Navier Stokes at outflow the condition of zero normal stresses $\tau_{ii}=0$ is used and the static pressure level is prescribed. In the region up- and downstream of the blade, periodic boundary conditions are used in circumferential direction.

RESULTS

The results of the numerical calculations of both methods are compared for the design-point and also for off-design points. The pressure distribution and velocity patterns are shown as well as integral parameters. The results are shown for various values of the flow coefficient $\varphi^* = \varphi/\varphi_{opt}$ and the pressure coefficient $\psi^* = \psi/\psi_{opt}$. Some of the calculated parameters are related to measurements.

PRESSURE DISTRIBUTION

At all examined operating points, there were only minor differences between the pressure distributions at the blade surface calculated with the viscous and the inviscid method. Therefore, only the pressure distribution along the blade profile at shroud calculated at optimum is shown (fig. 3). However, there are small differences at the leading-edge and close to the trailing- edge. The results of the viscous code show pressure peaks at the leading-edge that are significantly lower than the ones of the inviscid code. This may have two reasons. Firstly, the larger number of grid points at the leading edge and therefore a better resolution of the flow reduces the pressure peaks. Secondly, the viscous shear stresses are distributing energy and thus equalizing the flow which again reduces large gradients of velocity and pressure. Those effects may also give rise to the differences in results observed near the trailing-edge.

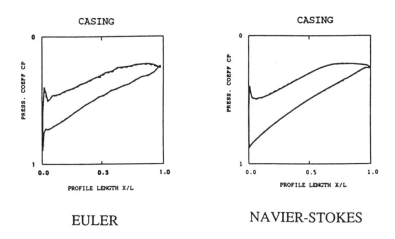

Fig. 3 Pressure distribution close to the shroud

FLOW PATTERNS

A comparison of flow patterns at best-efficiency point is shown for three quasi-stream surfaces (fig. 4) and for three meridional faces (fig. 5). Comparison of the velocity vectors at shroud, mid-span and hub shows qualitatively similar patterns. Going into detail, we find differences at the region near the leading edge. Whereas Euler shows a region of very low velocities and beginning of reverse flow at the pressure side, Navier-Stokes shows a regular flow in this region. The stagnation point calculated by Euler seems to be slightly more on the pressure side of the blade than the stagnation point calculated by Navier-Stokes. This could partly be an effect of the different mesh sizes, but viscous effects will also play a role in this area. These differences are most dominant at the shroud which will be considered in the following.

In the meridional view, the velocity vectors are shown close to the suction side, mid-pitch and close to the pressure side (fig. 5). Again, the flow patterns at mid-pitch and at the suction side are very similar for both Euler and Navier-Stokes. Alternatively, significant differences are shown at the pressure side in the region of the leading edge. The flow patterns in the rear part of the blading look similar. Also in this case, the flow vectors calculated by Euler are pointing outwards in this region, whereas Navier-Stokes shows regular velocitiy vectors at low speeds. In the following considerations, the flow at the pressure side will be evaluated as the largest discrepancies between the two computations occur here.

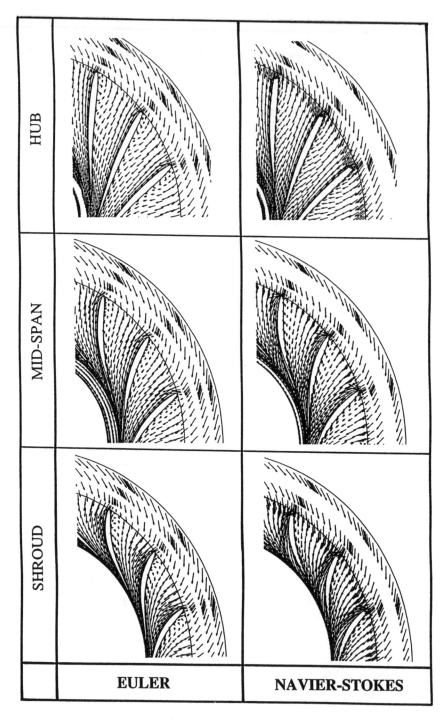

Fig. 4 Velocity vectors on quasi stream surfaces

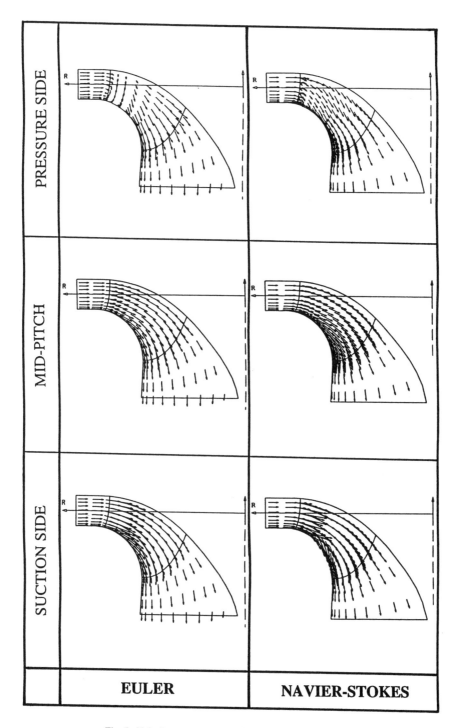

Fig. 5 Velocity vectors on meridional surfaces

The flow at quasi-stream surfaces at shroud, mid-span and hub is shown for different operating points: for the optimal energy-height (fig. 6) and at one operating point at increased turbine head (fig.8). At part load, recirculating flow in the region of the leading edge is shown. Again Euler is showing this effect more distinctively but perhaps somewhat overestimated. There are also differences in the vectors around the leading edge.
At full load, Euler predicts a region with reverse flow at the pressure side, whereas Navier-Stokes only shows low velocities as a result of initiating reverse flow. There are also small differences in the flow vectors at the suction side.
At the operation point of maximum head (fig. 8), significant differences between Euler and Navier-Stokes are demonstrated. Firstly, the region of recirculating flow at the pressure side in the front part of the blading is again overestimated by Euler. Secondly, at the suction side Navier-Stokes shows a three-dimensional separation downstream of the leading edge. This effect is not predicted by Euler. However, it can be discussed whether Navier-Stokes is overestimating this effect or not as the results of measurement did not indicate any particularly negative flow features.

Other interesting features appear at the meridional view at surfaces close to pressure side (fig. 7). Again Euler shows that the flow in the front part of the blade drives more outwards than in the Navier-Stokes computation. At full load, Euler shows a region of backward flow at the shroud near the leading edge, whereas Navier-Stokes shows low but almost regular velocities here. At maximum energy-height (fig. 9), both methods show backwards flow at the leading edge near the shroud.

PERFORMANCE CHARACTERISTICS

At the operating points at optimal turbine head, calculated and measured efficiencies are compared (fig. 10). For completeness, a fourth point at part load was added. Consequently, an Euler efficiency was defined as the difference between the turbine head and the mass-weighted kinetic energy of the tangential velocity at the outlet, normalized with the turbine head. Obviously, there must be significant differences between calculated and measured efficiencies as Euler does not consider any frictional losses and we only obtain the loss of momentum in the Euler-efficiency. The discrepancy between efficiency computed by Navier-Stokes and the measured efficiency is explained by the fact that in the calculation only the runner is considered. The measurements, however, correspond to the entire turbine including spiral casing, stay vanes, wicket gate and draft tube. Comparing the three efficiencies we find good agreement regarding the shape of the curves and the location of the best efficiency point. The quantitative predictions of the efficiency by the Navier-Stokes code leaves room for improvement, as the runner losses are overestimated relative to the losses of the entire turbine.

CONCLUSION

As an enhancement of the Euler calculations, Navier Stokes gives the possibility to detect regions with high losses in the flow field and enables us to predict regions of flow separation more accurately. Yet, also the results of Navier Stokes have to be interpreted carefully and the influence of mesh generation has to be considered. But Navier Stokes gives the possibility to select an optimal geometry by comparing different solutions with respect to viscous losses. Euler, however, shows the inviscid design parameters with good accuracy even at off-design points. For design purposes of turbine runners with an accelerating flow field Euler will continue to be the main design tool.
By using the capabilities of both Euler and Navier Stokes, design engineers are able to study three-dimensional flow phenomena in their full complexity and can finetune the technical solutions. Considering also three-dimensional viscous effects enables the engineer to minimize hydraulic losses and to predict the performance of a hydraulic turbine more reliably.

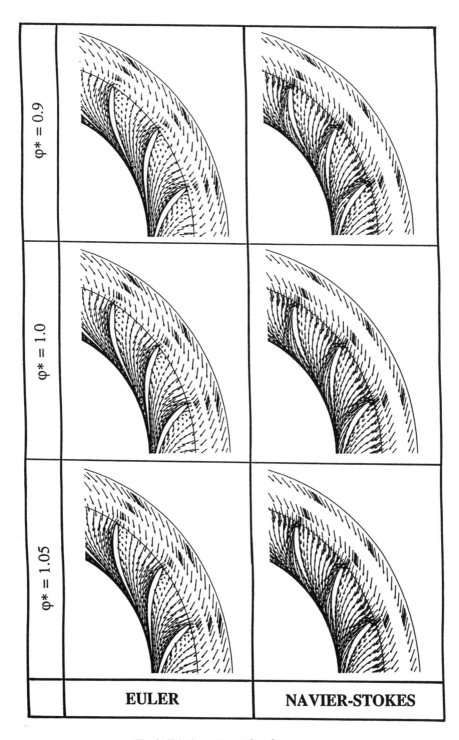

Fig. 6 Velocity vectors at shroud

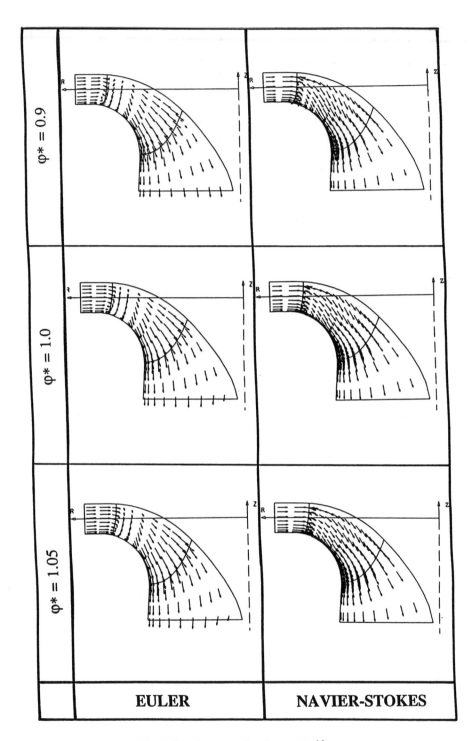

Fig. 7 Velocity vectors close to pressure side

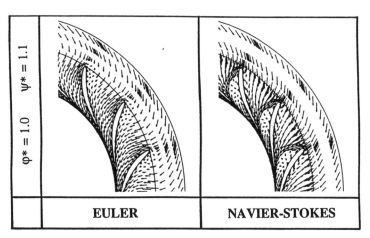

Fig. 8   Velocity vectors at shroud

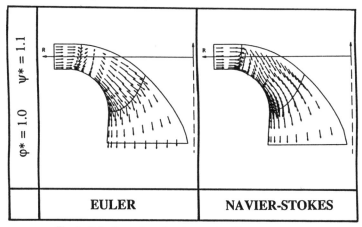

Fig. 9   Velocity vectors close to pressure side

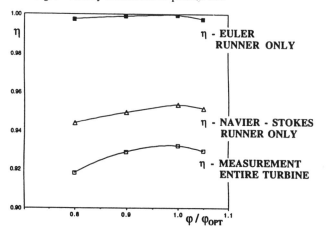

Fig. 10   Performance characteristics

NOMENCLATURE

t	time	u_x ⎤ cartesian		τ	viscous stress
p	pressure	u_y ⎬ velocity		ν	viscosity
ρ	density	u_z ⎦ components		η	efficiency
φ	flow coefficient	ψ	pressure coefficient		

REFERENCES

[1] ERIKSSON, L.E., RIZZI,A., „Numerical solutions of the steady incompressible Euler equations applied to water turbines ", 1984 AIAA Seattle, Washington.

[2] ERIKSSON, L.E., RIZZI,A., „Computation of inviscid incompressible flow with rotation ", 1985 Journal of Fluid Mech., Vol. 153.

[3] GOEDE, E., RHYMING, I.L.: „ 3-D computation of the flow in a Francis runner" Sulzer Technical Review 69 (1987), pp. 31-35, Sulzer Brothers Ltd., Winterthur.

[4] GOEDE, E., Cuenod, R., BACHMANN, P., „Theoretical and experimental investigation of the flow field around a Francis runner ", IAHR Symposium 1988, Trondheim.

[5] GOEDE, E., CUENOD, R., KECK, H., PESTALOZZI, J., „ 3-dimensional flow simulation in a Pump-turbine ", ASME Winter Annual Meeting 1989, San Francisco.

[6] GOEDE, E., SEBESTYEN, A., SCHACHENMANN, A., „ Navier-Stokes Flow Analysis for a Pump Impeller ", IAHR-Symposium 1992, Sao Paolo.

[7] KECK, H., GOEDE, E., PESTALOZZI, J., „Experience with 3-D Euler flow analysis as a practical design tool „ , IAHR Symposiun 1990 Belgrade.

[8] LAUNDER, B.E., SPALDING, D.B., „ The numerical computation of turbulent flows", Comp. Meth. Appl. Mech. Eng., vol 3, pp. 269-289, 1974.

[9] RODI, W., „ Turbulence models and their application in Hydraulics - A state of the art review ", 1984 IAHR stae of the art paper, Delft.

[10] VAVRA, M., „ Aerothermodynamics and flow in turbomachinery ", Wiley, 1960.

A SUBDOMAIN METHOD IN THREE-DIMENSIONAL NATURAL AND MIXED CONVECTION IN INTERNAL FLOWS

J. Salom, J. Cadafalch, A. Oliva, M. Costa

Laboratori de Termotècnia i Energètica
Dept. Màquines i Motors Tèrmics. Universitat Politècnica de Catalunya
Colom 11, 08222 Terrassa, Barcelona (Spain)

SUMMARY

In this work, a computational procedure to solve the incompressible three-dimensional equations which govern the heat transfer processes in regions divided into an arbitrary number of overlapping subdomains is developed. A pressure-based method to solve each subdomain is used. This study focuses on the analysis of two methods of transferring the information at the subdomain boundaries: a direct interpolation of the dependent variables, which is a non-conservative scheme, and a conservative interpolation of flux of mass at the subdomains boundaries. Several features have been analyzed in both natural and forced convection. An application of this method in a specific problem is used to illustrate the capability of this procedure.

INTRODUCTION

Thermal and fluid-dynamic phenomena involving natural and mixed convection are present in many technological applications such as: ventilation in buildings and enclosures, active and passive solar systems, electronic circuitry, etc. Typically, the numerical methods use a single grid to solve the features of the flow. This is sometimes very difficult or even impracticable when structured grids are used. In the last years and as consequence of this difficulty, some procedures have been developed which use the idea of decomposition of the domain. This approach consists in the division of the domain into a number of simple *blocks* or *subdomains*. The grids in each block can be created independently with an overlapping zone. This kind of method is called either *composite overlapping grids* or *overlaid grids*.

A number of works on the subject of domain decomposition have appeared in the recent literature [1][2][3]. The subdomain methods represent an attempt to significantly reduce the storage requirements of the full-domain direct methods [3]. Another advantage is that the subdomains resulting from decomposition provide natural building blocks for parallel computing schemes. With the advent of parallel processors in modern computers, each subdomain can be allocated to a different processor from which numerical simulation can proceed concurrently. Domain decomposition methods have a good prospect to take advantage of parallel computation capabilities [4].

MATHEMATICAL FORMULATION AND NUMERICAL SOLUTION

The fluid and heat transfer phenomena of a laminar fluid flow are described by the Navier-Stokes equations. The scalar components in a system of cartesian coordinates can be written in the compact form of the convection-diffusion equation.

$$\frac{\partial}{\partial t}(\rho\phi) + \frac{\partial}{\partial x}(\rho u\phi) + \frac{\partial}{\partial y}(\rho v\phi) + \frac{\partial}{\partial z}(\rho w\phi) = \frac{\partial}{\partial x}\left(\Gamma\frac{\partial\phi}{\partial x}\right) + \frac{\partial}{\partial y}\left(\Gamma\frac{\partial\phi}{\partial y}\right) + \frac{\partial}{\partial z}\left(\Gamma\frac{\partial\phi}{\partial z}\right) + S \quad (1)$$

where the dependent variable ϕ, the diffusion coefficient Γ and the source term S are given in table 1. All fluid properties are taken as constant except the influence of the density variations on the buoyancy term of the momentum equations (Boussinesq approximation).

Table 1 Dependent variables, diffusion coefficients and source terms

equation	ϕ	Γ	S
mass	1	0	0
x-momentum	u	μ	$-\frac{\partial p}{\partial x} + g_x \beta (T-T_o)$
y-momentum	v	μ	$-\frac{\partial p}{\partial y} + g_y \beta (T-T_o)$
z-momentum	w	μ	$-\frac{\partial p}{\partial z} + g_z \beta (T-T_o)$
energy	T	$\frac{k}{c_p}$	0

The governing equations of the flow (mass, momentum and energy) are solved at each subdomain in a segregated form using the SIMPLEC method proposed by Van Doormal and Raithby [5]. These equations are discretized using the Power-Law scheme proposed by Patankar [6]. The set of algebraic equations is solved using the SOR method.

SUBDOMAIN NUMERICAL PROCEDURE

The aim of this study is to develop a computational procedure to solve the incompressible Navier-Stokes equations written in three-dimensional form in domains divided into an arbitrary number of overlapping subdomains.

In the method presented in this work, the solution domain is first divided up into a number of overlapping regions. Every subdomain must have a parallepipedic shape and must be in correspondence between the cartesian axis. At each subdomain, an independent structured and staggered three-dimensional grid is used.. The generation system of the set of grids is supported by a Computer Assisted Design (CAD) program that allows to define complicated and irregular regions. It is possible to concentrate the grids, symmetrically or partially over its right or left side, using several stretching functions of different concentration factors.

The first difficulty is the task of determining the information flow from block to block. The location of where internal boundary information is to be obtained is not trivial, especially in cases in which an arbitrary number of overlapping subdomain grids is defined. In these cases, various grids may be available to supply information to some parts of an internal boundary; while for other parts, information may only be obtained from one neighboring block. An important feature in our subdomain algorithm is the order of the subdomain grids, so that higher-numbered grids cover up parts of lower-numbered grids. The points that lie underneath a higher-numbered subdomain grid and are unnecessary for the interpolation system are removed and do not intervene in the calculation. For this reason it is important in which order the subdomains are defined. For example, in Fig. 1.a, where two subdomain grids are shown, subdomain 2 cover up part of subdomain 1. In Fig. 1.b., the interpolated values of velocities, which are boundary conditions for subdomain 1, are indicated. The Control Volumes (CV) in the shaded zone are removed. In Fig. 1.c the interpolated values are shown. It is interesting to stand out that the physical boundaries coincide always with the calculation boundary, in the highest-numbered block.

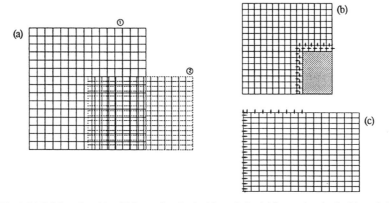

Fig. 1 (a) Subdomain grids; (b) Interpolated velocities sub. 1; (c) Interpolated velocities sub. 2

Interpolation procedures

There are various systems of transferring the information to a subdomain interface. This study analyzes two of these systems. Although the software is three-dimensional, in order to explain the interpolation methods a two-dimensional description is made.

Direct interpolation. A linear interpolation of all the independent variables is used in all directions. For instance, the update of the value of the x component of velocity in subdomain 1 (referenced u), Fig. 2.a, involves u_1, u_2, u_3, and u_4 belonging to subdomain 2

291

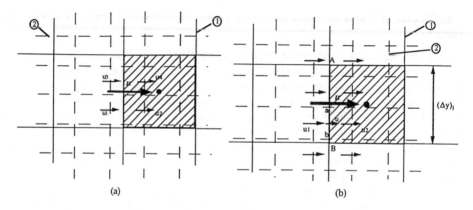

Fig. 2. Vertical overlap region. (a) Direct interpolation (b) Conservative interpolation

Conservative interpolation of mass fluxes. The update of the values of velocity variables in one of the blocks requires the calculation of the fluxes of mass through the boundaries of the block. Fig. 2.b shows an overlap region where the block drawn with solid lines is the one where variables are to be updated, and drawn with dotted lines is the neighbour block. First, we obtain the normal velocity component u on the sub-segment ab using a linear interpolation from velocity components u_1 and u_2, which are just left and right, located within block two. Next, the mass flux contribution is obtained by multiplying the values of normal velocity by the density and the sub-segment length. Note that all the velocities drawn in Fig. 2.b are involved in the calculation of mass flux. The velocity component at the CV boundary for block one, denoted U in Fig. 2.b, is given by: $u = \dfrac{\text{mflux}}{\rho \, (\Delta y)_1}$.

NUMERICAL RESULTS

The validation of the three-dimensional set of governing equations is a difficult task due to the lack of enough numerical three-dimensional well-established results to which compare our code. The situation which has received more attention in three-dimensional studies is the case of natural convection developed in an air-filled cubical enclosure, which is heated differentially at two vertical side walls. For this situation, our numerical results are in good agreement with those given by other authors [7].

Several features of the use of subdomains with the two interpolation systems described before are analyzed. The forced and natural convection in a cubical cavity is studied. In both of them air (Pr=0.71) is the working fluid. For natural convection, the case of a heated differentially cavity at two vertical side walls is studied, hereafter called *problem A*; the Rayleigh number is 10^4 in all of them. For forced convection, the application of a constant velocity at two vertical side walls of the cubical enclosure, u_o, is studied with different direction at each side, hereafter called *problem B*. The Reynolds number is 100 in this problem.

The first feature analyzed, is the influence of the interpolation system on cases where the density grids are the same in all the subdomains. In all cases there is no difference using the two systems because when the grids coincide the mass is conserved automatically at the subdomain boundaries. However, if we compare the results with the reference case (one subdomain) there are some differences. These differences are essentially due to the fluxes of momentum, which are not conserved with these methods of interpolation. Table 2 outlines the results in problem A for different cases using different numbers of subdomains.

Table 2. Comparison values at the symmetry plane in problem A between reference case and several cases using different number of subdomains. Grid density (40x40x40) in all cases.

	reference	Interpolation		
Subd.	1	2	3	4
U max	0.1972	0,1939	0.1975	0.1939
error	-	1.67%	0.15%	1.67%
Z	0.8375	0.8125	0.8375	0.8125
error	-	2.98%	0	2.98%
W max	0.2207	0.2204	0.2214	0.2204
error	-	0.13%	0.31%	0.13%
X	0.1125	0.1125	0.1125	0.1125
error	-	0	0	0
Nu mean	2.266	2.365	2.366	2.364
error	-	4.18%	4.41%	4.14%

Another important feature to be analyzed is the situation where the density grids are not maintained. The possibility of working with different density meshes is of most interest for the subdomain applications. In Fig. 3, W velocities at the symmetry plane are presented using different systems of interpolation and different density grids between subdomains. In this case we observe that conservative interpolation of mass flux is much better than direct interpolation. The results with direct interpolation show an unphysical behaviour.

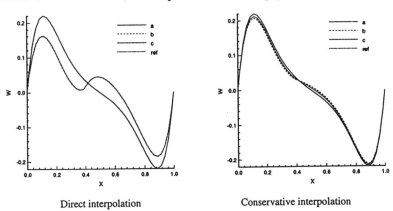

Direct interpolation   Conservative interpolation

Fig. 3 W velocities at symmetry plane in problem A using 2 subdomains. (a) Grids 24x20x40 and 24x20x40 (b) Grids 24x20x40 and 24x20x60   (c) Grids 24x20x40 and 24x20x80   (ref) Grid 40x20x40

In Fig. 4, the influence of interpolation boundary location is analyzed in problem B. This location has an important influence in the cases where mass is not conserved. If the scheme is conservative, the results have a little variation.

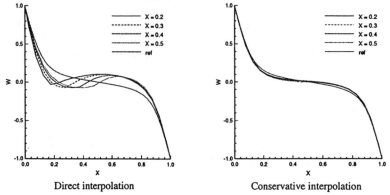

Fig. 4 W velocities at symmetry plane in problem B using 2 subdomains. X indicates the location of interpolation boundary. Grids 12x20x20 and 12x20x30. Reference grid: 20x20x20

The variations between the results applying mass conservation and the results of the reference case disappear when a conservative interpolation scheme of mass and momentum fluxes is implemented. This aspect is confirmed in Fig. 5 where a comparison in a two-dimensional simulation in problem A, between results using the three kinds of interpolation (direct, mass conservation and mass and momentum conservation), is showed.

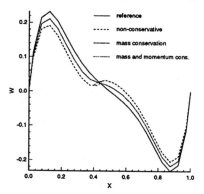

Fig. 5 W velocities in a two-dimensional simulation in problem A using 2 subdomains with different interpolation schemes. Grids 12x20x20 and 12x20x30. Reference grid: 20x20x20

Finally, an illustrative application of the subdomain method is presented. We study a three-dimensional simulation of natural convection in two reservoirs connected by a horizontal duct, Fig. 6.a, where the working fluid is air and the Rayleigh number of the problem is 10^4. All boundaries are insulated except the two vertical side walls of the reservoirs, which are isothermal. When x=0, T is T_h and when x=3L, T is T_c. Three subdomains have been employed: two for each reservoir (30x15x36 grid) and one for the duct (40x8x6). In Fig. 6.b and 6.c we present the velocity vectors and isotherms at the symmetry plane respectively and, in the Fig. 6.d, the isotherms at plane Z=0.5.

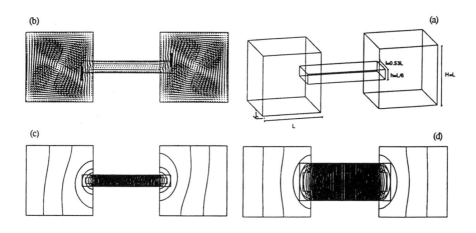

Fig. 6 (a) Physical model analyzed (b) Velocity vectors at symmetry plane (Y=0.5)
(c) Isotherms at symmetry plane (Y=0.5) (d) Isotherms at plane Z=0.5

NOMENCLATURE

c_p	specific heat	g	gravitational acceleration	k	thermal conductivity
L	reference length	mflux	flux of mass	Nu	Nusselt number
p	pressure	Pr	Prandtl number	T	temperature
$Re = \rho u_o L/\mu$	Reynolds number	u,v,w	velocities	T_o	reference temperature
$Ra = g\beta\rho^2 \Delta T L^3 c_p/\mu k$	Rayleigh number	U,V,W	dimensionless velocities	x,y,z	general coordinates
ΔT	temperature difference (T_h-T_c)	u_o	reference velocity	X,Y,Z	dimensionless coordinates
β	thermal expansion coefficient	μ	dynamic viscosity	ρ	density

REFERENCES

[1] G.Chesshire, W.D.Henshaw. "Composite Overlapping Meshes for the Solution of Partial Differential Equations", Journal of Computational Physics, Vol. 90, No. 1, pp. 1-64, 1990.
[2] J.A. Wright and W. Shyy, "A Pressure-Based Composite Grid Method for the Navier-Stokes Equations", Journal of Computational Physics, vol. 107, pp. 225-238, 1993.
[3] N.R.Reyes, A.Oliva, M.Costa, C.D.Pérez-Segarra, "Subdomain Method in both Natural and Forced Convection: A Special Attention to the Treatment of Irregular Geometries", Numerical Methods in Laminar and Turbulent Flow, Vol. VIII, Part. 1, pp. 424-435, Swansea, 1993.
[4] M.Wang, J.G. Georgiadis, "Parallel Computation of Forced Convection using Domain Decomposition", Numerical Heat Transfer, Part B, vol. 20, pp. 41-59, 1991.
[5] J.P. Van Doormal and G.D. Raithby, "Enhancements of the Simple Method for Predicting Incompressible Fluid Flows", Numerical Heat Transfer, vol. 7, pp. 147-163, 1984.
[6] S.V.Patankar, "Recent Developments in Computational Heat Transfer", Journal of Heat Transfer, Vol. 110, pp. 1037-1045, 1988.
[7] T. Fusegi, J.M. Hyun, K. Kuwahara and B. Farouk, "A Numerical Study of Three-Dimensional Natural Convection in a Differentially Heated Cubical Enclosure", Int. J. Heat Mass Transfer, vol. 34, pp. 1543-57, 1991.

Parallel multi-block computations of three-dimensional incompressible flows

M.L. Sawley, O. Byrde, J.-D. Reymond

Institut de Machines Hydrauliques et de Mécanique des Fluides
Ecole Polytechnique Fédérale de Lausanne
CH-1015 Lausanne, Switzerland

D. Cobut

Département de Mécanique, Université Catholique de Louvain
B-1348 Louvain-la-Neuve, Belgium

Abstract

Parallel computation has shown to provide considerable potential for the numerical simulation of complex three-dimensional flows. A number of studies have shown that CFD codes can be parallelized for efficient use on present-day parallel computer systems. However, of particular importance for industrial applications is the total time to solution, comprised not only of the resolution of the flow equations but also the pre- and post-processing phases. Results are presented of a study of the use of high-performance parallel computing to facilitate such numerical simulations. This study is being undertaken using a 256-processor Cray T3D system, within the framework of the joint Cray Research–EPFL Parallel Application Technology Program.

1 FLOW SOLVER

The parallel code used in this study is based on a multi-block code developed within IMHEF-EPFL for the numerical simulation of unsteady, turbulent, incompressible flows. This code solves the Reynolds-averaged Navier-Stokes equations on 3D structured and block-structured computational meshes.

1.1 Numerical Scheme

The numerical method employs a cell-centered finite volume discretization with an artificial compressibility method to couple the pressure and velocity fields. A high-order upwind spatial discretization scheme based on the approximate

Riemann solver of Roe is employed for the advection terms, while the diffusion terms are discretized using a central approximation. The time integration of the unsteady Navier-Stokes equations is performed using an implicit two-stage Runge-Kutta scheme. At each time step, a non-linear system resembling the equations for stationary flow is solved using an ADI method.

1.2 Parallelization

To date, only a subset of the original code has been parallelized; the parallel code can presently be used to compute steady, laminar, 3D incompressible flows. Two-dimensional flows are computed by replication in the degenerate third dimension. Parallelism is achieved by dividing the computational domain into a number of blocks (sub-domains), with the flow equations being solved in all blocks in parallel by assigning one block to each processor. Communication between processors is necessary to exchange data at the edges of neighbouring blocks. The communication overhead is minimized by data localization using two layers of ghost cells surrounding each block. A preliminary study [3] of a number of different parallel programming models (data parallel, message passing, implicit shared memory, data passing) has shown that message passing provides a good compromise between performance and portability. The PVM message passing library is employed in the parallel code to undertake the necessary communication between processors.

2 PERFORMANCE ANALYSIS

Studies have been undertaken to investigate the performance of the parallel code on both the Cray T3D system and a cluster of Hewlett-Packard 9000/720 workstations interconnected via Ethernet.

The present code uses an implicit numerical scheme to solve the flow equations within each block, combined with an explicit updating of the boundary values. It is therefore relevant to question if, due to convergence degradation, such a scheme is scalable to the large number of processors available on the Cray T3D.

As a test case, inviscid flow between the blades of the Durham low speed turbine cascade [2] has been considered. Computations have been performed for 3D flows using a mesh containing 120x52x64 cells. Figure 1 presents the convergence history as a function of the number of blocks employed. These results show that the convergence degradation for the present test case is not significant, resulting in a nearly linear speedup with increasing number of processors. The observation that the slowest convergence is obtained using 16 blocks is presumably due to the fact that the convergence rate is determined not only by the number of blocks, but also by the transient solutions within each block.

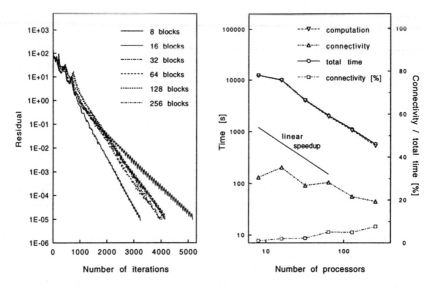

Figure 1. Convergence history and time to solution on the Cray T3D.

The time to solution (i.e. the wall clock time required to obtain a residual of 10^{-5}) and the relative importance of the connectivity overhead (i.e. the time spent in communication and synchronization between blocks relative to the total time) are also presented in Fig. 1. These results show that due to the very fast communication network of the Cray T3D, approximately linear speedup is obtained even when a large number of processors is employed. Such a conclusion can not always be drawn: 2D computations using a cluster of up to 24 workstations [1] have shown that the poor performance of the Ethernet interconnection network can result in a greater time to solution as the number of processors is increased!

3 PRE- AND POST-PROCESSING

It is well recognized that for the numerical simulation of complex 3D industrial flows involving an enormous quantity of data, the pre- and post- processing phases of the simulation procedure can necessitate a time (for data file manipulation and processing, I/O, etc.) often substantially longer than the cpu time required by the flow solver. This potential problem is further exacerbated if the flow solution is obtained using an efficient solver on a high-performance parallel system. For this reason, it is essential to incorporate the pre- and post-processing phases into the parallel environment.

3.1 Parallel mesh generation

The construction of a suitable block structured mesh for complex 3D geometries comprises three stages: surface definition (generally employing CAD techniques), block boundary determination, and mesh generation within each block.

Block structured meshes, while being structured within each block, are generally irregular at the block level. The task of generating the block boundaries is therefore equivalent to determining an appropriate unstructured mesh. While the block boundary mesh is currently determined manually, it is desirable that its construction be performed using automatic unstructured hexahedral mesh techniques; such an approach is presently under investigation. Once the block topology is determined, the structured mesh within each block is generated by transfinite interpolation in an independent – and thus parallel – manner. An elliptic smoothing procedure is then applied to avoid discontinuities in the gradient of mesh lines across the block boundaries. Such smoothing, which allows some movement of the block boundaries, necessitates communication between processors.

The use of parallel mesh generation – even in the above-described limited form – has been shown to provide a substantial reduction in mesh generation time. However, more significantly, since only the block boundary information need be imported into the parallel system a significant reduction in both disk storage and I/O time are also obtained.

3.2 On-line visualization

Due to the enormous quantity of data generated by an unsteady 3D flow solver each time step, it is not practical to store these data on disk for later "off-line" processing and visualization. To overcome this problem, an "on-line" visualization procedure has been developed, using the inter-process communication (IPC) library to interface to the commercial TECPLOT visualization package [4]. The transfer of both data from the parallel system to the graphics workstation and the sending of instructions back to the code running on the parallel system is undertaken using UNIX sockets (rather than message passing). This provides a fast and portable means of communication that is transparent to the user.

4 INDUSTRIAL APPLICATIONS

The parallel flow solver described above has been used to compute a number of different large-scale 3D flows; two specific examples, of interest to the turbomachinery and automotive industries, are briefly presented in the following sections.

4.1 Water turbine

Water turbines of the Francis type are widely used for hydroelectric power generation. The numerical simulation of the flow in a Francis turbine is conventionally undertaken by computing the flow separately in each component of the turbine: spiral casing, distributor, runner and draft tube. The coupling of the flow in these components is complicated both by the rotation of the runner relative to the other components and by the different periodicities of the distributor and the runner.

While it is planned to extend the present study to account for coupling of different components of the turbine, computations to date have concentrated on inviscid flow in the runner. Numerical simulations of the runner presently undertaken by industry generally assume that the flow is periodic; this allows only one inter-blade channel to be computed rather than the whole runner. Using the computational power of the Cray T3D, the flow in the entire runner can be computed in a reasonable time, which provides an initial assessment of the flow coupling between the runner and the distributor.

Figure 2 presents the results of a computation performed on 13 processors of the Cray T3D. For this computation, the periodicity of the flow at the exit of the distributor (24 blades for the present case) has been imposed as an input condition for the runner (13 blades), resulting in non-periodic flow in the runner. The results of this simulation has shown that the initial periodicity is quickly "lost" within the runner, with the flow at the entry of the diffuser section being essentially axisymmetric.

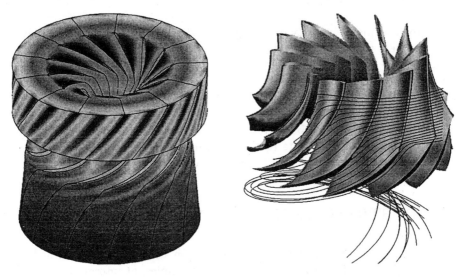

Figure 2. Local speed (left) and surface pressure and streamlines (right) for inviscid flow in the runner of a Francis turbine.

4.2 Formula 1 racing car

Racing cars provide a valuable test bed for automotive design in a number of different areas. While external aerodynamics are traditionally analysed experimentally via the use of windtunnel results, CFD has the potential to make a significant contribution to the reduction of design time and costs. Due to the complex geometry and flowfield, it is currently not possible to simulate the flow around an entire Formula 1 racing car. The aerodynamic properties of the car are however principally determined by certain critical regions, such as the air inlet, undertray, wheels and front and rear wings. While the flow in these regions is coupled, it is nevertheless useful to study each region separately to obtain a detailed understanding of the flow. The present study has concentrated on a rear wing consisting of a series of multi-element airfoils (see Fig. 3). Four different configurations have been computed, corresponding to the three angles of the central element (denoted by c1-c3) plus the removal of this element (denoted by c0).

The primary role of the rear wing is to produce downforce. Although significant drag also results, this is smaller than that produced by, for example, the wheels. Assuming that the flow is attached on the rear surface of each of the wing elements, inviscid computations should provide reasonable estimates of the downforce. It is also noted that even at maximum car speeds (up to 330 km/h), the flow can be considered to be incompressible. The numerical flow simulations performed to date have assumed zero angle of yaw; symmetry along the central vertical plane enables only half of the rear wing to be considered. For 3D flow computations, a mesh comprised of 121 blocks with a total of 3.8 million cells has been employed. Such computations required approximately 4 hours on 121 processors of the Cray T3D. The flow solution obtained for one of the rear wing configurations is presented in Fig. 3. These results clearly show the presence of outboard trailing vortices near the top of the side plates, as is observed at the race track.

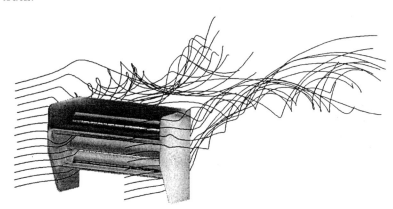

Figure 3. Surface pressure and streamlines for inviscid flow around the rear wing (configuration c2).

Since the flow over a large portion of the wing elements is seen to be essentially two dimensional, computations have also been performed using a 2D mesh with 32 blocks. It should be noted that since there is not the same number of mesh cells in each block, load imbalance occurs, as shown in Fig. 4. This problem can be alleviated by computing more than one block per processor, as has been undertaken on the workstation cluster.

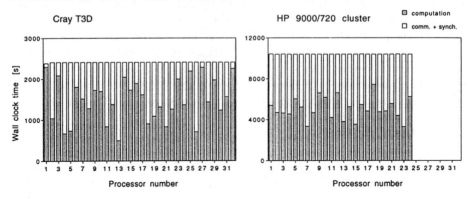

Figure 4. Wall clock time required for 2D flow computations on two different parallel computer systems.

A number of 2D flow simulations have been undertaken to investigate the characteristics of each wing configuration (in the symmetry plane). Computation of the pressure coefficient has allowed the contribution to the total wing downforce due to each element to be determined. The computed lift coefficients (normalized to the chord length of the largest element) are presented in Fig. 5 for each of the configurations considered. Figure 5 shows that a modification of the central element results in a change of the lift coefficient for each of the other elements, indicating that the flow around each component element is strongly coupled. For each configuration, approximately half of the total downforce is provided by the largest element.

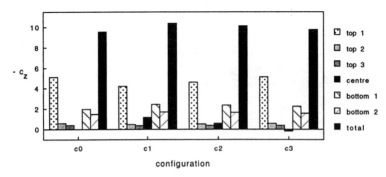

Figure 5. Component and total lift coefficient for the four different rear wing configurations considered.

302

5 CONCLUSION

The present study has shown that an existing 3D multi-block flow solver can be adapted to employ the enhanced capabilities of high-performance parallel computer systems. The use of the PVM message passing library has been seen to provide low communication overhead on the Cray T3D, as well as a high level of portability.

The use of the Cray T3D has enabled flow computations to be performed that are inaccessible to mono-processor computer systems. While workstation clusters can be considered as an alternative to supercomputers for small and medium size applications, the present study has demonstrated that large-scale applications can only be performed on massively parallel systems with the necessary memory requirements and computational power.

The present study has reinforced the fact that to solve certain industrial problems, the computation time required by the flow solver is not the limiting factor, since the pre- and post-processing phases generally require significantly more time. It has been shown that the integration of the pre- and post-processing phases into the parallel environment can produce a significant reduction in the overall time to solution.

6 ACKNOWLEDGEMENTS

The authors wish to thank Y. Marx for the numerous discussions regarding the original code and S.A. Williams for his aid with visualization. The geometry of the rear wing of the Formula 1 racing car was provided by PP Sauber AG. Financial support was provided by the Cray Research–EPFL Parallel Application Technology Program. One of the authors (D.C.) received additional support via the European Project ERASMUS.

REFERENCES,

[1] Byrde, O., Cobut, D., Reymond, J.-D., Sawley, M.L., 1995, Parallel multi-block computation of incompressible flow for industrial applications, Proceedings of Parallel CFD '95 (Pasadena), to appear.

[2] Gregory-Smith, D.G., 1994, Test case 3: Durham low speed turbine cascade, ERCOFTAC seminar and workshop on 3D turbomachinery flow prediction II (Val d'Isère) Part III, pp. 96-109.

[3] Sawley, M.L., Tegnér, J.K., 1995, A comparison of parallel programming models for multi-block flow computations, Journal of Computational Physics, accepted for publication.

[4] Williams, S.A., Sawley, M.L., Cobut, D., 1995, On-line visualization for scientific applications, EPFL Supercomputing Review, 7, 45-50.

On the treatment of fluid problems by methods of Clifford analysis

WOLFGANG SPRÖSSIG
TU - Bergakademie Freiberg
Fakultät für Mathematik und Informatik
Bernard-von-Cotta-Str.2
D-09596 Freiberg
Germany

KLAUS GÜRLEBECK
TU Chemnitz-Zwickau
Fakultät für Mathematik
09107 Chemnitz
Germany

1 Introduction

The study and application of methods of Clifford analysis and especially quaternionic analysis has recently pursued by a growing number of mathematicians and physicists (cf. [8], [1],[2]). Starting with fundamental papers of R.Fueter [3],[4] in the early thirties and his follower W.Nef [10] in 1944 then Rose's work [11] (cf. also [9]) on quaternion velocity potentials for axial-symmetric fluid flow were some of the first who applied analytic methods of (complex) quaternions representing solutions of partial differential equations. In this context should be mentioned the paper of Stein-Weiss [13] on generalized Cauchy-Riemann systems.

We developed Clifford operator methods for the investigation of linear and non-linear three-dimensional boundary value problems (cf. [6],[7]). In a unified Clifford operator calculus can be dicussed existence, uniquness, regularity, representation formulas of the solutions of these boundary value problems, " almost" a-priori estimates and even a suitable numerical analysis. In order to study three-dimensional problems in general domains it is sufficient to work in the special Clifford algebra of real quaternions $\mathbb{H}$. The letter $\mathbb{H}$ is used in honour to the discoverer of the skew-field of quaternions the Irish mathematician Sir W.R. Hamilton who in 1843 considered for the first time quaternions. To simplify matters suppose that our domains should be bounded by a smooth Liapunov surface. Mention that the theory is also working in n-dimensional domains which can also be unbounded.

The aim of this paper is to demonstrate the action of these Clifford operator methods considering a suitable example. Most of the applications in fluid mechanics only require analysis in the algebra of the real quaternions and its complexification. Therefore we will only introduce the quaternionic operator calculus. At first we have to explain some fundamental facts of quaternionic function theory and the corresponding operators.

2 Preliminaries

Let be $e_0 = 1, e_1, e_2, e_3$ the quaternionic units which fulfil the properties $e_i e_j + e_j e_i = -\delta_{ij}, i, j = 1, 2, 3$. Setting $e_1 e_2 = e_3, e_2 e_3 = e_1$ and $e_3 e_1 = e_2$ we finally get the skew-field of real quaternions $\mathbb{H}$. Each element $u \in \mathbb{H}$ can be represented by $u = \sum_{i=1}^{3} u_i e_i$. Let be $\underline{u} = \sum_{i=1}^{3} u_i e_i = Vec\, u$ the vector part and $u_0 = Sc\, u$ the scalar part of u. Furthermore $\overline{u} = u_0 - \underline{u}$ is the conjugated quaternion. Let be $G \subset \mathbb{R}^n$ a bounded domain. Then a $\mathbb{H}$-valued function $u \in C^1(G, \mathbb{H})$ is called $\mathbb{H}$-*(left)regular* in G iff $Du = 0$ in G, where $D = \sum_{i=1}^{3} \partial_i e_i$ is the so-called *Dirac operator*. Note that the space $C^1(G, \mathbb{H})$ is defined by

$$C^1(G, \mathbb{H}) = \left\{ \sum_{i=1}^{3} u_i e_i, u_i \in C^1(G) \right\}. \qquad (1)$$

The function spaces of $\mathbb{H}$-valued functions $L_2(G, \mathbb{H}), W_2^r(G, \mathbb{H}), \ldots$ are in the same way componentwise to understand. W_2^r are spaces of generalized differentiable functions in the sense of Sobolev-Slobotetzkij (see [14]). The space $\overset{\circ}{W}{}_2^1(G, \mathbb{H})$ is roughly speaking the Sobolev space of all $\mathbb{H}$-valued functions which componentwise belong to $W_2^1(G)$ and vanish on the boundary Γ. Furthermore let

$$\|u\|_2^2 := \int_G |u(x)|^2 dx \quad \text{and} \quad \|u\|_{2,1}^2 := \|u\|_2^2 + \sum_{i=1}^{3} \|\partial_i u\|_2^2.$$

It is easy to see that

$$Du = grad\, u_0 + rot\, \underline{u} - div\, \underline{u}.$$

Therefore $\mathbb{H}$-(left)regular functions are in case of $u_0 = 0$ just all sourceless and rotationless fields. Unfortunately is the quaternionic product of two $\mathbb{H}$-(left)regular functions u and v not again $\mathbb{H}$-(left)regular. Consider for instance $u = x_1 e_1 - x_2 e_2 \in ker D$. Using the generalized Leibniz rule

$$D(uv) = (Du)v - \overline{u}Dv + 2Sc(uD)v$$

we get $u^2 \notin ker\, D$. We find also $u^{-1} \notin ker\, D$ for $u \neq 0$.

Let Γ be a sufficiently smooth boundary and $e(x) = (4\pi)^{-1}(-x)|x|^{-3}$ with $x = x_1 e_1 + x_2 e_2 + x_3 e_3$. This function $e(x)$ is $\mathbb{H}$-(left)regular.

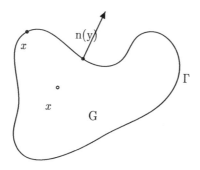

Using the function $e(x)$ we introduce the following integral operators namely the so-called *Teodorescu transform*

$$(T_G u)(x) := \int_G e(x-y)u(y)dy, \quad x \in \mathbb{R}^3,$$

the *Cauchy-type operator*

$$(\mathcal{F}_\Gamma u)(x) := \int_\Gamma e(x-y)n(y)u(y)d\Gamma_y, \quad x \notin \Gamma,$$

and a multidimensional *singular integral operator of Cauchy-type*

$$(S_\Gamma u)(x) := 2\int_\Gamma e(x-y)n(y)u(y)d\Gamma_y, \quad x \in \Gamma,$$

where $n(y) = \sum_{i=1}^3 e_i n_i(y)$ is the unit vector of the outer normal at the point y. The integral which defines the operator S_Γ is to understand in the sense of Cauchy's principal value. The operator $P_\Gamma := \frac{1}{2}(I + S_\Gamma)$ denotes the projection onto the space of all $\mathbb{H}$-valued functions which may be (left)regular extended into the domain G. $Q_\Gamma := \frac{1}{2}(I - S_\Gamma)$ denotes the projection onto the space of all $\mathbb{H}$-valued functions which may be (left)regular extended into the domain $R^3 \setminus \overline{G}$ and vanish at infinity. Immediately from [6] we get the following statements:

Lemma 1: *Let $u \in C^1(G, \mathbb{H}) \cap C(\overline{G}, \mathbb{H})$. Then we have the formulas*

(i) $\quad (\mathcal{F}_\Gamma u)(x) + T_G Du(x) = \begin{cases} u(x), & x \in G \\ 0, & x \in R^3 \setminus \overline{G} \end{cases}$ (Borel-Pompeiu formula)

(ii) $\quad (DT_G u)(x) = \begin{cases} u(x), & x \in G \\ 0, & x \in R^3 \setminus \overline{G} \end{cases}$

(iii) $\quad (D\mathcal{F}_\Gamma u(x) = 0$ in $G \cup (R^3 \setminus \overline{G})$

Lemma 2 *(Plemelj-Sokhotzkij's formulas): Let $u \in C^{0,\alpha}(G, \mathbb{H})$, $0 < \alpha < 1$. Then we have*

(i) $\lim\limits_{\substack{x \to \xi \in \Gamma \\ x \in G}} \mathcal{F}_\Gamma u)(x) = (P_\Gamma u)(\xi)$

(ii) $\lim\limits_{\substack{x \to \xi \in \Gamma \\ x \in R^n \setminus \overline{G}}} \mathcal{F}_\Gamma u)(x) = (-Q_\Gamma u)(\xi)$

for any $\xi \in \Gamma$.

Corollary: *Let $u \in C^{0,\alpha}(\Gamma, \mathbb{H})$. Then the relations*

(i) $(S_\Gamma^2 u)(\xi) = u(\xi)$ \qquad (ii) $(\mathcal{F}_\Gamma P_\Gamma u)(\xi) = \mathcal{F}_\Gamma u(\xi)$

(iii) $(P_\Gamma^2 u)(\xi) = (P_\Gamma u)(\xi)$ (iv) $(Q_\Gamma^2 u)(\xi) = (Q_\Gamma u)(\xi)$

are valid for any $\xi \in \Gamma$.

We remark that the operators $\mathcal{F}_\Gamma$, S_Γ, P_Γ and Q_Γ allow an extension to $L_2(G, \mathbb{H})$. The restriction of a $\mathbb{H}$-function u to a function defined on the boundary Γ is expressed by $tr_\Gamma u$. Let us now consider the Hilbert space $L_2(G, \mathbb{H})$ with an inner product $(u,v) = \int_G \bar{u} v \, dx \in \mathbb{H}$.

Lemma 3: *The Hilbert space $L_2(G, \mathbb{H})$ allows an orthogonal decomposition*

$$L_2(G, \mathbb{H}) = [\ker D \cap L_2(G, \mathbb{H})] \oplus D[\overset{\circ}{W}{}_2^1(G, \mathbb{H})].$$

Corollary 1: *There exist two orthoprojections $\mathcal{P}$ and $\mathcal{Q}$ with*

$$\mathcal{P} : L_2(G, \mathbb{H}) \to \ker D \cap L_2(G, \mathbb{H})$$
$$\mathcal{Q} : L_2(G, \mathbb{H}) \to D[\overset{\circ}{W}{}_2^1(G, \mathbb{H})],$$
$$\mathcal{Q} = I - \mathcal{P}.$$

Corollary 2: [12] *Let $f \in L_2(G, \mathbb{H})$. Then $f \in im\mathcal{Q}$ iff $tr_\Gamma T_G f = 0$.*

3 Nonlinear free convection problem

In this part we will show how the methods of Clifford analysis can be used for the treatment of boundary value problems of partial differential equations. For this reason we have chosen the time-harmonic case of the non-linear stream problem with free convection. The corresponding mathematical description reads as follows:

In the bounded domain G the following equations has to be valid:

$$-\Delta \underline{u} + a_1(\underline{u} \cdot grad)\underline{u} + f(\underline{u}) + a_2 grad\, p + a_3(-e_3)w = F(x)$$
$$div\, \underline{u} = 0$$
$$-\Delta w + a_4(\underline{u} \cdot grad)w = g.$$

On the boundary $\partial G = \Gamma$ the velocity $\underline{u}$ and the temperature w are identically zero. The coefficients can be given a well-defined physical meaning. If we denote by ρ the density of the fluid, by η the viscosity, by γ the Grashoff number, by κ the temperature conductivity and by m the Prandl number, then there is the following connection to the coefficients a_i (i=1,2,3,4):

$$a_1 = \frac{\rho}{\eta}, \quad a_2 = \frac{1}{\eta}, \quad a_3 = \frac{\gamma}{\eta}, \quad a_4 = \frac{m}{\kappa}$$

p denotes the pressure. Conditions for the function f are more technical. We will formulate them later.

In order to apply Clifford analysis we have to transform this problem. Using the results of Part 2 we get the following operator integral equation:

$$u = -a_1 T_G Q T_G [M(u) + a_3 e_3 w] - a_2 T_G Q p$$
$$0 = Sc\{ a_1 Q T_G [M(u) + a_3 e_3 w] + a_2 Q p \}$$
$$w = a_4 T_G Q T_G Sc(u D w) + T_G Q T_G g \quad ,$$

where $M(u) = (u \cdot grad)u + f(u) - F(x)$. Note that e_3 can be identified with the vector $(0,0,1)$. The term u means now $u := \underline{u}$.

Remark: The boundary conditions are fulfilled. In fact, the definition of the orthoprojection Q delivers immediately

$$tr_\Gamma T_G Q u = tr_\Gamma T_G D w, \qquad w \in \overset{\circ}{W}{}_2^1(G, \mathbb{H}) \ .$$

Borel-Pompeiu's formula yields now $tr_\Gamma T_G Q u = tr_\Gamma w - tr_\Gamma F_\Gamma w = 0$.

4 Iteration method

The following procedure reduces our problem to the more simple so-called *Stokes problems*:

$$u_n = -a_1 T_G Q T_G [M(u_{n-1}) + a_3 e_3 w_{n-1}] - a_2 T_G Q p_n \qquad (2)$$
$$0 = Sc\, a_1 Q T_G [M(u_{n-1}) + a_3 e_3 w_{n-1}] + a_2 Sc\, Q p_n \qquad (3)$$
$$w_n = a_4 T_G Q T_G\, Sc(u_n D w_n) + T_G Q T_G g. \qquad (4)$$

For the computation of w_n we use the "inner" iteration:

$$w_n^{(i)} = a_4 T_G Q T_G\, Sc(u_n D w_{n-1}^{(i-1)}) + T_G Q T_G g \quad (i = 1, 2, ...).$$

Theorem 1: *1. Let $u_n \in \overset{\circ}{W}{}_2^1(G, \mathbb{H})$. If $a_4 \neq 4$ and $\|u_n\| < 1/(a_4 C K)$ where C is the embedding constant from W_2^1 in L_2 and $K = a_1 \|T_G\|_{[im Q, W_2^1]} \|T_G\|_{[L_q, L_2]}$ then $(w_n^{(i)})$ converges in the sense of $W_2^1(G, \mathbb{H})$. The norm $\|w_n\|_{2,1}$ can be estimated.*

2. Let $F \in L_2(G, \mathbb{H}), g \in L_2(\Gamma, \mathbb{H}), \quad f: W_2^1(G, \mathbb{H}) \mapsto L_2(G, \mathbb{H})$ with $\|f(u) - f(v)\|_2 \leq L\|u - v\|_{2,1}, \; (L > 0)$ and $f(0) = 0$. Under the conditions

(i) $a_1 \|F\|_2 + a_3 K |d|^{-1} \|g\|_2 < 1/(16 K^2 C), \quad d = (4 - a_4)\kappa$

(ii) $\|g\| < (1 - 1/\sqrt{2}) a_2^{-1} d^2 / (32 K^3 C m)$

(iii) $a_4 < 4$

the iteration procedure (2)-(3)-(4) converges in $W_2^1(G, \mathbb{H}) \times W_2^1(G) \times L_2(G)$ to a unique solution

$$\{u, w, p\} \in \overset{\circ}{W}_2^1(G, \mathbb{H}) \times \overset{\circ}{W}_2^1(G) \times L_2(G)$$

(p is unique up to a constant) of the above formulated boundary value problem.

Remark: The smallness conditions (i) and (ii) can be realized for instance for fluids with a sufficient "big" viscosity number. Condition (iii) is easy to check.

Next we will explain how to realize the iteration procedure in practice. There are more or less two possibilities to do this within the Clifford operator calculus. At first we note that the solution of Stokes problem can be given explicitly. We have:

Theorem 2: With the abbreviations

$$h_n := M(u_{n-1}) + a_3 e_3 w_{n-1} \tag{5}$$

$$\mathcal{Q}' := I - Vec\,\mathcal{F}(tr_\Gamma T_G\, Vec\mathcal{F})^{-1} tr_\Gamma T_G, \quad (\mathcal{Q}'^2 = \mathcal{Q}') \tag{6}$$

we get the following representation of the solution $\{u_n, p_n\}$ of (2)-(3) :

$$u_n = -a_1 T_G \mathcal{Q}' Vec T_G\, h_n$$
$$p_n = \eta a_1\, Sc\mathcal{Q}' Vec T_G\, h_n, \quad (n = 1, 2, ...).$$

It remains to analyse the operators $\mathcal{Q}'$ and T_G. K.Gürlebeck developed in his habilitation paper [5](see also [6]) a discrete quaternionic function theory which guarantees also a discrete analogue to Borel-Pompeiu's formula. Furthermore he could also get discrete specimen of the operators T_G and $\mathcal{Q}$. The operator $\mathcal{Q}'$ is of the same structure like the operator $\mathcal{Q}$. Applying methods in [6] one can deduce a discrete analogue of the operator $\mathcal{Q}'$. Unfortunately it is not possible to introduce these discrete operators in detail. These constructions are very comprehensive and therefore we can only refer to our book [6].

The final theorem is related to the connection of the corresponding discrete solutions of the discretized analogue of (5)-(6) and the solution (5)-(6):

Theorem 3: Let G_h be the lattice to the domain G and h the corresponding meshwidth. If the right hand side $h_n \in L_p(G, \mathbb{H})$ $(p > 3)$ then we obtain for the difference of the solution u represented in (5) and the discretized solution u_h (=solution of the corresponding discretized problem)

$$\|u - u_h\|_{2,h} \longmapsto 0 \quad for \quad h \longmapsto 0 \quad ,$$

where

$$\|u\|_{2,h}^2 = \sum_{s \in G_h} |u(s)|^2 h^3.$$

References

[1] F.Brackx, R.Delanghe and F.Sommen, *Clifford analysis*, Pitman, Boston-London -Melbourne ,(1982).

[2] R.Delanghe,F.Sommen and V. Soucek, *Clifford algebra and spinor-valued functions*, Kluver Academic Publishers, Dordrecht ,(1992).

[3] R.Fueter, *Die Funktionentheorie der Differentialgleichungen $\Delta u = 0$ und $\Delta\Delta u = 0$ mit vier reellen Variablen*, Comm. Math. Helv. 7, (1935), 307-330.

[4] R.Fueter, *Reguläre Funktionen einer Quaternionenvariablen*, Math. Inst. d. Universität Zürich, (1949).

[5] K.Gürlebeck, *Grundlagen einer diskreten räumlich verallgemeinerten Funktionentheorie und ihrer Anwendungen*, Thesis, TU Karl-Marx-Stadt, (1988).

[6] K.Gürlebeck and W.Sprößig, *Quaternionic Analysis and Elliptic Boundary Value-Problems*, Birkhäuser-Verlag, Basel, (1990).

[7] K.Gürlebeck and W.Sprößig, *Clifford Analysis and Elliptic Boundary Value Problems*, In: R. Ablamowicz and P. Lounesto (eds.): Clifford algebras and Spinor structures, Kluver Academ. Publ., (1995), 325-334.

[8] D. Hestenes, *New Foundations for Classical Mechanics*, Reidel Publ. Co., Dordrecht/Boston, (1985).

[9] P. Lounesto and P. Bergh, *Axially Symmetric Vector Fields and Their Complex Potentials*, Complex Variables,Vol. 2, (1983), 139-150.

[10] W.Nef, *Die Funktionentheorie der partiellen Differentialgleichungen zweiter Ordnung(Hyperkomplexe Funktionentheorie)*, Bull. Soc. Fribourgoise Sc. Nat. 37, (1944), 348-370.

[11] A.Rose, *On the use of a complex (quaternion) velocity potential in three dimensions*, Comment. Math. Helv. 24,(1950), 135-148.

[12] W.Sprössig, *Representation formulae of solutions of boundary value problems with Clifford analysis methods* Preprint TU-Bergakademie Freiberg: Fakultät für Mathematik und Informatik ,Nr. 95-06 , 1-30.

[13] E.M.Stein and G.Weiss, *Generalization of the Cauchy-Riemann equations and representations of the rotation group*, Amer. J. Math. 90,(1968), 163-196.

[14] J.Wloka, Partielle Differentialgleichungen, B.G.Teubner, Stuttgart, (1982).

NUMERICAL SIMULATION OF FLOW AROUND A TRAIN

Masahiro Suzuki and Tatsuo Maeda
Railway Technical Research Institute
2-8-38 Hikari-cyo, Kokubunji-shi, Tokyo 185, JAPAN

Norio Arai
Tokyo University of Agriculture and Technology
2-24-16 Nakamachi, Koganei-shi, Tokyo 184, JAPAN

SUMMARY

It was recognized that the vibration amplitude of the tail car is greater, especially in the tunnel, than that of the other cars in a high-speed train, in which an aerodynamic force has some effects. However, few aerodynamical studies have been conducted and little knowledge has been gained. To clarify the aerodynamical effect on the train, a three-dimensional unsteady Navier-Stokes simulation was carried out. Unsteady flow separations on the rear nose, which cause fluctuations of the yawing moment of the tail car, were successfully obtained by the simulation. In the tunnel section, it is proved that the tunnel wall makes the flow separation asymmetric and that the expansion of the effective flow area along the rear nose causes a greater pressure fluctuation. We also proposed a modified shape of rear nose, which suppresses the flow separation and reduces the yawing moment fluctuation.

1. INTRODUCTION

Recently, the vibration of the tail car in a high-speed train has become an issue. From the vibration data obtained simultaneously on several cars in the same Shinkansen train (Japanese high-speed train), the followings were discovered[1].
   (a) The vibration amplitude of the tail car is greater than that of the other cars in a train.
   (b) In the tail car, the yawing vibration is more prominent than the lateral vibration.
   (c) The yawing vibration of the tail car in the tunnel section is more noticeable than that in the open section (non-tunnel section).
As the factors causing the phenomena, the followings are considered.
   (1) Track irregularity (Forced vibration)
   (2) Hunting motion (Self-excited vibration. A train winds horizontally on the track like a snake. Also called 'snaking motion'.)
   (3) Aerodynamic force
It was reported that there is a correlation between the vibration and the track irregularity in the open section, but no correlations in the tunnel section. An analysis of car body hunting characteristics and a vehicle dynamics simulation show that there is a tendency of greater vibration of the tail car when the train runs fast. As for the aerodynamic force, the effect of

Karman-like vortices on the vibration has been suggested. No mechanisms, however, have been clarified in detail. Since the factors (1) and (2) can not explain the greater vibration in the tunnel, the aerodynamic force seems to act essentially on the vibration of the tail car in the tunnel.

In this study, flows around train both in the open section and in the tunnel section are simulated by solving the three-dimensional unsteady Navier-Stokes equations. It is the purpose of this paper to clarify the effect of the unsteady aerodynamics force on the vibration of the tail car. Analyzing the flow structure, we also propose a modified shape, which decreases the yawing moment fluctuation.

2. NUMERICAL PROCEDURE

When a train enters a tunnel, a pressure wave is generated and it travels between both portals of the tunnel. When the train does not encounter the pressure wave, flow around the train becomes quasi-stationary and can be treated as an incompressible flow. We assume this situation and use the three-dimensional unsteady incompressible Navier-Stokes equation as the governing equation.

The numerical scheme is based on the well-known MAC method [2]. All spatial derivatives, except nonlinear terms, are represented by the second-order central-differences scheme. The nonlinear terms are represented by a third order upwind scheme [3], which enables resolving turbulent eddies without any turbulence models. The Euler implicit scheme is employed for time integration. The computations are done on a generalized coordinate system [4].

In fig. 1, the grid system is presented. Though the Shinkansen train is usually composed of 6-16 cars, our model has a length of 2.5 cars in this computation because of limitations of the computer memory. Convexities like bogies, pantograph and couplers are omitted in the grid. There are also no gaps between cars. A rear part of the length of one car is, however, regarded as a tail car. The numbers of grid points are 1.08 million in the open section and 0.94 million in the tunnel section.

The boundary conditions are as follows: $\mathbf{u}=(1,0,0)$ on the upstream boundary, the ground and the tunnel wall; $\mathbf{u}=(0,0,0)$ on the train. $\mathbf{u}$ is extrapolated on the outer boundary except the above boundaries. Pressure is determined by the extrapolation on all boundaries.

The Reynolds number is set at 10^5 based on the train height, which is the same order of a wind tunnel experiment.

Computations are done on the CRAY-YMP/4E at Railway Technical Research Institute.

3. RESULTS AND DISCUSSION

3.1 Comparison with experiments

To verify the simulation, computational results are compared with the wind tunnel experimental results and running test data available. Figure 2 shows the pressure distribution along the symmetric plane of the rear nose. Good agreements are obtained both in the tunnel section and in the open section. Surface flow patterns are compared between computational and wind experimental results (fig. 3). Though a flow visualization by the oil does not decisively yield the flow pattern in the separated region in the experiment, it is perceived that both computational and experimental results have some tendencies in respect of primary separations. The above agreement confirms verification of the simulation.

3.2 The yawing moment fluctuation of the tail car

The yawing moment is obtained by integrating the pressure and the shear stress on the tail car. The greater vibration in the tunnel section is successfully simulated (fig. 4). In fig. 5, the power spectrum analysis indicates that its value in the tunnel is about 10 times that in the open section.Though there is no specific peak in the open section, a peak around 0.1 non-dimensional frequency is noticed in the tunnel section.

3.3 Wake structures and surface pressure fluctuations

Using the averaged field data, wake structures behind the train are investigated. Figure 6 demonstrates total pressure distributions on different cross sections. In the open section, a pair of trailing vortex is clearly observed. In the meanwhile, no specific pairs of trailing vortex are recognized in the tunnel section. Figure 7 exhibits total pressure distributions on a horizontal surface at a half height of the train. In contrast with the symmetric flow separation in the open section, flows separate asymmetrically in the tunnel section. The flow separation occurs earlier on the tunnel wall side of the tail car than on the tunnel center side. This is due to more rapid expansion of the effective flow area on the tunnel wall side and the difference of the boundary layer between both sides.

The above mentioned separation patterns are easily imagined to be unsteady because of the shape of the train, which has moderate curves without any abrupt changes. Unsteady separation patterns and pressure distributions are scrutinized by making animations. The separation pattern moves from upstream to downstream with the surface pressure fluctuation. Typical flow patterns are displayed in figs. 8 and 9. Surface streamlines evidence separations, reattachments, secondary separations and so on. Surface pressure distributions manifest pressure variations corresponding to the surface streamline patterns. These pressure distributions are asymmetric not only in the tunnel section, but also in the open section. The asymmetry of the pressure distribution generates the yawing moment. In the tunnel section, the augmentation of the effective flow area along the rear nose causes a greater pressure fluctuation. In addition to this, the pressure distribution with significant asymmetry in the tunnel section results in further increase of the yawing moment fluctuation.

3.4 A countermeasure for the yawing moment fluctuation

Taking the above discussion into consideration, a shape, which decreases the yawing moment fluctuation, is proposed here. The main cause of yawing moment fluctuations is the unsteadiness of separation patterns. Two possible countermeasures for the yawing moment fluctuation are deliberated. One is to fix the separation point. The other is to suppress the separation on the side of the tail car. An option of the former is to attach some devices like fin at a point, where the separation is likely to arise. There are several additional matters to be cautioned in this option. Unsteady flow reattachments, which may generate pressure fluctuations, should be also avoided. Because of the characteristic of the train, the tail car of the upbound train will be the first car of the downbound train. Special care will be taken so that the apparatus on the first car generates no aerodynamic noise. We could not help saying that it is hard to design a such gadget meeting these conditions within the rolling stock gauge (car clearance). Consequently, the latter way of decreasing the yawing moment fluctuation, which suppresses separations, is deliberated here. To suppress the separation on the side of the train, the shape of the side should be flat. In the tunnel section, the abrupt expansion of the flow area with a great pressure change is ascribed to separations with greater pressure fluctuations. Therefore, the change rate of the cross sectional area of the rear nose should be decreased. A shape of rear nose among candidates, which satisfy above requirements, is illustrated in fig. 10. Sides of the nose are flat. The change rate of the cross sectional area is moderate except at the tip. To prove the effect, the flow around a train with the modified rear nose in the tunnel section is calculated, where the original shape is used as the front nose for the sake of eliminating the influence of the shape of the front nose. The power spectrum of the yawing moment is drawn in fig. 5. A peak value of the modified shape is successfully diminished and comparable to that of the original shape in the open section.

4. CONCLUSIONS

Computations of unsteady flows around a train were conducted. Yawing moment fluctuations of the tail car were successfully simulated. With the analysis of flow structures, it was proved that the cause of yawing moment fluctuations is unsteady separations with the pressure fluctuations. In the tunnel section, the asymmetric separation pattern and the pressure fluctuation with a sudden expansion of the effective flow area cause a greater yawing moment variation. On the basis of the above consideration, a modified shape, which suppresses the yawing moment fluctuation, was presented. It was confirmed that the proposed shape makes the yawing moment fluctuation considerably lower in the tunnel section.

While the present computation was focused on flow separations from the rear nose, the turbulent boundary layer developed on middle cars and the vibration of the train itself occur in the real situation. In future studies, the flow induced vibration with turbulent layer should be considered.

REFERENCES

[1] FUJIMOTO, H., MIYAMOTO, M., SHIMAMOTO, Y. : "Lateral vibration of a Shinkansen and its decreasing measure", *RTRI Report*, 9-1 (1995) pp.19-24.

[2] HARLOW, F. H., WELCH, J. E. : "Numerical calculation of time-dependent viscous incompressible flow of fluid with free surface", *Phys. Fluids*, 8 (1965) pp.2182-2189.

[3] KAWAMURA, K., TAKAMI, H., KUWAHARA, K. : "Computation of high Reynolds number flow around a circular cylinder with surface roughness", *Fluid Dyn. Res.*, 1 (1986) pp.145-162.

[4] THOMPSON, J. F., WARSI, Z. U. A., MASTIN, C. W. : "*Numerical grid generation*", North-Holland, Amsterdam, 1985.

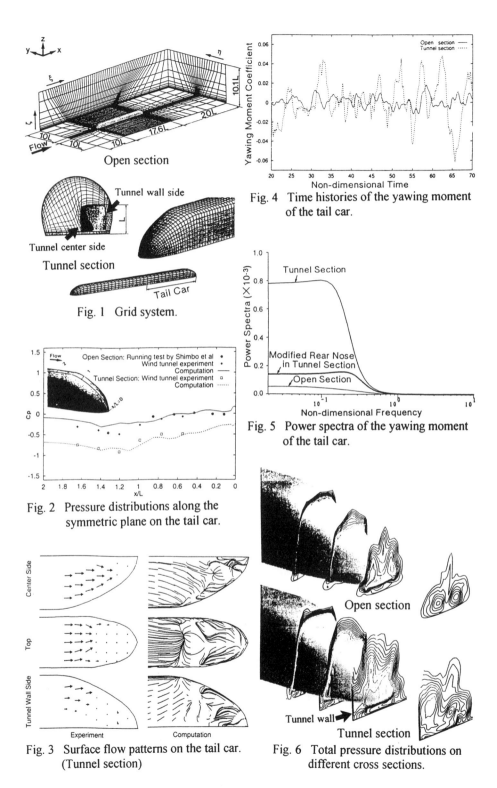

Fig. 1 Grid system.

Fig. 2 Pressure distributions along the symmetric plane on the tail car.

Fig. 3 Surface flow patterns on the tail car. (Tunnel section)

Fig. 4 Time histories of the yawing moment of the tail car.

Fig. 5 Power spectra of the yawing moment of the tail car.

Fig. 6 Total pressure distributions on different cross sections.

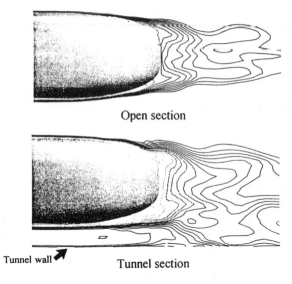

Open section

Tunnel wall Tunnel section

Fig. 7 Total pressure distributions on a horizontal surface at a half height of the train.

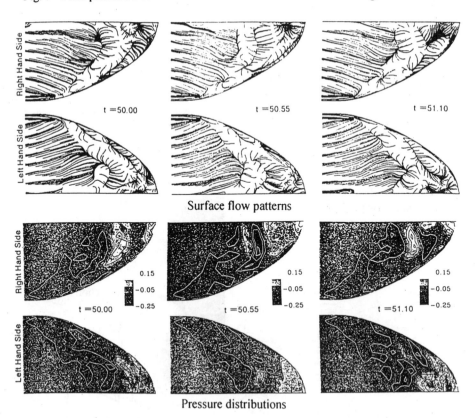

Surface flow patterns

Pressure distributions

Fig. 8 Surface flow patterns and pressure distributions on the tail car. (open section)

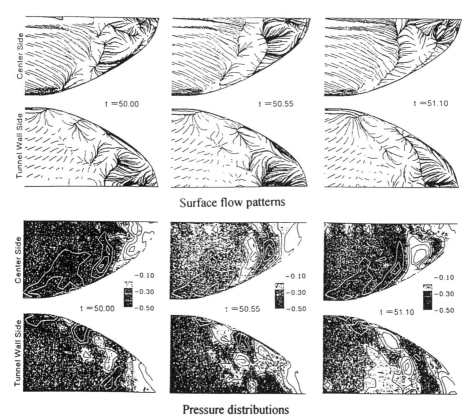

Fig. 9 Surface flow patterns and pressure distributions on the tail car. (tunnel section)

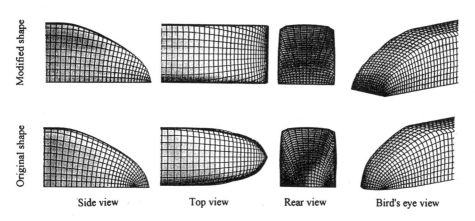

Fig. 10 A modified nose shape, which decreases yawing moment fluctuations

Parallelization of a 3D code for simulation of turbulent channel flow on the Cray T3D

L.J.P. Timmermans and F.T.M. Nieuwstadt
Delft University of Technology
Laboratory for Aero- and Hydrodynamics
Rotterdamseweg 145
2628 AL Delft, The Netherlands

Abstract

The development of massively parallel processing machines may make it possible to simulate in the near future flow problems which by far exceed present computational capacity. In this paper the parallelization of a code for both Direct Numerical Simulation and Large-Eddy Simulation of turbulent channel flow is discussed. The code is ported to a Cray T3D. The parallel programming tool used is the data parallel programming model CRAFT. The parallelized code scales very well up to 256 processors. To increase performance several optimization techniques have also been applied.

1 Introduction

A flow is generally called turbulent when it exhibits chaotic fluctuations in both space and time. Most flows occurring in engineering applications are turbulent. Although turbulent flows have been investigated for more than a century, no general approach to the solution of problems in turbulence exists.

Despite the apparent chaos in turbulence, detailed analysis has shown that turbulent flows are not completely random in space and time. It appears that turbulent flows contain spatial structures that develop in time. These structures are called coherent as they consist of spatially correlated flow properties. A possible method of analysis is to explain the dynamics of turbulence in terms of the development of these coherent structures as a function of time.

At the moment one of the more powerful tools to obtain detailed information on turbulent flows is a numerical simulation. The Navier–Stokes equations, which govern turbulent flow, can be solved numerically. The main advantage of a numerical simulation is that it gives detailed information on the three-dimensional structure of the flow. The numerical approach has been made increasingly feasible due to the fast growth in computer power over the last ten years. Especially the development of massively parallel processing (MPP) machines may make it possible to simulate in the near future flow problems which presently by far exceed computational capacity and which are real challenges from a physical point of view.

This paper describes the parallelization of a code for the simulation of three-dimensional turbulent channel flow on the first generation Cray MPP system: the Cray T3D. In section 2 the basics of numerical simulation techniques such as Direct Numerical Simulation and Large-Eddy Simulation are briefly discussed. Section 3 contains details of the computational techniques applied. Section 4 deals with the parallel implementation of the code. Firstly, attention is given to the Cray T3D: the hardware structure and the parallel programming paradigms available on this machine. Secondly, details and results of the parallelization are given. Finally, the main results and conclusions are summarized in section 5.

2 Numerical simulation of turbulent flow

Although turbulence is far from being completely understood from a physical point of view, the equations for turbulent flow have been known for a long time and for an incompressible Newtonian fluid these are the Navier–Stokes equations [1]. They consist of conservation laws for mass and momentum which in vector notation read

$$\nabla \cdot \mathbf{u} = 0, \qquad \frac{\partial \mathbf{u}}{\partial t} + \nabla \cdot (\mathbf{uu}) = -\nabla p + \nu \nabla^2 \mathbf{u}. \qquad (1)$$

Here $\mathbf{u}(x,y,z,t)$ is the velocity vector, $p(x,y,z,t)$ the kinematic pressure (pressure divided by density) and ν the kinematic viscosity. An important parameter is the dimensionless Reynolds number $Re = \mathcal{UL}/\nu$ where $\mathcal{L}, \mathcal{U}$ are the length and velocity scales of the largest turbulent flow motions. Turbulence by definition can only exist for large Re, so that in turbulent flows the non-linear advective term dominates the viscous term.

Direct Numerical Simulation
In a Direct Numerical Simulation (DNS) of turbulent flow the Navier–Stokes equations are solved directly, resolving all the turbulent scales. Unfortunately, DNS is limited by severe computational restraints. Since all the details of the flow need to be resolved, the distance between the grid points should at least be of the order of the smallest length scale, whereas the computational domain should be large enough to capture the largest scales.

Consider a DNS of a turbulent channel flow. The height of the channel is denoted by L. The number of grid points needed to resolve all the scales can be derived to be

$$N_{\text{DNS}} \sim \left(\frac{L}{\mathcal{L}}\right)^3 Re^{\frac{9}{4}}. \qquad (2)$$

For fully turbulent flow, for which $Re \gg 1$, the number of grid points needed may become very large, in fact far beyond reach of today's supercomputers. Consequently, DNS at the moment is only feasible for turbulent flows at modest Reynolds numbers which are much smaller than those occurring in engineering applications. For a more thorough analysis of DNS the reader is referred to e.g. [11], [12], [13].

Large-Eddy Simulation
An alternative method for the numerical simulation of turbulence is Large-Eddy Simulation (LES). Conceptually a spatial filter is applied to the governing equations to remove the small scales. The influence of the unresolved small scale motions appears in the form of new unknown stress terms that have to be modelled. Since in LES only the larger scales are computed explicitly, the computational requirements so stringent on the DNS procedure are relaxed.

Consider again a simulation of turbulent channel flow. The filter width of the LES is denoted by $\mathcal{L}_f$. The number of grid points needed for LES then becomes

$$N_{\text{LES}} \sim \left(\frac{L}{\mathcal{L}_f}\right)^3. \tag{3}$$

Thus, for LES there is in principle no dependency on Re. A more detailed analysis of the computational costs of both DNS and LES is presented in [9].

A commonly used filter in LES is the top-hat filter [10]. The filter length $\mathcal{L}_f$ is typically taken to be two grid spacings. This means that precisely those scales are filtered out that cannot be represented on the numerical grid. Applying the filtering operation to the velocity and pressure variables in the Navier–Stokes equations defines a decomposition of these variables into a resolved part and an unresolved subgrid scale (SGS) part.

As already mentioned, in the Navier–Stokes equations the subgrid part leads to new stress terms also known as the SGS stress τ. In order to model this stress it is usually decomposed into an isotropic and a non-isotropic part. The isotropic part can be combined with the resolved pressure. The non-isotropic part remains to be modelled. In analogy to the molecular theory this is done by relating the stress to the deformation rate of the resolved velocity field with the aid of an eddy viscosity ν_t

$$\tau = -\nu_t \overline{S}, \tag{4}$$

with $\overline{S} = (\nabla \overline{u} + (\nabla \overline{u})^T)$ the strain rate of the resolved velocity field. The so-called Smagorinsky model is chosen to specify ν_t. In this model the eddy viscosity is related to the deformation rate of the resolved velocity field. For a detailed discussion on the validity of the Smagorinsky model the reader is referred to [5], [8], [11].

3 Computational techniques

This section deals with the numerical techniques used to solve the governing equations. These are written in the following general form in order to incorporate the features specific to LES

$$\nabla \cdot \mathbf{u} = 0, \quad \frac{\partial \mathbf{u}}{\partial t} = -\nabla p - \nabla \cdot (\mathbf{u}\mathbf{u}) + \nabla \cdot \left(\nu \left(\nabla \mathbf{u} + (\nabla \mathbf{u})^T\right)\right), \quad (5)$$

with $\mathbf{u}$ either the velocity field (DNS) or the resolved velocity field (LES), p the pressure including the isotropic part of the SGS stresses in the case of LES, and ν the viscosity including for LES the eddy viscosity ν_t.

A second-order accurate explicit Adams-Bashforth method is chosen for time integration, see e.g. [6]. Application of this method to the system (5) gives

$$\nabla \cdot \mathbf{u}^{n+1} = 0, \quad (6)$$

$$\frac{\mathbf{u}^{n+1} - \mathbf{u}^n}{\Delta t} = -\nabla p^{n+1} + \frac{3}{2}(-\mathcal{A}\mathbf{u} + \mathcal{D}\mathbf{u})^n - \frac{1}{2}(-\mathcal{A}\mathbf{u} + \mathcal{D}\mathbf{u})^{n-1}, \quad (7)$$

with $\mathcal{A}\mathbf{u} = \nabla \cdot (\mathbf{u}\mathbf{u})$ and $\mathcal{D}\mathbf{u} = \nabla \cdot (\nu(\nabla \mathbf{u} + (\nabla \mathbf{u})^T))$. The above system forms a coupled set of equations. An easy to implement decoupling procedure is provided for by the pressure correction method [7]. This scheme proceeds as follows:

- Calculate an intermediate velocity field $\mathbf{u}^*$ from

$$\mathbf{u}^* = \mathbf{u}^n + \frac{3}{2}\Delta t(-\mathcal{A}\mathbf{u} + \mathcal{D}\mathbf{u})^n - \frac{1}{2}\Delta t(-\mathcal{A}\mathbf{u} + \mathcal{D}\mathbf{u})^{n-1}. \quad (8)$$

- Solve for the pressure from the Poisson equation

$$\nabla^2 p^{n+1} = \frac{\nabla \cdot \mathbf{u}^*}{\Delta t}. \quad (9)$$

- Calculate the velocity from

$$\mathbf{u}^{n+1} = \mathbf{u}^* - \Delta t \nabla p^{n+1}. \quad (10)$$

Finally, a second-order accurate finite volume technique is used to discretize the equations in space. For details on the numerical techniques see [14].

4 Implementation on the Cray T3D

The Cray T3D is a distributed memory MIMD (Multiple Instruction Multiple Data) parallel machine consisting of processing elements (PE's) interconnected by a high bandwidth, low latency network based on a 3D torus topology. For an extensive description of the hardware configuration see [3].

It has become clear that despite the in theory impressive peak performance of MPP systems, in practice it is a real challenge to extract a significantly high percentage of peak performance. The hardware alone does not determine performance. Of great importance also are software, compilers etc. The available parallel programming paradigms by which the Cray T3D can be programmed are message passing and data parallel CRAFT (Cray Research Adaptive Fortran) [4]. CRAFT consists of Fortran 77 plus a number of additional compiler directives. This model focuses mainly on distribution of data and work across PE's in order to achieve:

- Data sharing: in conjunction with large data sets it is necessary to share large arrays due to memory limitations. By default data are private, each PE has its own unique copy.

- Work sharing: distributing data is not very useful unless the work can also be distributed. This is achieved through the use of shared loops.

Although memory is physically distributed, using the data parallel approach it is logically shared: any PE can address the memory of any other PE. Of course, local memory accesses are faster than remote ones.

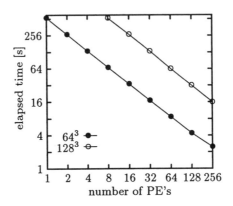

Figure 1 Elapsed time of the initial code as function of the number of PE's for 64 and 128 grid points in each direction.

First results
The use of the data parallel model enables the straightforward porting of codes to the Cray T3D. For this reason the turbulence code was parallelized at first instance using CRAFT. Details on the implementation of the CRAFT directives in the code are found in [14]. Figure 1 shows the elapsed time of the code as function of the number of PE's for a fixed number of time steps. The simulation was performed for resolutions of 64 and 128 grid points in each spatial direction. For the larger resolution a minimum of 8 PE's is needed due to memory limitation. It is clear that performance of the code scales very well up to 256 PE's.

Although this result may seem surprising at first sight it can be easily understood. The computational model given in equations (8)–(10) contains a lot of inherent parallelism:

- The implementation of the righthand side of the equations on the finite volume grid involves evaluation of 3-points finite difference stencils in each direction. This means that there is only communication between neighbouring points. In the parallel implementation there is only communication between sets of data residing on PE boundaries.

- Equation (9) also involves the solution of a discrete Poisson equation. The solution of the system is obtained by a fast Poisson solver, which involves Fast Fourier Transforms and the solution of a tridiagonal system using the Thomas algorithm [7]. To reduce communication the data distribution is adjusted in such a way that the solution algorithm works on private data only [14].

Performance optimization
The data parallel approach places a large burden on the compiler. This is reflected in a relatively small percentage of peak performance. A way to increase the performance of CRAFT significantly is by combining the model with the fast explicit task shared memory communication routines, and by using subroutines that work on private data only. This programming model is fast, but more complicated.

The latter approach is taken here. The objective is to make full use of the potential locality of shared data. Although data may actually reside on a particular PE the compiler does not always recognise this and will treat array references as remote. Memory access times can thus be slower than necessary.

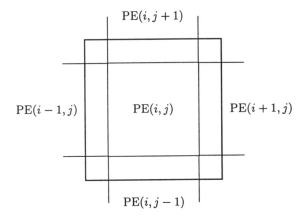

Figure 2 Domain of the private array containing the elements of a shared array residing on PE(i, j) including the boundaries of neighbouring PE's (in a 2D topology of PE's).

Several techniques, presented in [2], are used with which local addressing can be achieved. Here, one technique is briefly discussed that works particularly well for the present code. By making use of intrinsic functions that determine the lowest and highest index of a shared array on a particular PE, the parts of arrays residing on that PE are copied into local arrays that are declared to be private, i.e. each PE has its own unique copy of (different) local elements. Since in the finite volume context evaluation of the discrete equations involves communication between array element boundaries, these boundaries must be included in the private arrays, see Figure 2. These arrays are then passed on to the subroutines in which the evaluations are performed, ensuring local memory access.

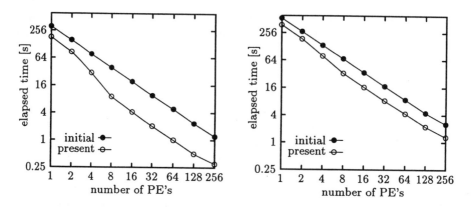

Figure 3 Comparison of elapsed time of the initial and present code as function of the number of PE's for 64 grid points in each direction. Left: momentum equation. Right: overall code.

Figure 3 (left) shows the considerable increase in performance gained by applying the above technique in the implementation of the momentum equation (8), the most expensive part of the code. The resolution is 64 grid points in each direction. For the overall code the performance increase is shown in Figure 3 (right).

5 Conclusions

In this paper the parallelisation on the Cray T3D of a code for both DNS and LES of turbulent channel flow is described. The programming model used is the data parallel CRAFT which allows a straightforward porting of the code. The computational techniques involved possess a lot of inherent parallelism. As a consequence, the initial implementation shows very good scaling of performance up to 256 processors. The overall percentage of peak performance achieved is relatively small. This is mainly due to the fact that the compiler does not make optimal use of the potential locality of shared data. A considerable increase in performance can be obtained using optimization techniques that ensure more local memory access. This research is still very much ongoing, however the first results indicate that in the near future it may well be possible that MPP machines can play a significant role in the numerical simulation of highly turbulent flows.

Acknowledgments

The authors would like to thank Cray Research B.V., The Netherlands, for intensive support and for giving access to Cray T3D systems.

References

[1] G.K. Batchelor(1967), *An introduction to fluid dynamics* (Cambridge University Press, Cambridge).
[2] S. Booth, P. MacCallum, N. MacDonald and A. Simpson (1995), Performance optimization on the Cray T3D, *EPCC Report*, EPCC, Edinburgh.
[3] Cray Research Inc. (1993), Cray T3D: technical summary, *Technical Report*, Cray Research Inc.
[4] Cray Research Inc. (1994), Cray MPP Fortran reference manual, *Reference manual SG-2504 6.1.*
[5] J.H. Ferziger (1983), Higher-level simulations of turbulent flows, in: *Computational methods for turbulent, transonic and viscous flows (ed. J.A. Essers)* (Hemisphere Publications, Washington), pp. 93-182.
[6] C.W. Gear (1971), *Numerical initial value problems in ordinary differential equations* (Prentice-Hall, Englewood Cliffs).
[7] C. Hirsch (1988), *Numerical computation of internal and external flows. Vols. 1 & 2* (John Wiley & Sons, Chicester, New York.
[8] P.J. Mason (1994), Large-eddy simulation: A critical review of the technique, *Q. J. R. Meteorol. Soc.* **120** pp. 1-26.
[9] F.T.M. Nieuwstadt, J.G.M. Eggels, R.J.A. Janssen and M.B.J.M. Pourquié (1994), Direct and large-eddy simulations of turbulence in fluids, *Future Generations Computer Systems* **10** pp. 189-205.
[10] U. Piomelli, P. Moin and J. Ferziger (1988), Model consistency in large-eddy simulation of turbulent channel flow, *Phys. Fluids* **31** pp. 1884-1891.
[11] W.C. Reynolds (1989), The potential and limitations of direct and Large Eddy simulations, in: *Whither turbulence? Turbulence at the crossroads (ed. J.L. Lumley)*, (Springer-Verlag, Berlin), pp. 313-349.
[12] R.S. Rogallo and P. Moin (1984), Numerical simulation of turbulent flows, *Ann. Rev. Fluid Mech.* **16** pp. 99-137.
[13] U. Schumann and R. Friedrich (1987), On direct and large eddy simulation of turbulence, in: *Advances in turbulence (eds. G. Comte-Bellot and J. Mathieu)* (Springer-Verlag, Berlin), pp. 88-104.
[14] L.J.P. Timmermans (1995), Numerical solution of turbulent flow on the Cray T3D, *TRACS Report*, EPCC, Edinburgh.

A NEW APPROACH TO MODELLING AN UNSTEADY FREE SURFACE IN BOUNDARY INTEGRAL METHODS WITH APPLICATION TO BUBBLE-STRUCTURE INTERACTIONS

R. P. Tong
School of Mathematics and Statistics
The University of Birmingham
Birmingham
UK

SUMMARY

The interaction of transient cavities in fluid flows with rigid boundaries is investigated numerically. The class of fluid motions considered here can be described by a velocity potential and a newly-developed fully three-dimensional boundary integral method is employed to compute the evolution of the free surface. Particular attention is given to the approximation of the surface velocities by radial basis function methods thus avoiding some of the limitations of polynomial approximation. The efficiency of the algorithm is demonstrated through computed examples which show the interaction of buoyancy with the Bjerknes effect of a rigid boundary for the growth and collapse of transient cavities in a fluid.

1 INTRODUCTION

Many computational problems in engineering involve following the evolution in time of an unsteady free surface. These include the interaction of water waves with marine structures as well as the dynamics of cavitation and explosion bubbles near some submerged vessel. The boundary integral method has often been applied to flows of this type because it leads to a saving in computational effort when information for the flow inside the fluid domain is not required at every time step. Since the work of Longuet-Higgins and Cokelet [8] on water waves in two dimensions, the method has been extended to axisymmetric flows [3] and more recently fully three-dimensional flows [4], [7].

Although a large number of applications have appeared there are still unresolved numerical problems. This paper addresses the aspect of surface representation and the computation of surface velocities in the fully three-dimensional case. The motivation for the work is the dynamics of gas bubbles in a liquid and the mathematical formulation is first described. This is followed by a brief account of radial basis function (*RBF*) interpolation methods and their utilisation in the algorithm for the construction of the free surface velocities. Finally, examples of bubble growth and collapse are given to illustrate the numerical scheme.

2 MATHEMATICAL FORMULATION

The particular application used to illustrate the computational method is that of the growth and collapse of a bubble near a rigid boundary under the action of gravity. This may be a cavitation bubble containing only vapour with a constant internal pressure p_c, or an explosion bubble which includes, in addition, a non-condensible gas producing an internal pressure of,

$$p_b = p_c + p_0 \left(\frac{V_0}{V}\right)^\lambda,$$

where the subscript 0 denotes an initial quantity, λ is the ratio of specific heats and V is the bubble volume.

For bubbles of this type which expand rapidly to a maximum volume after inception and then collapse, it is natural to scale lengths by the maximum radius, R_m, attained by a spherical bubble in an infinite fluid given the same initial condition. From this it follows that time can be scaled by $R_m(\rho/\Delta p)^{1/2}$ and pressure by $\Delta p = p_\infty - p_c$ where p_∞ is the hydrostatic pressure at the depth at which inception of the motion occurs. Cartesian coordinates are chosen with gravity acting vertically downwards in the negative z direction. The bubble motions considered are well-described by potential flow. Let Ω be the fluid domain with boundary $\partial\Omega \equiv S \cup \Sigma$, where S is the free surface of the bubble and Σ represents the solid boundaries. Then the velocity potential $\Phi(\boldsymbol{x},t)$, $\boldsymbol{x} = (x,y,z)$, satisfies Laplace's equation,

$$\nabla^2 \Phi = 0, \quad \text{in } \Omega,$$

together with the kinematic and dynamic free surface boundary conditions,

$$\frac{D\boldsymbol{x}}{Dt} = \nabla\Phi, \quad \boldsymbol{x} \in S,$$

$$\frac{D\Phi}{Dt} = \frac{1}{2}|\nabla\Phi|^2 - \alpha\left(\frac{V_0}{V}\right)^\lambda - \delta^2(z - z_0) + 1,$$

where D/Dt is the Lagrangian derivative. The buoyancy parameter is defined as, $\delta = (\rho g R_m/\Delta p)^{1/2}$ and the strength of an explosion is given by $\alpha = p_0/\Delta p$, so that $\alpha = 0$ for a cavitation bubble. The normal component of the velocity is assumed known on the rigid boundaries Σ.

To complete the formulation an initial condition is required to give the free surface shape along with Φ and $\nabla\Phi$ at $t = 0$ on this surface. The conditions used are those given in [1], with a spherical bubble of radius R_0 where,

$$R_0 = 0.1, \qquad \Phi_0 = -2.25806976,$$

with an initial radial velocity of 25.806976 for a cavitation bubble and

$$R_0 = 0.1651, \qquad \Phi_0 = 0,$$

with an initial velocity zero for an explosion bubble with $\alpha = 100$, $\lambda = 1.4$.

The mathematical problem defined above is solved in the following way. Given the values of the potential function Φ_i, at N nodes on the bubble surface together with $\partial\Phi/\partial n$ on the solid boundaries the Laplace equation can be expressed as a system of integral equations of the form,

$$c(\boldsymbol{x})\Phi(\boldsymbol{x}) = \int_{\partial\Omega} \left(G \frac{\partial\Phi}{\partial n'}(\boldsymbol{x}') - \Phi(\boldsymbol{x}') \frac{\partial G}{\partial n'} \right) dS', \qquad \boldsymbol{x}, \boldsymbol{x}' \in \partial\Omega,$$

where $c(\boldsymbol{x}) = 1/2$ for a smooth surface and

$$G(\boldsymbol{x}, \boldsymbol{x}') = \frac{1}{4\pi} \frac{1}{|\boldsymbol{x} - \boldsymbol{x}'|}.$$

The integration is carried out by applying a 7-point quadrature rule to a triangulation of the surface where Φ and $\partial\Phi/\partial n$ are represented by quadratic shape functions. Duffy's transformation [11] is used for the singular cases. The system is then solved by the iterative solver GMRES [10].

With Φ and $\partial\Phi/\partial n$ known at the surface nodes the surface velocity is constructed. A fourth order Adams-Bashforth-Moulton predictor-corrector method is then used to integrate the free surface boundary conditions in time.

3 SURFACE REPRESENTATION

A major difficulty in the implementation of the numerical scheme lies in the construction of the surface velocity at each time step which is dependent on the technique used to represent the free surface. Small errors at this stage are transmitted to the solution at successive time steps and can lead to instability and a breakdown of the algorithm due to a loss of smoothness. The chosen approximation method must therefore maintain both the accuracy of the solution and smoothness of the free surface. An element-based approach is rejected here because of the work needed to achieve continuity of derivatives throughout a surface patch comprising several elements when derivative information is not available *a priori*. An alternative is to use polynomials fitted to a particular node and its neighbours [4]. However, if interpolation is used then certain point arrangements will give a rank deficient matrix for the coefficients, for example, the starting shape of a sphere with a quadratic polynomial interpolant in three variables [5]. If, instead, a least squares fit is used to avoid this problem the solution is easily degraded and numerical tests with this method produced relatively large errors.

These difficulties are avoided by using local RBF interpolants at each node. A RBF interpolant to a set of scattered data f_i defined at m points $\boldsymbol{x}_i \in \mathcal{D} \subset \mathcal{R}^n$ can be expressed as,

$$s(\boldsymbol{x}) = \sum_{i=1}^{m} a_i \psi(|\boldsymbol{x} - \boldsymbol{x}_i|) + \sum_{i=1}^{K} b_i q_i(\boldsymbol{x}).$$

$\{q_i\}$ is a basis for the space $\mathcal{Q}_k$ of polynomials of order not exceeding k. In the present work the radial function ψ is chosen as the multiquadric,

$$\psi(|\boldsymbol{x} - \boldsymbol{x}_i|) = \sqrt{|\boldsymbol{x} - \boldsymbol{x}_i|^2 + c^2},$$

where c is a constant. In addition to the interpolation conditions it is required that,

$$\sum_{j=1}^{m} a_j q_i(\boldsymbol{x}_j) = 0, \quad \text{for} \quad 1 \le i \le K = \binom{k+n-1}{n}.$$

Error bounds have been obtained in [12] for RBF interpolation, for functions fulfilling a condition on their Fourier transform. Numerical experiments indicate that the multiquadric gives a better approximation to the data than interpolation or least squares fitting by a quadratic polynomial in many cases [2]. Further, the constant c can be viewed as a tensioning parameter and as c becomes larger the multiquadric approaches a polynomial fit [6].

The present method here uses multiquadric interpolation to approximate the bubble surface and its velocity at each node, given the values Φ_i and $\partial \Phi_i / \partial n$, $i = 1, 2, \ldots, N$. Starting from a distribution of nodes on the surface of a sphere resulting from refinements of an icosahedron, each node is associated with its immediate neighbours. The coordinate axes are then rotated so that the z-axis is in the direction of the vector product $\boldsymbol{a} \times \boldsymbol{b}$ where $\boldsymbol{a}$ and $\boldsymbol{b}$ are vectors formed by linking pairs of the neighbour nodes. This orientation of the axes ensures that the projections of these nodes onto the xy-plane are in general all distinct. A multiquadric is interpolated to the z values associated with the projections of the nodes to define a surface patch as $z \approx s(x, y)$ with $s(x_i, y_i) = z_i$. The unit normal to the surface can then be computed as,

$$\hat{\boldsymbol{n}} = \frac{(s_x, s_y, -1)}{(s_x^2 + s_y^2 + 1)^{1/2}}.$$

To complete the specification of the surface velocity at a particular node, the tangential derivative of the potential function must be approximated. This is achieved by applying a local trivariate multiquadric interpolant to the values $\Phi(\boldsymbol{x})$, $\boldsymbol{x} \in S$, giving the required tangential component as,

$$\nabla\Phi \cdot \hat{t} = \nabla\Phi - (\nabla\Phi \cdot \hat{n})\hat{n}.$$

Numerical investigations carried out for the present application show that this approach is more accurate for derivative estimation than the surface adapted forms, such as the spherical multiquadric of [9], when the surface is not spherical.

4 RESULTS

A simple test of the accuracy of the algorithm is the pulsation of a spherical explosion bubble in an infinite fluid in the absence of body forces. The motion is described by the Rayleigh-Plesset equation for the bubble radius $R(t)$,

$$R\ddot{R} + \frac{3}{2}\dot{R}^2 = \alpha\left(\frac{R_0}{R}\right)^{3\lambda} - 1.$$

This can be solved by a standard ODE solver to any desired accuracy. The numerical scheme described in this paper gives good accuracy for a discretization of the bubble surface with $N = 42$ nodes and more than six periods can be easily computed. When $N = 162$, the match with the ODE solution is almost exact, but the surface begins to lose smoothness after the second period. This is acceptable because the purpose of the present study is to investigate the deformation of a bubble during the first period of its motion.

Comparison can also be made with the results of the axisymmetric boundary integral method for bubble growth and collapse near a rigid wall [1]. Here the fully three-dimensional method presented in this paper agrees with the axisymmetric case for $\gamma = 1.5$ to within 1% of the bubble period for $N = 362$, where γ is the non-dimensional distance from the wall.

A simple illustration of the numerical method without axisymmetry is the growth and collapse of a cavitation bubble near a vertical rigid wall with a buoyancy force acting parallel to the wall. In this situation there is an interaction between the Bjerknes effect of attraction horizontally towards the rigid boundary and the upward force of buoyancy. For $\gamma = 4.0$, $\delta = 0.25$ (fig.(1)) the buoyancy force is dominant. The bubble remains nearly spherical in shape for most of the period of motion and during the collapse phase its centroid moves upward at an angle of $77°$ to the horizontal (directed towards the wall). By the time $t = 1.899$ a jet has begun to form on the underside of the bubble which broadens and moves upward in the direction of the centroid translation already established.

The case $\gamma = 1.1$, $\delta = 0.25$ shows the effect of a stronger Bjerknes force, but with buoyancy still significant. The closeness of the bubble to the wall causes a flattening of the near side by $t = 1.483$ (see fig.(2)). The appearance of the bubble at this time is similar to cases computed axisymmetrically where a jet penetrates the bubble in the direction of the wall. However, in the case illustrated, the effect of buoyancy causes a distortion of the shape and a small jet appears which moves vertically to pinch off the upper end of the bubble.

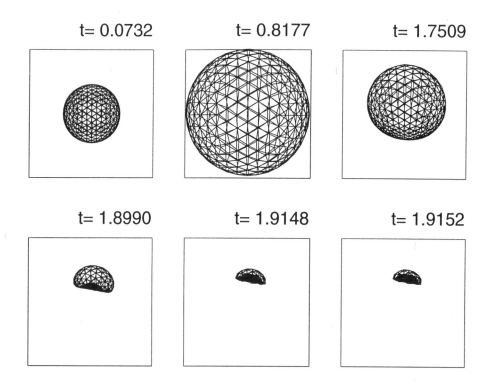

Figure 1 The growth and collapse of a cavitation bubble for $\gamma = 4.0$, $\delta = 0.25$. The right hand side of the frame is parallel to the wall at a distance 3.0.

The numerical method thus successfully models the evolution of a free surface in the context of bubble dynamics. The use of *RBF* approximations for the surface velocities avoids some of the drawbacks of polynomial fitting while improving accuracy. Further work is planned to extend the method to other applications and to validate the results obtained by experimental observations.

Funding for this work from the Defence Research Agency is gratefully acknowledged.

References

[1] J. P. Best and A. Kucera. A numerical investigation of non-spherical rebounding bubbles. *J. Fluid Mech.*, 245:137–154, 1992.

[2] J. R. Blake, J. M. Boulton-Stone, and R. P. Tong. Boundary integral methods for rising, bursting and collapsing bubbles. In H. Power, editor, *BE Applications in Fluid Mechanics*, pages 31–72. Computational Mechanics Publications, 1995.

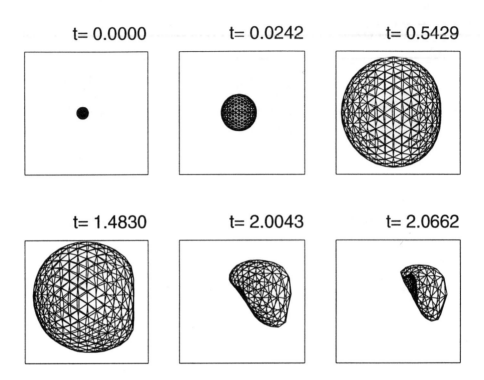

Figure 2 The growth and collapse of a cavitation bubble for $\gamma = 1.1$, $\delta = 0.25$. The right hand side of the frame shows the position of the wall

[3] J. R. Blake, B. B. Taib, and G. Doherty. Transient cavities near boundaries. Part 1. Rigid boundary. *J. Fluid Mech.*, 170:479–497, 1986.

[4] G. L. Chahine and T. O. Perdue. Simulation of the three-dimensional behavior of an unsteady large bubble near a structure. *3rd International Colloquium on Bubbles and Drops, Monterey*, 1988.

[5] C. de Boor and A. Ron. Computational aspects of polynomial interpolation in several variables. *Math. Comp.*, 198:705–727, 1992.

[6] M. Eck. MQ-curves are curves in tension. In T. Lyche and L. L. Schumaker, editors, *Mathematical Methods in Computer Aided Design*, volume II, pages 217–228, 1992.

[7] P. J. Harris. A numerical model for determining the motion of a bubble close to a fixed rigid structure in a fluid. *Int. J. Num. Meth. Eng.*, 33:1813–1822, 1992.

[8] M. S. Longuet-Higgins and E. D. Cokelet. The deformation of steep surface waves on water: I. A numerical method of computation. *Proc. R. Soc. Lond. A*, 350:1–26, 1976.

[9] H. Pottmann and M. Eck. Modified multiquadric methods for scattered data interpolation over a sphere. *Comp. Aid. Geom. Des.*, 7:313–321, 1990.

[10] Y. Saad and M. H. Schultz. GMRES: a generalized minimal residual algorithm for solving nonsymmetric linear systems. *SIAM J. Sci. Stst. Comput.*, 7:856–869, 1986.

[11] C. Schwab and W. L. Wendland. On numerical cubatures of singular surface integrals in boundary element methods. *Num Math*, 62:343–369, 1992.

[12] Z-M Wu and R. Schaback. Local error estimates for radial basis function interpolation of scattered data. *IMA J. Num. Anal.*, 13:13–27, 1993.

Numerical experimentation of the first transition to unsteadiness of air free convection in a differentially heated cubic cavity with non active adiabatic walls

E. Tric, G. Labrosse and M. Betrouni [1]

CNRS - LIMSI, Université Paris–Sud, 91405 Orsay Cedex, France

Abstract

The transition to unsteadiness of air free convection flows in differentially heated cubic cavity (two opposite vertical faces have fixed temperature, the other ones are adiabatic) is observed in (pseudo-spectral) numerical experiments. It presents a hysteretic behaviour characterized by two critical Rayleigh number values, Ra_{c1} and Ra_{c2}, found to lie in the respective ranges $[3.3, 3.4] \times 10^7$ and $[3.1, 3.2] \times 10^7$. Those values are about six times smaller than the corresponding $Ra_{2D,c}$ value for the square cavity. Preliminary results are given about the pre-transitional and transitional flows, their symmetry properties in particular, and some first comparisons between the 2D and the mid-plan 3D flows.

1 Introduction

In the past two decades, thermal convection in differentially (on two opposite vertical faces) heated cavity has been studied extensively, both numerically and experimentally, for its wide variety of technical applications. The only configuration which we are interested in, with this contribution, has *adiabatic non active walls*, in contrast with the case where at least one non active wall is thermally conductive. This latter situation mixes the dynamics we are investigating with something else (a Rayleigh-Bénard behaviour for instance) and, consequently, its analysis deserves to be delayed, somehow.

Of course, the numerical experiments have been mainly carried out for the 2D (square, rectangular) case which can be considered as well known to-day. In particular, the transition to unsteadiness is largely documented [1, 2, 3, 4, 5] with a threshold value $[Ra_{2D,c} = (1.82 \pm 0.01) \times 10^8, [3]]$ but the mechanism not quite identified.

Relevant 2D experimental data are out of reach, may be for ever, at least until the role of the third direction be understood, while, now, 3D accurate numerical experiments are quite conceivable. To our knowledge, nothing, neither experimental nor numerical, has been published to-day about the first transition to unsteadiness in the differentially heated cubic (and parallellepipedic, a fortiori) cavity. Let us only point out the early numerical works, as the very first

---

[1] on leave from Institut de Physique, USTHB, El Alia, BP 32, Alger, Algeria

description [6] of the 3D flow structure, or the exploration in liquid metal [7], or, at last, the first attempts [8, 9] made with large Rayleigh number values, up to $10^{7.5}$ (i.e. not far from the onset of unsteadiness!). The transition quoted in [10, 11, 12] concerns the non adiabatic case.

The present investigation is focused on the numerical analysis of the first transition in the differentially heated cube and on the comparison of the 2D and 3D mechanisms. In particular, the additional symmetry, introduced by the third direction, is seen to be clearly broken in this transition whose hysteretic character is not observed in 2D: good indications of a typically 3D effect.

2 The mathematical model and the numerical method

The usual dimensionless Boussinesq equations are:

$$\frac{\partial \mathbf{v}}{\partial t} + (\mathbf{v} \cdot \nabla)\mathbf{v} = -\nabla p + Pr \nabla^2 \mathbf{v} + Ra\, Pr\, T\, \widehat{\mathbf{e}}_z$$

$$\nabla \cdot \mathbf{v} = 0$$

$$\frac{\partial T}{\partial t} + (\mathbf{v} \cdot \nabla)T = \nabla^2 T$$

where:
- the unit vector $\widehat{\mathbf{e}}_z$ points the upward vertical direction, $\widehat{\mathbf{e}}_x$ and $\widehat{\mathbf{e}}_y$ being the unit vectors in the horizontal plane;
- the lengths, the velocity $\mathbf{v} = u\widehat{\mathbf{e}}_x + v\widehat{\mathbf{e}}_y + w\widehat{\mathbf{e}}_z$ and temperature T are respectively scaled by the cube edge H, the thermal diffusion velocity $\frac{\kappa}{H}$ and the imposed temperature difference ΔT, the other scales being derived from these ones;
- Ra and Pr (fixed at 0.71) are the Rayleigh and Prandtl numbers.

No-slip boundary conditions are imposed on all the faces of the cube. The two vertical thermally active faces are defined by $T(x = \pm\frac{1}{2}, y, z) = \mp\frac{1}{2}$ the other ones being adiabatic: $\frac{\partial T}{\partial n} = 0$, at $(x, y = \pm\frac{1}{2}, z)$ and $(x, y, z = \pm\frac{1}{2})$, $\frac{\partial}{\partial n}$ standing for the appropriate normal derivative.

A Chebyshev Gauss-Lobatto collocation method has been used to evaluate the fields spatial derivatives. The velocity and pressure have been uncoupled by a "projection - diffusion" approach, recently proposed in [13]. It is an unconditionally stable direct solver of the unsteady Stokes problem. No additional time scheme is required to split the operators. The pressure is first evaluated, from a Darcy system, to ensure the numerical value of the divergence of an intermediate field which becomes, in its turn, the source of a diffusion (and advection, in Navier-Stokes) equation to be solved for the velocity field. The obtained divergence improves itself exponentially as the nodes number increases, as expected from the numerical divergence of any analytically given (regular) velocity field.

These equations have been time integrated by a classical second order finite differences scheme (Crank-Nicolson for the diffusion terms and Adams-Bashforth for any explicit evaluation).

All the maxima quoted, from now on, for the velocity divergence (scaled by the largest modulus of the velocity), on one hand, and for the velocity components, on the other hand, have been looked for respectively on the nodes, and by a second order Lagrangian interpolation. Their exponential convergence with the nodes number has been checked. For instance the 3D divergence for the $Ra = 10^6$ flow is reduced by 10 doubling the nodes number in each direction.

3 Results

3.1 The Steady Flows

We come up, to the validation of our numerical results, to the characterization of the 3D pre-transitional flows and the comparison with the 2D flows.

Convergence to steadiness is declared when the following criterion is satisfied: $\frac{|\phi_n - \phi_{n-1}|}{|\phi_n|\delta t} \leq 10^{-1}$ for all ϕ, where ϕ_n stands for the maximum absolute value, found on the nodes at time $n\delta t$, of one of the physical fields (the velocity components and temperature).

Table 1: Compared $Ra = 10^6$ 3D air flow results following [11] (velocity data are scaled according to [15]).

Authors	Grid	u_{max}	w_{max}	S
This study	61^3	0.7748	0.2559	0.9101
	81^3	0.7645	0.2581	0.9095
[10]	62^3	0.8910	0.2588	-
[11]	120^3	0.8575	0.2585	0.9103

All the flows mentioned in this subsection have been obtained (with a 51^3 grid and $\delta t = 4 \times 10^{-6}$) from steady flows converged at a lower Ra value. The highest Ra value for which the steadiness has been declared is 3.3×10^7.

An excellent agreement has been observed between our 2D results and the benchmark data given in [14].

In the absence of 3D benchmark data, we adopt the approach proposed by the only published attempt to compare 3D results [11] which quotes, surprisingly, data which are not so relevant (as shown hereinafter) to the specific 3D flow structure. Table 1 adds our $Ra = 10^6$ results to those (taken from [11]) of previous studies. All the tabulated quantities concern the mid-plan $y = 0$ in which the flow looks much like the 2D flow: $u_{max} = max[u(0, 0, z)]$, $w_{max} = max[w(x, 0, 0)]$ and the central stratification $S = \frac{\partial T}{\partial z}|_{(0,0,0)}$. As suggested in [15] for the 2D flow, the velocity data, in this table, have been respectively scaled by $(RaPr)^{\frac{1}{3}}$ and $\sqrt{RaPr}$. To comment the disagreement between the u_{max} data requires the knowledge of the effective order of accuracy of the schemes implemented with unequal meshing by [10, 11].

To characterise the 3D structure of the pre-transitional flows, Table 2 gives the following results: in the first two columns come the 3D data, (a) the maximum, in the cavity, of each velocity component and of the divergence (denoted d_{3D}), and (b) the maximum, in the mid-plan ($y = 0$), of the u and w components (v cancels there) and of the (scaled, pointwise) quantity $\frac{\partial v}{\partial y}|_{(x,0,z)}$ (denoted d_{mp}); (c) the last column brings reference 2D data, at the same Ra values, namely the maximum of each of the velocity components and of the divergence (denoted d_{2D}), both maxima found in the square, of course. Each of these extrema has its own location.

As already noted by [8, 9], the three-dimensionality is not quantitatively dominant: for instance,

Table 2: Characteristic data from the steady 3D flows and the corresponding 2D flows.

Rayleigh number	3D		2D
	cavity	mid-plan	
10^7	$u = 383.7$	$u = 383.7$	$u = 387.2$
	$w = 767.7$	$w = 698.7$	$w = 697.9$
	$v = 83.4$		
	$d_{3D} = 0.12$	$d_{mp} = 0.20$	$d_{2D} = 0.14$
3.0×10^7	$u = 647.1$	$u = 647.1$	$u = 650.8$
	$w = 1347.$	$w = 1210.$	$w = 1217.$
	$v = 171.9$		
	$d_{3D} = 0.36$	$d_{mp} = 1.58$	$d_{2D} = 0.26$
3.1×10^7	$u = 657.3$	$u = 657.3$	$u = 660.8$
	$w = 1370.$	$w = 1230.$	$w = 1238.$
	$v = 175.3$		
	$d_{3D} = 0.37$	$d_{mp} = 1.61$	$d_{2D} = 0.26$
3.2×10^7	$u = 667.3$	$u = 667.3$	$u = 670.6$
	$w = 1392.$	$w = 1251.$	$w = 1258.$
	$v = 178.7$		
	$d_{3D} = 0.38$	$d_{mp} = 1.65$	$d_{2D} = 0.26$
3.3×10^7	$u = 677.1$	$u = 677.1$	$u = 680.3$
	$w = 1415.$	$w = 1271.$	$w = 1277.$
	$v = 182.$		
	$d_{3D} = 0.38$	$d_{mp} = 1.67$	$d_{2D} = 0.27$

the maximum of v, the "depth" ("third" direction) component, does not exceed 13% of the maximum of the vertical component w. On the other hand, the mid-plan contains the maximum of the horizontal u component, and does not differ so much from the 2D case, as far as the velocity component extrema are concerned (although systematically smaller in the mid-plan than in the 2D case). So the square cavity can be considered, to some extent, as an approximation to (under) evaluate the thermal transfer properties of the cube.

The significant departure from zero of the tabulated d_{mp}, compared to d_{2D}, indicates clearly that the mid-plan flow must be somewhat different from the 2D flow, unless the 3D structure be so particular as to lead to cancel $\frac{\partial v}{\partial y}|_{(x,0,z)}$ everywhere in the mid-plan. Indeed, in addition to the 2D centro-symmetry (CS), w.r.t. the axis $x = z = 0$, the 3D flows have the "depth" as an other symmetry direction, with the mid-plan (MP) symmetry of the u and w components and, from the $\nabla.\mathbf{v} = 0$ constraint, a MP antisymmetric v component.

An other typical 3D effect comes from Table 2 which confirms the already mentioned [11] dependence of the velocity vertical component on the depth, its maximum being located near the end walls, at a y-distance which scales with $Ra^{-\frac{1}{4}}$. Figure 1 shows the maximum (denoted $w_{max}(y)$), in each nodal y-plan, of $w(x, y, z)$.

3.2 The Unsteadiness

(1) A different time evolution appears, without convergence to steadiness, starting from the converged steady flow at $Ra = 3.3 \times 10^7$ to get (with 51^3 and 61^3 grids) the $Ra = 3.4 \times 10^7$ and $Ra = 3.5 \times 10^7$ flows. For this latter case, figure 2 presents the time trace of the u component, at some given point, on a duration of 0.88 units of diffusif time, with a 0.35 units transient stage preceding a permanent time behaviour. The former case presents an analogous time evolution, but must be experimented on a large enough duration to be declared as unsteady.

(2) Starting from the unsteady flow at $Ra = 3.5 \times 10^7$ to go down to the $Ra = (3.4, 3.3, 2, 3.1) \times 10^7$ flows, a converged steady state has been recovered only with the last smaller Ra value. An hysteretic behaviour characterizes this transition. On figure 3 are plotted the relative amplitudes (scaled by the time average of the signal) of the velocity components oscillations, measured at a given point, for these unsteady flows, as a function of the Rayleigh number. These data fit a linear law which gives $Ra = 3.15 \times 10^7$ as the starting value of the hysteretic branch.

All the unsteady flows have in common to break the MP symmetry, while keeping the 2D CS property, as can be seen clearly on figure 4 where are plotted the contour lines, in the mid-plan, of an instantaneous $v(x, y = 0, z)$ component for the $Ra = 3.5 \times 10^7$ case. The MP symmetry breaking is localized in the two "hydraulic jumps" regions. Everyone, because of the 2D CS property, is crossed, in both directions, by the flow, a very weak forward flow across a major part of this plan being necessary to ensure the balance between the net forward and backward flows. The MP symmetry is recovered by the steady flow obtained at $Ra = 3.1 \times 10^7$.

4 Conclusion

Preliminary numerical results are reported on the onset of unsteadiness of air convection flows in the thermally driven cubic cavity. In contrast with the square cavity case, an hysteresis characterises this transition which occurs at Ra values significantly smaller than the corresponding transition level of the 2D case. Moreover, the symmetry w.r.t. the third direction is precisely the only one broken by this transition. These are good indications of a specifically 3D transition mechanism.

Acknowledgements

The authors wish to thank the IDRIS (CNRS computing center) scientific committee for providing support through computing hours given on the Cray-C94, and the University PARIS-SUD computing service for opening the access to the Cray-YMP of CRI. We are pleased to acknowledge Mr A. Batoul and Miss C. Sambourg for their valuable help.

Bibliography

[1] LE QUÉRÉ, P., ALZIARY de ROQUEFORT, Th.,"Transition to unsteady natural convection of air in vertical differentially heated cavities: Influence of thermal boundary conditions on the horizontal walls", *Proc. 8th Int. Heat Transfer Conf., San Francisco, (1986) pp. 1533-1538.*

[2] PAOLUCCI, S., CHENOWETH, D. R.,"Transition to chaos in a differentially heated vertical cavity", *J. Fluid Mech., 201 (1989) pp. 379-410.*

[3] LE QUÉRÉ, P., BEHNIA, M.,"From onset of unsteadiness to chaos in a differentially heated square cavity", *Rapport interne du LIMSI, 94-02 (1994).*

[4] RAVI, M. R., HENKES, R. A. W. M., HOOGENDOORN, C. J.,"On the high-Rayleigh-number structure of steady laminar natural-convection flow in a square enclosure", *J. Fluid Mech., 262 (1994) pp. 325-351.*

[5] JANSSEN, R. J. A., HENKES, R. A. W. M.,"Influence of Prandtl number on instability mechanisms and transition in a differentially heated square cavity", *J. Fluid Mech., 290 (1995) pp. 319-344.*

[6] MALLINSON, G. D., DE VAHL DAVIS, G., JONES, I.P.,"Three-dimensional natural convection in a box: a numerical study", *J. Fluid. Mech., 83 (1977) pp. 1-31.*

[7] VISKANTA, R., KIM, D. M., GAU, G.,"Three-dimensional natural convection heat transfer of a liquid metal in a cavity", *Int. J. Heat Mass Transfer, 29 (1986) pp. 475-485.*

[8] HALDENWANG, P.,"Résolution tridimensionnelle des équations de Navier-Stokes par methodes spectrales Tchebycheff: Application à la convection naturelle", *Thèse d'Etat, Université de Provence (1984).*

[9] HALDENWANG, P., LABROSSE, G.,"2-D and 3-D spectral Chebyshev solutions for free convection at high Rayleigh number", *Proc. 6th Int. Symposium on finite element methods in flow problems, (1986) pp. 261-266*

[10] FUSEGI, T., HYUN. J. M., KUWAHARA, K.,"Three-dimensional numerical simulation of periodic natural convection in a differentially heated cubical enclosure", *Appl. Sc. Res., 49 (1992) pp. 271-282.*

[11] JANSSEN, R. J. A., HENKES, R. A. W. M., HOOGENDOORN, C. J., "Transition to time-periodicity of a natural-convection flow in a 3D differentially heated cavity", *Int. J. Heat Mass Transfer, 36 (1993) pp. 2927-2940.*

[12] HENRY, D., BUFFAT, M.,"Two and three-dimensional numerical simulations of the transition to oscillatory convection in low-Prandtl fluid", *Report Lab. de Méc. des Fluides et d'Acoust. (ECL) (1991).*

[13] BATOUL, A., KHALLOUF, H., LABROSSE, G., "Une méthode de résolution directe (pseudo-spectrale) du problème de Stokes 2D/3D instationnaire. Application à la cavité entraînée carrée.", *C.R. Acad. Sci. Paris, t.319, Série II (1994)* pp. *1455-1461*.

[14] LE QUÉRÉ, P.,"Accurate solutions to the square thermally driven cavity at high Rayleigh number", *Computers & Fluids, 20 (1991) pp. 29-41*.

[15] HENKES, R. A. W. M.,"Natural-convection boundary layer", *PhD-thesis, Delft University of Technology, (1990)*.

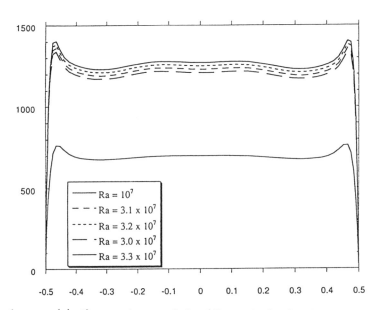

Figure 1: $w_{max}(y)$, the maximum of the (dimensionless) velocity vertical component in each nodal y-plane, as a function of y, for the steady flows obtained at $Ra = (1, 3, 3.1, 3.2, 3.3) \times 10^7$.

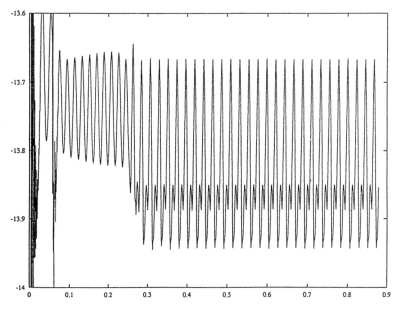

Figure 2: Time history of the u component at one given point of the 51^3 grid, at $Ra = 3.5 \times 10^7$.

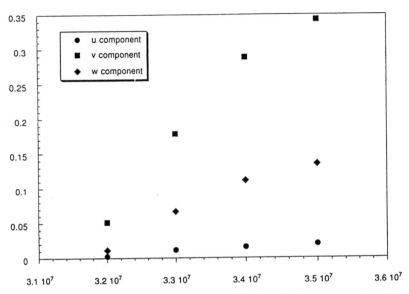

Figure 3: Dependence on the Ra value of the amplitude of the temporal oscillations of u, v and w components.

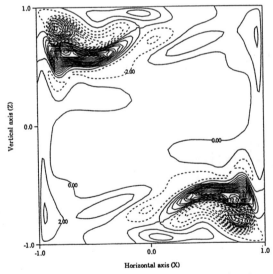

Figure 4: Instantaneous contour lines of the v component in the mid-plane ($y = 0$), at $Ra = 3.5 \times 10^7$, with a 51^3 grid. The continous and dashed lines correspond to positive and negative values respectively.

COMPUTATION OF UNSTEADY 3-D TRANSONIC FLOWS DUE TO FLUCTUATING BACK-PRESSURE USING k-ε TURBULENCE CLOSURE

I. Vallet, G.A. Gerolymos
LEMFI, URA CNRS 1504, Bldg. 511, Université Pierre-et-Marie-Curie, 91405 Orsay, Paris, France
P. Ott, A. Bölcs
LTT–DGM, Ecole Polytechnique Fédérale de Lausanne, CH-1015 Lausanne, Switzerland

Abstract

A 3-D Navier-Stokes solver is applied to the computation of unsteady nozzle flows due to fluctuating back-pressure and validated by comparison with experimental measurements. The flow is modelled by the 3-D compressible Navier-Stokes equations using the Launder-Sharma near-wall $k - \epsilon$ turbulence model. The mean-flow and turbulence-transport equations are integrated in time using a 1-order implicit scheme with 3-order MUSCL upwind-biased Van Leer flux-vector-splitting. The configuration studied is a 3-D nozzle, with thick sidewall boundary-layers. The 3-D effect is produced by the shock-wave/boundary-layer interaction at the corners of the nozzle, where an important 3-D recirculating zone appears. The flow is unsteady due to back-pressure fluctuation, produced experimentally by a rotating rod of elliptical cross-section. The unsteady computations were run for a back-pressure fluctuation-frequency of 180 Hz. The computed and measured results compare satisfactorily. An analysis of the 3-D steady and unsteady flow is undertaken.

Introduction

As early as 1984 Liou and Coakley[1] investigated the unsteady 2-D transonic flow in diffusers using the 2-D Navier-Stokes equations with $k - \omega^2$ turbulence closure, using grids where the first grid-point away from the wall was at $n_w^+ \cong 5$. Both forced and natural flow oscillations were studied. The validity of 2-equation closures for unsteady internal transonic flows has only been tested in 2-D. Carefully investigation of existing experimental data for transonic shock-wave/boundary-layer interaction[2] suggests that there invariably exists nonnegligible 3-D contamination of nominally 2-D configurations. Therefore assessing the validity and drawbacks of models is only possible by 3-D computations. With this objective, the unsteady 3-D transonic flow in a Laval nozzle, due to fluctuating back-pressure, using the Navier-Stokes equations with the Launder-Sharma[3] near-wall turbulence closure, is studied in this work. The numerical method used is based on the steady flow solver developed by Gerolymos and Vallet[2]. The configuration studied has been investigated experimentally by Ott et al.[4].

Flow Model and Computational Method

The flow is modelled by the compressible Favre-Reynolds-averaged 3-D Navier-Stokes equations, with the Launder-Sharma[3] near-wall $k - \varepsilon$ closure, written in tensor-invariant form[5]

$$\frac{\partial \bar{\rho}}{\partial t} + \operatorname{div}[\bar{\rho}\tilde{\vec{V}}] = 0 \quad ; \quad \frac{\partial \bar{\rho}\tilde{\vec{V}}}{\partial t} + \operatorname{div}\left[\bar{\rho}\tilde{\vec{V}} \otimes \tilde{\vec{V}} + \bar{p}\boldsymbol{I} - \bar{\boldsymbol{\tau}} + \bar{\rho}\widetilde{\vec{V}'' \otimes \vec{V}''}\right] = 0$$

$$\frac{\partial(\bar{\rho}h_t - \bar{p})}{\partial t} + \operatorname{div}\left[\bar{\rho}\tilde{\vec{V}}h_t - \tilde{\vec{V}} \cdot (\bar{\boldsymbol{\tau}} - \bar{\rho}\widetilde{\vec{V}'' \otimes \vec{V}''}) + (\bar{\vec{q}} + \bar{\rho}\widetilde{e''\vec{V}''})\right] = -\left(P_k - \bar{\rho}\varepsilon^* - 2\mu[\operatorname{grad}\sqrt{k}]^2\right)$$

$$\frac{\partial \bar{\rho}k}{\partial t} + \operatorname{div}\left[\bar{\rho}\tilde{\vec{V}}k - (\mu + \frac{\mu_T}{\sigma_k})\operatorname{grad}k\right] = \left(P_k - \bar{\rho}\varepsilon^* - 2\mu[\operatorname{grad}\sqrt{k}]^2\right) \quad (1)$$

$$\frac{\partial \bar{\rho}\varepsilon^*}{\partial t} + \operatorname{div}\left[\bar{\rho}\tilde{\vec{V}}\varepsilon^* - (\mu + \frac{\mu_T}{\sigma_\varepsilon})\operatorname{grad}\varepsilon^*\right] = \left(C_{\varepsilon 1}P_k\frac{\varepsilon^*}{k} - C_{\varepsilon 2}f_{\varepsilon 2}\bar{\rho}\frac{\varepsilon^{*2}}{k} + 2\frac{\mu\mu_T}{\bar{\rho}}[\nabla^2 \tilde{\vec{V}}]^2\right)$$

$$\bar{p} = \bar{\rho} R_g \tilde{T} = \bar{\rho}\frac{\gamma-1}{\gamma}\tilde{h} = \bar{\rho}(\gamma-1)\tilde{e}$$

$$-\bar{\rho}\widetilde{\vec{V}''\otimes\vec{V}''} \cong \mu_T\left(2\tilde{\boldsymbol{D}} - \frac{2}{3}\mathrm{div}\tilde{\vec{V}}\boldsymbol{I}\right) - \frac{2}{3}\bar{\rho}k\boldsymbol{I} \quad ; \quad \widetilde{\bar{\rho}e''\vec{V}''} \cong -k_T\mathrm{grad}\tilde{T} \qquad (2)$$

where t is the time, $\tilde{\vec{V}}$ the velocity, $\bar{\rho}$ the density, $\bar{p}$ the pressure, $\tilde{T}$ the temperature, $\tilde{h}$ the enthalpy, $\gamma = 1.4$ the isentropic exponent, $R_g = 287.04 \text{ m}^2 \text{ s}^{-1} \text{ K}^{-1}$ the gas-constant for air, $h_t = \tilde{h} + \frac{1}{2}\tilde{V}^2$ the total enthalpy of the mean flow (which is different from $\tilde{h}_t = h_t + k$), k the turbulence-kinetic-energy, ε^* the isotropic part of the turbulence-kinetic-energy dissipation-rate ($\varepsilon^* = \varepsilon - 2\nu[\mathrm{grad}\sqrt{k}]^2$, where ε is the dissipation-rate), $\bar{\tau}$ the viscous-stress-tensor, $\bar{q}$ the heat-flux-vector, $\tilde{\boldsymbol{D}}$ the rate-of-deformation-tensor, $\boldsymbol{I}$ the identity-tensor, μ the molecular dynamic viscosity, μ_T the eddy viscosity, k the molecular heat conductivity, k_T the eddy heat conductivity, $\tilde{}$ denotes Favre-averaging, and $\bar{}$ denotes nonweighted-averaging.

The mean-flow and turbulence-transport equations are discretized in space, on a structured grid, using a 3-order upwind-biased MUSCL scheme with Van Leer flux-vector-splitting and Van Albada limiters. The resulting semi-discrete scheme is integrated in time using a 1-order implicit procedure. The mean-flow and turbulence-transport equations are integrated simultaneously. The Jacobian-flux-matrix is computed using a 1-order space-discretization, to reduce bandwidth, and factored. The resulting linear systems are solved by LU decomposition. The time-step is based on a combined convective (Courant) and viscous (von Neumann) criterion[2]. For the unsteady computations presented in this work $CFL_{max} \cong 500$.

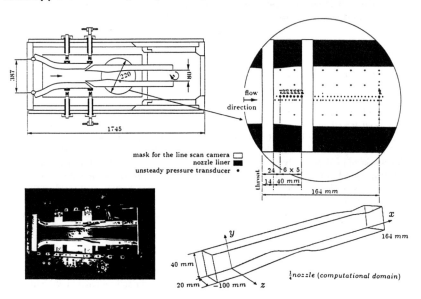

Figure 1: Experimental set-up and computational domain

In this work $\frac{1}{4}$ of the nozzle was discretized. At the upstream boundary a reservoir condition was applied (since the flow is choked and no perturbation reaches the upstream boundary). This condition as well as the no-slip wall boundary-condition and the symmetry-conditions at the symmetry-plane are identical with those used for steady flow computations[2]. At the outflow boundary the pressure signal is imposed, described as the sum of 12 time-harmonics obtained from the measurements

$$\bar{p}_2(t) = {}^0\bar{p}_2 + \sum_{m=1}^{12}\left[|{}^m\bar{p}_2|\cos(2\pi mft + {}^m\varphi_2)\right] \Longrightarrow \frac{\partial\bar{p}_2}{\partial t} = -\sum_{m=1}^{12}\left[2\pi mf|{}^m\bar{p}_2|\sin(2\pi mft + {}^m\varphi_2)\right] \qquad (3)$$

where ${}^0\bar{p}_2$, is the time-mean outflow pressure, $|{}^m\bar{p}_2|$ and ${}^m\varphi_2$ the amplitudes and phase-angles of the Fourier-coefficients, and m the harmonic number. Because experimental measurements were available only at the wall it was assumed that pressure is homogeneous on the exit-plane. The outflow boundary-condition imposes the pressure (and its time-derivative in the implicit phase of the scheme) and extrapolates all other variables ($\bar{\rho}$, $\tilde{V}$, k, ε^*)[2]. In order to achieve high time-steps it is indispensible to apply boundary-conditions both implicitly and explicitly.

Experimental Set-up

The experimental set-up has been described in detail by Ott et al.[4]. A Laval-nozzle was equipped with nozzle-liners giving a converging-diverging section (Fig. 1). On the nozzle-liners the boundary-layer was cut off before the throat (Fig. 1). On the sidewalls the boundary-layers were not modified. These relatively thick sidewall-boundary-layers interact with the shock-wave at the nozzle-corners and produce the significant 3-D effects observed. At the exit section of the nozzle a cylindrical rod with elliptical cross-section is rotated by a hydraulic motor. The rod is situated 480 mm downstream of the throat. The blockage and wake-losses of the rod induce a shock-wave in the test-section, which oscillates with the rotation of the rod. The rotation-frequency of the rod can be varied in the range 0–180 Hz. Steady-state pressure-taps are mounted on the nozzle-liners and on the sidewall, at the symmetry-planes of the nozzle (Fig. 1). Unsteady pressures are measured using 8 unsteady pressure transducers[4], mounted on the sidewall y-symmetry-plane, 1 located 164 mm downstream of the throat and 7 spaced 5 mm apart in the shock-wave neighbourhood (Fig. 1). The downstream unsteady transducer records the unsteady pressure signal that is applied as a downstream boundary condition for the computations (Eq. 3).

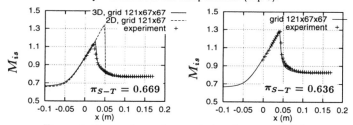

Figure 2: Comparison of computed and measured M_{is_w}.

Table 1: Steady computations summary

π_{S-T}	$N_i(N_x)$	$N_j(N_y)$	$N_k(N_z)$	y_w^+	z_w^+	r_y	r_z	Mpoints*	CPU-h†
0.6360	121	67	67	0.480	0.500	1.150	1.131	0.52	11
0.6525	101	53	53	1.520	1.536	1.152	1.149	0.27	3
	121	57	49	0.500	0.500	1.182	1.194	0.32	7
	121	67	67	0.500	0.500	1.149	1.131	0.52	11
0.6690	121	67	67	0.500	0.500	1.149	1.131	0.52	11

*1 Mpoint= 1024^2 points ; †CRAY-C98

Steady Computations

Steady pressure-measurements were available for 2 different static-to-total ($\pi_{S-T} = \bar{p}_2/p_{t_1}$) nozzle-pressure-ratios, corresponding to the nonrotating rod horizontal ($\pi_{S-T} = 0.6360$) or vertical ($\pi_{S-T} = 0.6690$). Steady 3-D computations were run for both these cases using a (121×67×67) grid, with $y_w^+ \cong z_w^+ \cong 0.5$ (Table 1)). The computational grids used in this study consist of planes normal to the x-axis, equidistant x-wise. The grid is geometrically stretched (with ratio r_y and r_z) in the y and z directions respectively[2]. At inflow the experimentally determined conditions were applied (inflow-conditions are denoted by $(\cdot)_1$ and outflow-conditions by $(\cdot)_2$)

$$p_{t_1} = 168600 \text{ Pa} \quad ; \quad T_{t_1} = 323 \text{ K} \quad ; \quad \delta_{y_1} = 0.5 \text{ mm} \quad ; \quad \delta_{z_1} = 5 \text{ mm}$$

$$T_{u_1} = 3.5\% \quad ; \quad \ell_{T_1} = 0.05 \text{ mm} \tag{4}$$

where δ_{y_1} is the boundary-layer thickness at inflow on the nozzle-liners, δ_{z_1} the boundary-layer thickness at inflow on the sidewall, p_{t_1} the inflow-total-pressure outside the boundary-layers, T_{t_1} the inflow-total-temperature outside the boundary-layers, $T_{u_1} = \sqrt{\frac{2}{3}k_1}\tilde{V}_1^{-1}$ the turbulence intensity at inflow, and $\ell_{T_1} = k_1^{\frac{3}{2}}\varepsilon_1^{*-1}$ the turbulence length-scale at inflow. The value of T_{t_1} is an average value, since during the measurements T_{t_1} varied in the range 313–333 K. The boundary-layer thicknesses at inflow were approximately chosen to represent the thick sidewall-boundary-layers and the effect of removing the boundary-layers on the liners (Fig. 1).

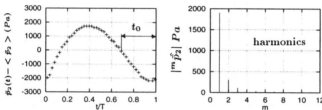

Figure 3: Measured outflow unsteady pressure signal

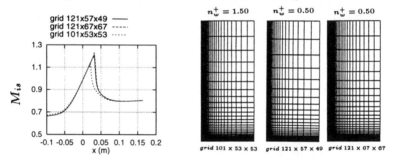

Figure 4: Grid influence on steady flow results
The comparison of computed and measured isentropic-wall-Mach-number

$$M_{is} = \sqrt{\frac{2}{\gamma - 1}\left[\left(\frac{p_{t_1}}{\bar{p}}\right)^{\frac{\gamma-1}{\gamma}} - 1\right]} \tag{5}$$

distributions at the y-symmetry-plane on the sidewall (Fig. 2) shows quite satisfactory agreement. In the same plot (Fig. 2) is included the 2-D result obtained with the 3-D code run in pseudo-2-D mode on the same grid ($\pi_{S-T} = 0.6690$). It is evident that 3-D effects are important. It is impossible to obtain 2-D results for the $\pi_{S-T} = 0.6360$ case, because this pressure-ratio is lower than the shock-at-exit pressure-ratio.

Unsteady Flow Results

Unsteady computations were initialized from a converged steady-state flowfield, corresponding to the arithmetic mean π_{S-T} between the 2 extreme positions of the rod. It was assumed that this arithmetic mean is a good approximation to the time-average unsteady flow $^0\pi_{S-T}$ (only dynamic unsteady signals were available in the experiment). The computations were performed for a rod rotation frequency $f = 180$ Hz. The exit of the computational grid coincides with the downstream unsteady pressure probe location ($x = 164$ mm). The computations were started at a negative time ($t = -t_0 \cong -1.775$ ms) corresponding to the nearest to the origin zero-crossing of the experimental signal (Fig. 3). It will be observed that the measured signal contains several harmonics. It is necessary to use the exact measured signal in order to obtain accurate results.

In order to reduce computing-time-requirements it was attempted to use coarser grids. Increasing n_w^+ is tempting because the minimum time-step is encountered in the nearest-to-the-wall cells, and scales with n_w^{+2}[2]. However the influence of n_w^+ on the results is known to be significant, especially in 3-D flows with corner-separation[2]. This is illustrated by the comparison of steady-state results using different grids (Table 1) for the mean π_{S-T} (Fig. 4). The grid with $n_w^+ \cong 1.5$ greatly underpredicts the shock-wave strength, and fails to predict the corner separation. It was however found possible, while retaining $n_w^+ \cong \frac{1}{2}$, to reduce the y-wise and z-wise grid-points, using a 121×57×49 grid (Table 1), which gives adequate steady-state results (Fig. 4) and contains ~40% less points than the 0.52 Mpoints grid used for the steady computations.

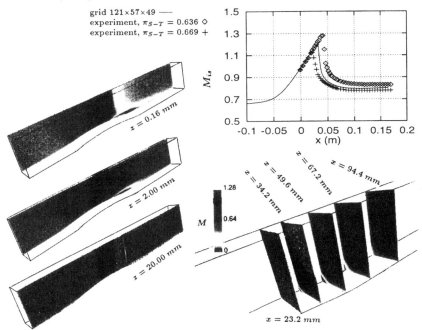

Figure 5: Initial steady-state flowfield

Time-integration was performed with a $\Delta t \cong 0.36$ μs, corresponding to $CFL_{max} \cong 500$. The rather substantial computational-times required did not permit a CFL_{max}-influence study. It was prefered to use this low $CFL_{max} \cong 500$ rather than risk computational divergence. There are 15408 instants/period corresponding to ~ 140 CPU-h/period on a CRAY-C98 computer. The fact that $\Delta t \ll T$ ($T = f^{-1} = 5.555\ldots$ ms is the oscillation period) justifies the use of an implicit scheme which is only 1-order-accurate in time (higher time-accuracy will presumably be required at high frequencies).

Starting at $t = -t_0 \cong -1.775$ ms from the mean steady-state flowfield (Fig. 5) the computations were continued until $t = 2T$ (corresponding to ~ 335 CPU-h on a CRAY-C98 computer). Unsteady pressures were computed as $\bar{p}(t) - <\bar{p}>$

$$<\bar{p}> \equiv {}^0\bar{p} = f \int_T^{2T} \bar{p}(t) dt \qquad (6)$$

Comparisons of computed and measured unsteady pressures (Fig. 6) show remarkable agreement between computation and experiment. The pressure signal propagates upstream and induces an oscillation of the shock-wave. There is a slight error at the shock-wave-foot, which can be attributed to the associated error in shock-wave position. An even more interesting comparison is given by the x-wise distributions of the first 3 harmonics of pressure

$$^m\tilde{p} = 2mf \int_T^{2T} \tilde{p}(t)e^{-i2\pi mft}\, dt \qquad (7)$$

The agreement in amplitude distribution is excellent (Fig. 7). The error in the real-part distribution of $^1\tilde{p}$ at the shock-wave foot corresponds to the phase-error observed in the time-plots. The time-plots illustrate the periodic convergence of the computations. It is noteworthy that the unsteady results agree even better with experiment than the steady results.

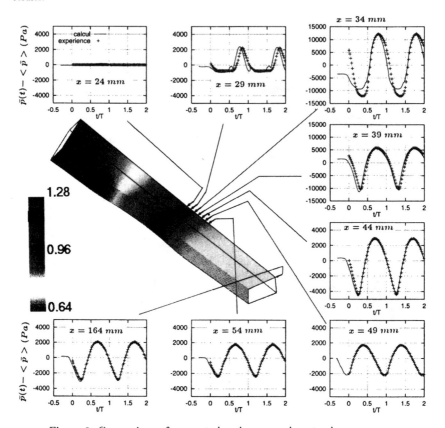

Figure 6: Comparison of computed and measured unsteady pressures

Discussion

The computed and measured results compare remarkably well. The question is naturally raised: why is the agreement of unsteady computation with experiment so good, when a standard turbulence model, with well-known drawbacks[5] was used? The computations reported were completely predictive. The measured steady and unsteady outflow pressure was applied as boundary-condition in the computations. No adjustment whatsoever was made, and the standard model constants[3] were used (the authors firmly believe that tempering with model constants is an inadmissible practice). The answer to this question is twofold. On the one hand the steady flow prediction is in this case quite satisfactory, due to the rather low M_{shock} ($\sim$1.2–1.3) and to the strongly 3-D nature of the flow (3-D flows are easier to predict than 2-D flows, even if computing-time-requirements are greater). On the other hand unsteady predictions are invariably better than steady predictions for driven flows such as those encountered in flutter and forced response analyses. It is therefore believed that improvement of unsteady predictions in such cases hinges upon bettering the prediction of the underlying steady flow.

The grid-influence study conducted for this configuration agrees with previous results[2] that n_w^+ is the most important grid-quality parameter. Even if using $n_w^+ = 0.5 - 0.75$ is time-consuming for unsteady computations, it is indispensible to do so for the quality of steady and unsteady results.

It is clear, despite the satisfactory results obtained in this study, that experimental validation was only partial. Even if unsteady pressures were exceptionally well predicted (and unsteady pressures are very important for practical applications) there was no way of assessing the prediction of several difficult flow characteristics, such as velocity profiles and wall-shear-stresses. There is need for experimental data that would permit such validation in realistic 3-D configurations. It is also important to develop a validation data-base for higher frequencies, for which spectral interaction between the driving unsteady phenomenon and turbulence is possible.

From a numerical and modeling point-of-view the necessary improvements include the use of near-wall Reynolds-stress closure[6] for better steady-flow prediction, and the extension of the implicit scheme to 2-order-time-accuracy[7] for better prediction at high frequencies.

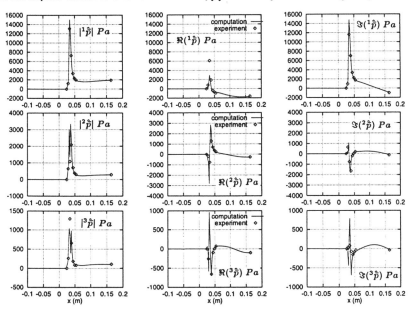

Figure 7: Comparison of computed and measured pressure-harmonics distributions

References

[1] Liou, M.S., Coakley, T.J.: *AIAA J.* **22** (1984) 1139–1145.
[2] Gerolymos, G.A., Vallet, I.: "Implicit Computation of the 3-D Compressible Navier-Stokes Equations using $k - \varepsilon$ Turbulence Closure", *AIAA J.*, **34** (1996) to appear in May.
[3] Launder, B.E., Sharma, B.I.: *Lett. Heat Mass Transf.* **1** (1974) 131–138.
[4] Ott, P., Bölcs, A., Fransson, T.H.: *ASME J. Turbom.* **117** (1995) 206–114.
[5] Gerolymos, G.A.: *AIAA J.* **28** (1990) 1707–1717.
[6] Vallet, I., Gerolymos, G.A.: "Computation of Shock-Wave/Boundary-Layer Interaction using Near-Wall Reynolds-Stress Modeling", 2. *ERCOFTAC Turbulence in Compressible Flows Workshop*, 1995, Paris.
[7] Anderson, W.K., Thomas, J.L., Rumsey C.L.: *AIAA J.* **27** (1989) 673–674.

Acknowledgments: The computations presented in this work were run at the Institut pour le Développement des Ressources en Informatique Scientifique (IDRIS), where computer ressources were made available by the Comité Scientifique.

Thermoconvection Simulation and Control in Melting Glass Furnaces

Nicolas Vanandruel
Université Catholique de Louvain ; CESAME – Applied Mechanics Division
Bâtiment Euler ; Av. G. Lemaître, 4–6; B-1348 Louvain-la-Neuve ; Belgium

Abstract

The major part of flat glass is currently produced in open-hearth furnaces in which the glass ingredients are flame-heated from above to form a melt. The charge reacts, is mixed by thermoconvection and diffusion and the produced glass is withdrawn at the opposite end of the furnace.

The understanding of various transport processes and chemical reactions occuring in a glass tank is an enormous task. The over-all problem involves radiant heat transfer from the flame-fired gas volume, melting thermodynamics, chemical reactions in the glass during the convection transport, ...

Traditionnally, industrial measurement and laboratory modelling were the sources of understanding and prediction for the melting glass processes. These methods suffer from severe drawbacks: restrictions to the measurements applicability (high temperature, low velocity, ...), poor economical and technical testing feasibility and difficult fulfilment of the long list of similarities.

On the other hand, mathematical modelling has recently become an interesting source of reproducible and unexpensive informations on the structure of the thermoconvective currents in the furnace.

The spatial discretization used for this study is a standard finite element method, based on a commercial software, with different interpolation order for temperature and velocities. Rather than presenting the results of new computational techniques or new computer architecture performances, this paper will therefore focus on novel applications of an available technique. Two recent points of interest will be presented:

– The "numerical" furnace becomes the source of information for the control and identification tools. The information between numerical modelling and identification/control techniques combine space and time representations.

– An example of numerical discovery in the furnace flow structure is what has been called the "Bénard cells". These are unexpected (and mostly unwanted) recirculation cells in the width direction. Critical locations of Bénard cells may result in the appearance of periodic defaults in the produced glass.

1 Continuous and discrete mathematical model

The equations set to be solved for the evaluation of the thermoconvective currents is the classical Boussinesq system (incompressible Navier-Stokes coupled with the energy equation):

$$\rho \frac{D\mathbf{v}}{Dt} = \nabla.(\mu \nabla \mathbf{v}) - \nabla p + \rho \mathbf{g}(1 - \alpha(T - T_o)) \quad (1.1)$$

$$\nabla.\mathbf{v} = 0 \quad (1.2)$$

$$\rho c \frac{DT}{Dt} = \nabla.(k\nabla T) \quad (1.3)$$

where $\mathbf{v}$, p and T stand for velocity, pressure and temperature distributions. Realistic molten glass simulation have to consider both μ and k, the dynamic viscosity and thermal conductivity, to be function of the temperature. The variation of k with the temperature is the result of the so-called Rosseland approximation for the radiative heat transfer that leads to an effective conductivity [1].

Characterizing the set of parameters usually encountered for molten glass simulation, the values of some dimensionless numbers are the following:
- Reynolds number, Re $\sim$ 1, creeping flow is observed and no numerical difficulties result from the momentum equation,
- Péclet number, Pé $\sim$ 300, the distribution of temperature is dominated by convection. Spatial conjunction of no-slip boundaries and modification of the flow direction leads to important temperature gradients,
- Richardson number, Ri $\sim$ 1500, the structure of the thermoconvective currents are essentially due to the temperature profile imposed at the tank surface (resulting from the fuel and burners distribution),
- Rayleigh number, Ra $\sim 10^7$, buoyancy effects dominate the stabilisation of viscous and thermal dissipation. Unsteady thermoconvection may appear if the geometry or the process parameters of the furnace is badly selected.

Associated with proper boundary conditions, the Boussinesq set of equations (Eqs. 1.1–1.3) is an adequate sub-model for the evaluation of the thermoconvective currents in the furnace. This sub-model may be coupled with the evaluation of thermodynamic properties of the glass, interaction with crown, flames or refractories ... Thermoconvective currents are however a crucial physical phenomenenon acting on the quality of the produced glass.

The non linearity of the system 1.1–1.3 is mainly due to the convective dominant nature of the energy equation. Other (but of less importance) non linearities result from the aboved mentionned variation of viscosity and thermal conductivity with temperature. Typically, the latter can be of the type [1]

$$k = k_{\text{cond}} + k_{\text{rad}} = k_{\text{cond}} + \sum_{i=0}^{3...4} a_i T^i \quad \text{with} \quad \frac{k_{\text{cond}}}{k_{\text{rad}}} \sim 0.01. \quad (1.4)$$

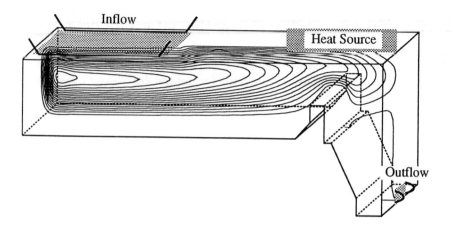

Figure 1 Geometry of a 10 × 5 × 1 [m] furnace with the circulation pattern in the (assumed) plane of symmetry. This benchmark furnace is an operating furnace running at 175 tons per day.

Discretizations

Spatial discretisation is achieved through the decomposition of the computationnal domain (by symmetry, only half the furnace is considered) into finite elements. A commercial code, Polyflow [2], was used as a toolbox for this study. A quadratic interpolation for the integration of energy equation coupled with linear-velocity and constant-pressure interpolation leads to affordable CPU cost [5]. This particular choice of low-order interpolation for velocity may result in spurious pressure modes that can be, however, easily detected.

Resolution of the Newton-Raphson linearized simultaneous set of equations is conducted by a direct elimination based on a frontal method. The typical "long box" shape of a furnace (length dimensions significantly excess width and height dimensions) is an important advantage for this solver choice as proper degree ordering maintains the frontwidth within CPU time and memory requirements. Large frontal width may however take full benefit of vector peak performance, more than 95% of the CPU time being spent during the frontal elimination.

The mesh used for the simulation of the flow in the furnace depicted in Figure 1 contains 5800 elements. This spatial discretisation leads to 50000 degrees of freedom for temperature, and 30000 for velocity and pressure. The resulting frontwidth is 1100, so that vector peak performance is reached on a Convex C3820. Typical CPU time for one resolution is 45 min. The start-from-zero simulation considers evolution on the Péclet number; each of the succession of 16 Péclet-steps containing an average of 4 Newton-Raphson iterations to ensure convergence and allow further stepping. Total CPU time reaches then globally 2 days for that single processor vector computer.

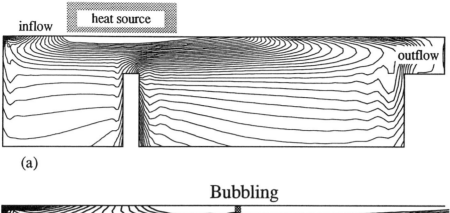

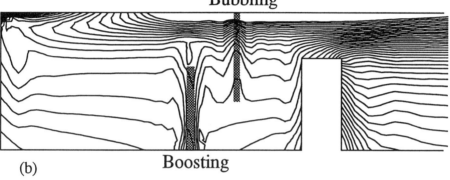

Figure 2 (a) Temperature distribution in the symmetry plane without boosting nor bubbling. (b) Local temperature distribution with boosting and bubbling. Increment between isotherms measures a 10K difference. Note the scale difference.

2 Industrial geometry and modification to natural convection

The glass flow structure results from the combined effect of the furnace geometry and the temperature profile imposed at the surface by the repartition of fuel oil burned. The furnace geometry is generally rather simple but the glass industry has included some technological tools to increase the quality of the produced glass. A first technique is the electrical boosting: the molten glass is heated by the Joule effect resulting from the electrical current induced by immerged electrodes. Another effect is the forced bubbling due to a train of air bubbles injected in the furnaces through nozzles. These bubbles create an intense mixing effect. Bubbling and boosting (Figure 2 for a different furnace geometry) may induce local and important modification of the flow structure and contribute to the complexity of the recirculation pattern.

3 Controlled numerical furnace and modelling of the furnace control

Numerical simulation of industrial processes may result from different modelling strategies. The numerical simulation of thermoconvective currents produces quantitative information on heat transfer within the batch. On the other hand, identification methods consider a set of measurements time series in order to build a simple black box input/output model.

For glass furnaces, the temperature profile at the bottom of the furnace is an adequate ouput variable for which variations (in time) results from combined actions of the fuel rate, glass pull rate, crown temperatures,...

The model to be calculated is then, for a succession of sampling time t_i

$$T_{\text{bottom}}(t_i) = \mathcal{L}(T_{\text{bottom}}(t_{j<i}), \text{fuel}, \text{pull}, T_{\text{crown}}, \ldots) \qquad (3.5)$$

where $\mathcal{L}$ stands generally for an algebraic linear low-order operator.
Some of the input variables may be considered as real inputs to the system while others are treated as perturbation sources.

The interest of this identification procedure is the small computing effort and the closeness of the $\mathcal{L}$ operator with industrial measurements.
On the other hand, the choice of relevant input variables, the degree of the operator and the time stepping are difficult problems.

Exchanges between thermoconvective modelling and identification procedures can be of two types :
– rather than using industrial measurements, the identification procedures may consider, as input variables, the results of the numerical simulation. This identification of a numerical furnace allows variation of process parameters that may not, for economical or technical reasons, be conducted on industrial sites [6].
– the values of some identification parameters may be considered as boundary conditions for the numerical thermoconvective simulation. Moreover, evolution of the technological equipments –like refractory corrosion– during the furnace lifetime (up to 20 years) results in variations of input parameters for the numerical simulation. These variations may be measured by an adaptative identification procedure.

4 Bénard cells

The distribution of the temperature in the vicinity of the boundaries may be such that hot glass flows beneath colder regions. This temperature "encapsulation" may result, if viscous and thermal diffusion does not stabilize the buoyancy term, in lateral recirculation. By analogy with the Bénard instabilities [4] and even if the surface tension does not strongly interact in this phenomenon, these cells have been described as "Bénard cells".

Quality of the molten glass is function of the residence time of the particles leaving the furnace (and of the homogeneity of the residence time distribution [3]). Secondary recirculation cells like the Bénard cells results in large differences in the residence time whether or not a particle has been captured in the lateral cell. Moreover, appearance of Bénard cells at the surface of the bath is the origin of great redox state difference along the trajectories of molten glass particles.

The Figure 3 illustrates the appearance of Bénard cell in the previously described furnace. The temperature distribution in the symmetry plane -i- shows the temperature encapsulation (increment between isotherms measures 10K difference): highest temperature, at the surface, is measured at the end of the furnace while the flow structure (Figure 1) leads to convective transport of hot glass beneath the batch.

The cross plane [a-a'] -k- details the tangential velocity field within the cell region. Vertical velocity profile at the surface is the result of molten glass inflow from the batch. The cross plane [b-b'] -l- shows the axial velocity distribution before the Bénard cell region (the flow structure is essentially composed by two layers).

Ackowledgements

This research is supported by the Walloon Regional Government under the "Convention de Recherche" ST 2576.

References

[1] Ungan, A., Viskanta, R.: "State-of-Art Numerical Simulation of Glass Melting Furnaces", Ceram. Eng. Sci. Proc.,9 [3-4] (1988) pp. 203–220.

[2] Crochet, M.J., Debbaut, B., Keunings, R., Marchal, J.M.: "Polyflow: a finite element program for continuous polymer flows" In K.O. Brien editor "Computer modelling for extrusion and other continuous processes", pp. 25–50, Hanser, Munich 1992.

[3] Beerkens, R., van der Heijden, T, Muijsenberg, E.: "Possibilities of Glass Tank Modeling for the Prediction of Quality of Melting Processes", Ceram. Eng. Sci. Proc.,14 [3-4] (1993) pp. 139–160.

[4] Koschmieder, E.L.: "Bénard Cells and Taylor Vortices", Cambridge Monographs on Mechanics and Applied Mathematics, Cambridge University Press, Cambridge 1993.

[5] Vanandruel, N.: "Simulation of Glass Processes with Polyflow, a finite element program", Chimica Chronica, New Series, 23 [2-3] (1994) pp. 95–100.

[6] Vanandruel, N., de Haas, B., Wertz, V.: "Modelling and Control of a Glass Melting Tank via Numerical Simulation" Proceedings of the XVII International Congress on Glass, Beijing (China), October 1995, in press.

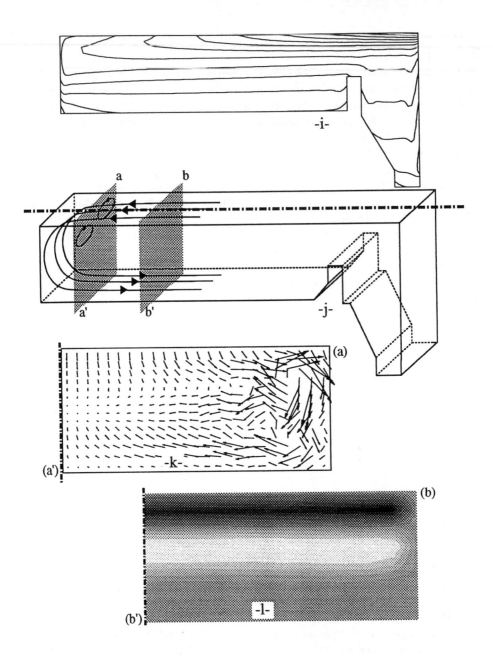

Figure 3 Bénard cells in glass melting furnace, -i- temperature distribution in the symmetry plane, -j- geometry of the furnace and location of the cross planes, -k- tangentiel velocity pattern in the [a-a'] plane, -l- axial velocity distribution in the [b-b'] plane.

Computational Methods for Large-Eddy Simulation in 2D Complex Geometries

Peter R Voke and Zhiyin Yang
Department of Mechanical Engineering
University of Surrey
Guildford GU2 5XH U.K.

Abstract

A method is described for the solution of the discrete Poisson equation in three dimensions, when one of the dimensions of the problem is periodic but the other two are of complex form. By utilising a direct Fourier method for the periodic dimension, the problem is reduced to a set of Helmholtz problems whose diagonal dominance increases with increasing wavenumber. The iterative solution of such problems on complex domains is shown to be highly efficient for the higher wavenumbers, and the use of multigrid acceleration for the lower wavenumber problems results in an effective overall solution procedure. Suitable line relaxation procedures and multigrid acceleration methods are described. The conditions under which the multigrid acceleration is beneficial are investigated, and the wavenumber range for which it should be activated is identified quantitatively. The method has applications in computational fluid dynamics and computational electromagnetics.

1 Introduction

The discrete Poisson equation occurs in a number of computational contexts such as computational fluid dynamics and computational electromagnetics. For example, a common strategy in the numerical solution or simulation of incompressible fluid flow is to extract an equation for the pressure p by taking the divergence of the Navier-Stokes equations:

$$\partial_t u_i = \partial_i \Sigma_{ij} . \tag{1.1}$$

Here the fluid stress Σ_{ij} is given by

$$\Sigma_{ij} = (-p\delta_{ij} - u_i u_j + 2\nu s_{ij}) , \tag{1.2}$$

where p is the physical pressure divided by the constant density, u_i is the vector velocity and ν and s_{ij} are respectively the kinematic viscosity and fluid strain-rate tensor. An implicit equation for the pressure is obtained by taking the divergence of (1.1),

$$\partial_i \partial_t u_i = \partial_i \partial_j \Sigma_{ij} = 0 \,, \qquad (1.3)$$

utilising the mass conservation law for the incompressible fluid,

$$\partial_i u_i = 0 \,. \qquad (1.4)$$

Equation (1.3) can be recast as a Poisson equation for the pressure,

$$\nabla^2 p = \partial_i H_i \,, \qquad (1.5)$$

by extracting the pressure part of Σ_{ij}, H_i being the divergence of the remainder of the fluid stress. The solution of this equation is then utilised to 'correct' the explicitly or implicitly advanced velocity field, integrated forward in time from (1.1) by omitting the unknown pressure term.

The above approach is common in computations of incompressible fluid flow, various forms being known as projection or pressure correction methods. In direct simulations of turbulence, for example, very fast fully explicit or mildly implicit methods of advancing the velocity are utilised because of accuracy requirements, stability restrictions and the great expense of three-dimensional calculations continued for many tens of thousands of time steps. Comparably efficient discrete Poisson solvers are required for solving (1.5) if this part of a simulation is not to overwhelm the computational cost and make the simulation uneconomic. Typically, practitioners of this computational art aim to find discrete Poisson solvers that take no more time to solve (1.5) than is taken to advance (1.1) in time to second or third order. Naturally, very simple geometries allowing the use of direct (spectral or cyclic reduction) methods have been the first recourse: see Schumann and Sweet [1], Hockney [2] and Schwarztrauber [3]. It is often assumed that iterative solution methods cannot compete without an unacceptably high residual threshold, leading to loss of the conservation properties of the algorithm for mass and energy.

Here we consider the problem of solving the discrete Poisson equation in a three-dimensional domain with one periodic dimension and the other two dimensions arbitrarily complex. The method utilises a direct Fourier method for the periodic dimension, in combination with a highly optimised iterative method in the other two dimensions, and it is applicable to three-dimensional simulations of incompressible turbulence in which one dimension only is statistically homogeneous.

2 Description of the Problem

We consider a three-dimensional computational domain in which one dimension, z, is homogeneous and only solutions periodic in z are sought. The other two dimensions x and y may involve a geometry of arbitrary complexity provided a general structured coordinate system exists. In this paper we concentrate on fairly simple geometric boundaries and general orthogonal coordinates (x, y) for the sake of clarity in the presentation.

We suppose that the discretisation of the physical problem leads to a discrete Poisson equation. The differential operators in x and y are represented by linear difference operators, and in z either a discrete Fourier expansion may be used or linear difference operators. In the former case the Poisson equation (1.5) will become

$$\Delta^2 \tilde{p} - k_z^2 \tilde{p} = \tilde{r} , \qquad (2.6)$$

where $\tilde{p}$ is the discrete Fourier representation of the field p in z, and

$$\Delta^2 = \Delta_x \Delta_x + \Delta_y \Delta_y \qquad (2.7)$$

is the two-dimensional discrete Laplacian based on linear difference operators Δ_x and Δ_y operating on a staggered mesh. The tildes will be omitted hereafter.

If linear differences are used in the z dimension also, an equation of the form (2.6) can still be obtained by discrete Fourier transformation of the Poisson equation in z. It is simple to show that this results in an equation formally identical to (2.6), provided the z mesh is uniform, though with

$$k_z^2 = \frac{4}{\delta z^2} \sin^2(k_z'/2) , \qquad (2.8)$$

where k' is the discrete Fourier wavenumber. Note that the coefficient of p in (2.6) is still negative definite. Similar forms are obtained for higher-order differencing in z.

The two-dimensional discrete Laplacian is a five-point star, with the sum of the coefficients zero. We indicate the spatial discretisation in x and y by subscripts m and n respectively:

$$\Delta^2 p_{m,n} = a_1 p_{m-1,n} + a_2 p_{m,n-1} + a_3 p_{m+1,n} + a_4 p_{m,n+1} - \sum_{d=1}^{4} a_d p_{m,n} . \qquad (2.9)$$

This form is appropriate not only on an even Cartesian mesh, where the coefficients a_d will possess simple spatial symmetries, but also in a general orthogonal coordinate system which has a diagonal metric tensor, though in this case the coefficients a_d will be more arbitrary. Additional terms arise from the off-diagonal components of the metric tensor in the case of completely general curvilinear coordinates, but apart from this slight additional complication the principles are much the same.

3 The Iterative Solution

The equations (2.6) are decoupled for the various values of k_z^2. We introduce an iterative procedure for the solution of this series of discrete Helmholtz problems:

$$a_1\, p_{m-1,n} + a_2\, p_{m,n-1} + a_3\, p_{m+1,n} + a_4\, p_{m,n+1} - \left(\sum_{d=1}^{4} a_d + k_z^2\right) p_{m,n} = r \,. \quad (3.10)$$

The equations are solved by black-white striped line-relaxation, each line equation taking the form:

$$a_1\, p_{m-1,n}^{i+1} + a_3\, p_{m+1,n}^{i+1} - \left(\sum_{d=1}^{4} a_d + k_z^2\right) p_{m,n}^{i+1} = r - a_2\, p_{m,n-1}^{i} - a_4\, p_{m,n+1}^{i} \,, \quad (3.11)$$

where the line-relaxation iteration number is indicated by the superscript i.

A similar equation is easily constructed for striped line-relaxation in the y rather than the x direction. It is known that the optimal procedure for the use of relaxation in the two directions depends on the form of the mesh. If the mesh is predominantly stretched in the y direction, resulting in the coefficients a_1 and a_3 being consistently larger in absolute magnitude than a_2 and a_4, (3.11) should be used alone; conversely, if a_2 and a_4 are consistently larger than a_1 and a_3, the equivalent y line-relaxation equations only should be solved. More typically on a real general mesh, the relationship between the coefficients will vary across the mesh, and then alternation of the direction of line-relaxation is preferable. Extremely stretched meshes should be avoided since they lead to poorly conditioned problems and possible failure of convergence depending on the rounding accuracy of the machine used for the computations. The tridiagonal system (3.11) is normally LU-decomposed to allow very rapid solution by forward and backward substitution.

This scheme can be extended easily to geometries in which a solid cylinder (or several cylinders) is present in the x-y plane, provided its surfaces are coincident with x-y coordinate lines and it is 2D in the z dimension. The Poisson weights of the solver (in each z plane) are altered to produce the equivalent of the homogeneous Neumann boundary conditions on the cylinder surface, just as on the external boundaries. Inside the cylinder, where no pressure solution is meaningful, the Poisson star is replaced by a trivial unit matrix. This strategy results in a pressure solution for a complex domain in the x-y plane that is just as efficient in general as that in a simple domain.

For all nonzero k_z the diagonal dominance of the matrix system (3.11) is enhanced by the presence of the negative definite term $-k_z^2$ in the diagonal coefficients. This property results in the rapid convergence of the line procedure for k_z larger than a minimum value which is found to be quite small. However, there is a zero wavenumber problem which typically will converge very slowly, and it is found that this single plane, together with a few other very-low-k_z

planes, completely dominates the computation time for the solution of the overall three-dimensional discrete Poisson equation unless further steps are taken to accelerate convergence. Multigrid acceleration [4] is therefore introduced to mitigate the slow convergence of the very-low-wavenumber problems.

4 Multigrid Procedure

The multigrid procedure restricts the residuals from the initial grid (grid level 1) line-relaxation solution onto grid level 2, which is coarser in both x and y by a factor of two. A single relaxation sweep is performed at this level, and the resulting residual problem is again restricted onto a two-times coarser grid, level 3. Line-relaxation proceeds at this level until the maximum residual falls below 10% of the maximum residual found on level 1, and the solution is prolonged to level 2 where it is used to correct the level 2 solution. Two line-relaxation sweeps are performed on level 2 to smooth the local interpolation errors, and the solution is again prolonged to level 1 to correct the fine grid solution at that level. Again two line-relaxation sweeps are performed, or more if the maximum residual is then close to the desired target. If the target is not achieved, the multigrid V-cycle is repeated, with the residual threshold on level 3 being reset to 10% of the final maximum residual encountered at level 1 for each cycle.

Restriction is achieved by straightforward averaging over the four fine-level cells contributing to the coarse-level cell, and the prolongation utilises bilinear smoothing. More sophisticated methods than those described (including more multigrid levels, higher-order smoothing, and variations in the restriction procedures and setting of thresholds for the residuals at the various levels) have been investigated by numerical experimentation, with negative or very marginal positive effects on the efficiency of the procedure judged by convergence per cpu second. The procedure, programmed in Fortran, is found to be robust on a number of platforms, having been originally developed on a small system and subsequently tested on superworkstations, Cray Y-MP and Cray T3D.

The simple multigrid algorithm described reduces the maximum residual on the original fine grid by an order of magnitude for each V-cycle. While the computational expense is obviously greater than the raw line-relaxation used for the higher wavenumber problems, it represents a marked saving for the low wavenumber planes. We have found that the imposition of multigrid cycling for non-dimensional wavenumbers k_z greater than about 0.4 is generally counterproductive, resulting in convergence rates per floating-point operation or per cpu second that are inferior to those achieved by the raw relaxation method, owing to the greater diagonal dominance of the high-wavenumber problems and the necessary expense of the multigrid injections and interpolations. The activation of the multigrid procedure is therefore made conditional on the value of the wavenumber, being used only for

$$k_z^2 < 0.16 \,. \tag{4.12}$$

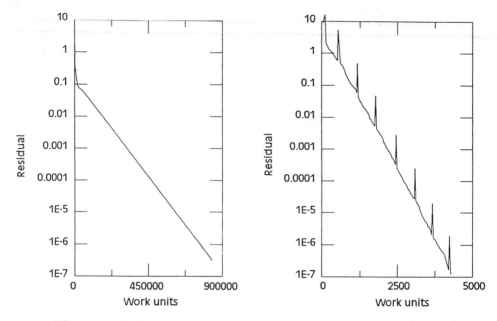

Figure 1 Reduction of the maximum residual of the two-dimensional Poisson equation (2.6) with $k_z^2 = 0$. Left, striped line relaxation; right, striped line-relaxation with multigrid acceleration. (The residuals on the coarse mesh levels are also shown.)

This value is relative to the weightings occuring in the x-y equation, and applies when the central weighting is equal to 4:

$$\sum_{d=1}^{4} a_d = 4 \ . \tag{4.13}$$

It has been found to be appropriate for grids typically from 16^3 up to 128^3 and has been confirmed on a number of machines; nevertheless it is machine- and mesh-dependent and needs to be optimised by numerical experimentation. It is affected be the relative efficiency of the multigrid acceleration on each architecture, and is therefore a function of vectorisation, caching and other machine-specific properties for the coarse mesh phases of the multigrid cycle.

Figure 1 shows the performance of a pressure solution for the $k_z = 0$ plane in a $264 \times 72 \times 64$ simulation of transitional fluid flow over a curved surface. The simulation is carried out using curvilinear orthogonal body-fitted coordinates and is reported elsewhere [5]. The figure shows the multigrid solution with the residuals from all levels included, as a function of a work unit proportional to the number of floating point operations performed for the solution in the x-y plane (1 work unit = flops/264/72). Since the form of interpolation used results in residuals on the coarse levels being numerically higher, but the work units expended are low, the excursions to the coarser multigrid levels appear as regular spikes.

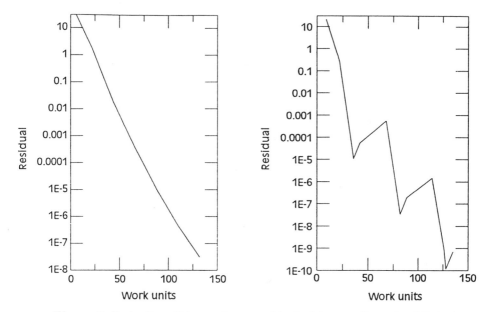

Figure 2 Reduction of the maximum residual of the two-dimensional Helmholtz equation (2.6) with $k_z^2 = 2$. Left, striped line relaxation; right, striped line-relaxation with multigrid acceleration. (The residuals on the coarse mesh levels are also shown.)

The effect of removing the multigrid acceleration is also displayed, showing very much slower convergence as expected. Figure 2 compares the performances for a higher wavenumber plane; here there is little benefit apparent from multigrid acceleration, and in practice the poorer vectorisation, greater complexity, and time overheads from subprogram calls make the multigrid solution more expensive.

5 Conclusions

The combined Fourier and striped line-relaxation method with multigrid acceleration for selected wavenumbers is found to be extremely efficient on a number of machines, for residual thresholds down to 10^{-8} of the initial error. It has been tested as part of a fluid dynamics code used for large-eddy simulation of transitional and turbulent flows and is found to perform on problems involving general orthogonal coordinates in two dimensions as well as direct pressure solvers previously applied to problems in Cartesian coordinates based on the cyclic reduction algorithm of Schwarztrauber [6] and applied to simulation of turbulence first by Gavrilakis [7].

The efficiency of our iterative solver does depend on the physics of the flow being simulated as well as on the chosen level of residual reduction. If the pressure field varies little from one time step of the simulation to the next, as is often the case when large-scale mass fluctuations are not imposed through the external boundaries of the domain, the pressure solution from one time step represents a good first approximation for the solution of the next. Under these conditions the iterative solution may be extremely fast, and the whole solution, including the discrete Fourier transform, may take as little as 12% of the total computation time to reduce the residual to an acceptable level.

Under more general circumstances in which the solution error is substantially regenerated each time step, the reduction of the three-dimensional discrete Poisson residuals by 7 or 8 orders of magnitude takes up to 65% of the total cpu time. However, this is still an acceptable balance of work and represents a powerful technique for the solution of the pressure field in simulations involving arbitrary geometry in two out of the three dimensions.

Work is underway utilising the method in domains that are multiply connected in the x-y plane, and using general non-orthogonal coordinates.

References

[1] SCHUMANN, U. and SWEET, R.A.: 'A direct method for the solution of Poisson's equation with Neumann conditions on a staggered grid of arbitrary size', *J. Comp. Phys.*, **20**, 171–182 (1976).

[2] HOCKNEY, R.W.: 'Computers, compilers and Poisson-solvers', in *Computers, Fast Elliptic Solvers and Applications*, (U. Schumann, ed.) Advance Publications, London (1978).

[3] SCHWARZTRAUBER P.N.: 'Fast Poisson solvers', *MAA Studies in Mathematics: Volume 24, Studies in Numerical Analysis*, (G.H. Golub, ed.) Math. Assoc. Am. (1984).

[4] BRANDT, A.: 'Multi-level adaptive solutions to boundary-value problems', *Math. Comp.*, **31**, 333–390 (1977).

[5] YANG, Z and VOKE P.R.: 'Large-eddy simulation of boundary layer transition on a flat plate with semi-circular leading edge', *Tenth Symp. on Turbulent Shear Flows*, Volume 2, paper 11-13. Penn. State Univ. (1995).

[6] SCHARZTRAUBER, P. N.: 'A direct method for the discrete solution of separable elliptic equations', *SIAM J. Num. Anal.*, **11**, 1136–1150 (1974).

[7] GAVRILAKIS, S.: 'Numerical simulation of low-Reynolds-number turbulent flow through a square duct', *J. Fluid Mech.*, **244**, 101–129 (1992).

Acknowledgements

This work has been performed in connection with research supported by the U.K. EPSRC and Rolls Royce plc, to whom we are grateful.

THREE DIMENSIONAL NUMERICAL SIMULATION OF THE FLOW IN A FOUR-STROKE I.C. ENGINE

E. von Lavante and J. Yao
Department of Mechanical Engineering, University of Essen,
Schuetzenbahn 70, 45127 Essen, Germany

SUMMARY

An implicit upwind finite volume algorithm according to Liou's AUSM (Advection Upstream Splitting Method) Flux Vector Splitting is developed in this work. The numerical method is capable of simulating three-dimensional unsteady compressible flows in Internal Combustion Engines using the multi-domain approach. The work focuses on the validation of the program in conjunction with the LES (Large Eddy Simulation) of the complicated turbulent exhaust flow. The computational domain consists of a manifold, a valve, an air port, cylinder wall and a moving piston. The computational grid is divided into five blocks to fit the complicated geometry.

1. INTRODUCTION

Numerical simulation is an effective method for research and design of piston engines. Many efforts have been made for the computation of flow motions in I.C. engines. The pioneering work was reported by Gosman, et al [1]. There, the implicit time differencing and first-order upwind spatial differencing were used on a 30×30 computational mesh. A standard $k - \epsilon$ model was applied in conjunction with the logarithmic-law-of-the-wall treatment at the solid boundaries. Amsden, et al [2] reported their KIVA program developed specifically for the simulating the reacting and turbulent flows in I. C. engines. It was a time-marching finite-difference approach that useed an acoustic subcycling method in order to overcome time step size limitation due to the CFL criterion. Taghavi et al [3] have simulated the intake and cylinder flow using the KIVA program with the $k - \epsilon$ turbulence model proposed by El Tahry (1983). Recently further works on this subject have been reported in the literature. Most of them, however, concentrated on the flow simulation in the cylinder field with simple models or simplified geometries only.

The configuration of the valve-port combination plays an important role on the exhaust flow, particularly at larger valve lifts. In this work, the exhaust flow on the combined flow field of the cylinder, port and valve are simulated numerically. The implicit upwind finite volume approach is employed to solve the filtered governing equations. Liou's AUSM Flux Vector Splitting method is used to calculate both steady and time accurate parts.

Turbulence modeling is an important component of simulation accuracy. The turbulence models based on statistical methods, such as $k - \epsilon$ model, utilize or imply Reynolds averaging of the governing equations and require modeling of the correlations of fluctuating quantities that represent turbulence. LES approach, however, separates the field variables of the turbulent flow into large-scale parts and small-scale quantities. The large-scale variables are calculated directly through solving the filtered governing equations, while the unresolved small-scale quantities need to be modeled in terms of resolved variables. This is considered to be more adequate to the cycle-to-cycle variant engine-like applications [8]. In the present work, the LES of the exhaust flow is considered in conjunction with the implicit upwind finite volume approach.

2 MATHEMATICAL FORMULATION

The flow is assumed to be unsteady, viscous and compressible. It is modeled by using the three-dimensional, time-dependent full Navier-Stokes equations describing the conservation of mass, momentum and energy of the flow. Their divergence form in the body-fitted curvilinear coordinates is:

$$\frac{\partial Q}{\partial \tau} + \frac{\partial (F - F_v)}{\partial \xi} + \frac{\partial (G - G_v)}{\partial \eta} + \frac{\partial (H - H_v)}{\partial \zeta} = 0, \qquad (2.1)$$

where $Q = J^{-1}(\rho \ \rho u \ \rho v \ \rho w \ e)^T$ is the vector of conserved unknown variables. J is the Jacobian of the coordinates transformation from physical (x, y, z, t) to computational (ξ, η, ζ, τ) space. It is given by

$$J^{-1} = x_\xi(y_\eta z_\zeta - y_\zeta z_\eta) - y_\xi(x_\eta z_\zeta - x_\zeta z_\eta) + z_\xi(x_\eta y_\zeta - y_\eta x_\zeta). \qquad (2.2)$$

The inviscid flux F and viscous flux F_v in ξ direction, for example, are given by

$$F = \frac{1}{J} \left\{ \begin{array}{c} \rho U_\xi \\ \rho U_\xi u + p\xi_x \\ \rho U_\xi v + p\xi_y \\ \rho U_\xi w + p\xi_z \\ (e+p)U_\xi - \xi_t p \end{array} \right\}, \qquad F_v = \frac{1}{JRe_\infty} \left\{ \begin{array}{c} 0 \\ \xi_x \sigma_{xx} + \xi_y \sigma_{xy} + \xi_z \sigma_{xz} \\ \xi_x \sigma_{yx} + \xi_y \sigma_{yy} + \xi_z \sigma_{yz} \\ \xi_x \sigma_{zx} + \xi_y \sigma_{zy} + \xi_z \sigma_{zz} \\ \xi_x \Omega_x + \xi_y \Omega_y + \xi_z \Omega_z \end{array} \right\},$$

with

$$\Omega_i = \sigma_{ij} u_j + q_i, \qquad \sigma_{ij} = -\frac{2}{3}\mu \frac{\partial u_k}{\partial x_k}\delta_{ij} + \mu(\frac{\partial u_i}{\partial x_j} + \frac{\partial u_j}{\partial x_i}), \qquad q_i = K\frac{\partial T}{\partial x_i}.$$

σ_{ij} is the components of the shear-stress tensor and q_i is the heat-flux vector with the help of Fourier's law. The contravariant velocity relative to the grid motion, $U_\xi = \xi_t + \xi_x u + \xi_y v + \xi_z w$, is in the direction normal to constant ξ.

For the turbulence calculations, the large-scale variables are defined through employing a convolution integral with a filter function G of the type

$$\bar{f}(x,t) = \int_\Delta G(x-\xi)f(\xi,t)d\xi \ . \tag{2.3}$$

The large-scale field equations are operationally simpler for the compressible flow when the variables are recast in terms of Favre-filtered quantities [9]. It is defined as a density-weighted variable by the ratio $\tilde{f} = \overline{\rho f}/\bar{\rho}$. Then each variable can be decomposed into $f = \tilde{f} + f'$. The large-scale filtered quantity $\tilde{f}$ appears explicitly in the governing equations. The small-scale filtered-out quantity f' and their correlations with the resolved large-scale motion are modeled with a subgrid model. To model the small scale turbulence, the shear stress in the governing equations includes not only the viscosity shear stress but also the subgrid-scale turbulence shear stress τ_{ij} [7] :

$$\overline{u'_i u'_j} = \tau_{ij} + \frac{1}{3}\delta_{ij}\overline{u'_i u'_j} \tag{2.4}$$

In the computation, the SGS eddy viscosity model proposed by Smagorinsky (1963) is used,

$$\tau_{ij} = -2\nu_t S_{ij} \ , \tag{2.5}$$

with

$$S_{ij} = \frac{1}{2}(\frac{\partial \tilde{u}_i}{\partial x_j} + \frac{\partial \tilde{u}_j}{\partial x_i}) \ , \qquad \nu_t = (C_s \Delta)^2 \sqrt{2S_{mn}S_{mn}} \ ,$$

where S_{ij} is the large scale strain rate and ν_t is the subgrid effective eddy viscosity. $C_s = 0.1$ is a model constant and $\Delta = (\Delta x \cdot \Delta y \cdot \Delta z)^{1/3}$ is the filter width. The term $\frac{1}{3}\delta_{ij}\overline{u'_i u'_j}$ is added to the pressure term and needs not to be calculated explicitly. In addition, the filtered temperature $\tilde{T}$ and the total energy $\tilde{e}$ are related to the filtered variables are defined by

$$\tilde{T} = \gamma \frac{\bar{p}}{\bar{\rho}} \ , \qquad \tilde{e} = \frac{\bar{p}}{\gamma - 1} + \frac{\bar{\rho}}{2}\tilde{u}_i \tilde{u}_i \ .$$

3 NUMERICAL METHODOLOGY

The filtered governing equations are implicitly integrated in time for both steady and time accurate calculations in conjunction with the upwind finite-volume scheme. The approximate factorization algorithm in the delta form, described by Beam and Warming [5], is as follows:

$$[I + h\delta_\xi^+(A^- - A_v^-) + h\delta_\xi^-(A^+ - A_v^+)]$$
$$[I + h\delta_\eta^+(B^- - B_v^-) + h\delta_\eta^-(B^+ - B_v^+)] \tag{3.6}$$
$$[I + h\delta_\zeta^+(C^- - C_v^-) + h\delta_\zeta^-(C^+ - C_v^+)]\Delta Q^n = h \cdot RHS \ ,$$

where δ^+ and δ^- are forward and backward difference operators. Let $U_\xi^\pm$ and $p^\pm$ represent the contravariant velocity and static pressure split according to Liou's AUSM Flux Vector Splitting, the split Jacobian matrix of the inviscid flux can be written as a sum of the convective terms and pressure terms as follows:

$$A^\pm = A_c(U_\xi^\pm) + A_p(p^\pm), \qquad (3.7)$$

with

$$A_c(U_\xi^\pm) = Q \otimes \frac{\partial U_\xi^\pm}{\partial Q} + U_\xi^\pm I, \qquad A_p(p^\pm) = P \otimes \frac{\partial p^\pm}{\partial Q} + p^\pm \frac{\partial P}{\partial Q},$$

where $P = J^{-1}(0 \ \xi_x \ \xi_y \ \xi_z \ U_\xi - \xi_t)^T$. The Jacobian matrices of the viscous flux vectors A_v, B_v and C_v are centrally split [4]. The residual RHS is given by

$$\frac{\partial Q}{\partial t^*} = -\frac{\partial (F - F_v)}{\partial \xi} - \frac{\partial (G - G_v)}{\partial \eta} - \frac{\partial (H - H_v)}{\partial \zeta} = RHS. \qquad (3.8)$$

It can also be calculated by using Liou's Flux Vector Splitting method. The inviscid flux value at the cell interface are estimated through

$$F_{i+1/2}^{Liou} = \frac{1}{2} \left[\tilde{U} (\Phi_r + \Phi_l) - |\tilde{U}| (\Phi_r - \Phi_l) \right]_{i+1/2} + (\tilde{p} \, \Xi)_{i+1/2}, \qquad (3.9)$$

where r, l denote the right and left states of the cell interface. Φ and Ξ are the vectors relative to the convective and pressure term, respectively. The convective term of the inviscid flux at the cell interface is a weighted average via the convective velocity $\tilde{U}$, while the pressure term is that via the intermediate pressure $\tilde{p}$. The reconstruction of the cell-centered variables to the cell-interface locations is done using the MUSCL type interpolation to effect a second-order accurate scheme. More details of the description can be found in [6].

4 COMPUTATIONAL GRID

In order to investigate the influence of the valve-port configuration on the exhaust flow from the cylinder, the computational domain in the present work consists of a manifold, a valve, an air port, cylinder wall and a moving piston, shown in Fig. 1. The valve stem is omitted for the sake of simplicity during the construction of the computational grid, without the change of the effective cross-sectional area of the main flow. The valve lift is 7 mm, at which the gas motion in the manifold has a significant effect on the exhaust flow. Due to the geometrical complexities of the domain, a multi-zone method has to be used. Moreover, special fine grids in the near-wall regions are applied, so that as high as possible resolution can be obtained, while relative coarse grids are used on the field far from the solid wall. The normal grid spacing at the surfaces around the valve is on the order of $y^+ = 5$. This is sufficient to properly resolve the turbulent boundary layer.

The computational grids used in this work are built using an in-house program, which is based on Poisson's differential equation. The fine grids at the important solid boundaries are orthogonal. No cell face of the final grid is shared by more than two cells at a time. The total number of the grid points of the 5-block computational grid is 217,600. The velocities of the grid motion in the cylinder field are computed from the moving piston. The grid singularity emerges on the centerline due to the radiation of the grid. It is treated as a special boundary condition. There, small cells including the singular points are introduced. These cells have non-vanishing cell faces, the cell-centered variables may be calculated by

$$\frac{\partial}{\partial t}\int_\Omega \vec{Q}d\Omega = \oint_S \overline{\overline{F}} \cdot d\vec{S}, \qquad (4.10)$$

where Ω represents the volume, S the surfaces of the small cell.

5 NUMERICAL RESULTS

The computations are performed on the 5-block three-dimensional grid, as shown in Fig. 1. The velocity of the moving grids are calculated from the rotational speed of the model engine. The bore of the cylinder is 100 mm and the ratio of the cylinder bore to the piston stroke is 1.0. The Reynolds number based on the cylinder bore is $Re = 1.12 \cdot 10^6$, in this case.

The numerical solutions are obtained using the implicit solver mentioned above. The local CFL number can be larger than 50. The exhaust flow is caused by the upward movement of the piston. The valve stays at its position in order to isolate the influence of the piston moving on the exhaust flow. Fig. 2-4 give the velocity fields around the valve head at three different CA. The flow behavior on this region is very sensitive to the movement of the piston. At the CA of 190^0, as shown in Fig. 2, no separated flow occurs on the flow field due to the lower Mach number. Flow separations come at the CA of 200^0 on the valve corners and near the solid walls behind the valve seat, as shown in Fig. 3. At the CA of 220^0 some recirculations are enlarged due to the acceleration of the main flow, for example, near the valve given in Fig. 4. The recirculations in the manifold diminish and even vanish. This appearance is thought to be the result of the increase of the local pressure. The computed velocity vectors and Mach number contours of the exhaust flow at the CA of 220^0 are given in Fig. 5 and Fig. 6, respectively. Flow features such as separated recirculations, corner flow expansions and subsequent recompressions are evident. A considerable acceleration of the flow occurs at the gap between the poppet valve and the seat. The flow inside the cylinder is characterized by a low Mach number, and small gradients of the pressure and density, in this case. The simulations show that exhaust flow is extremely sensitive to geometric details of the valve and valve seat.

6 CONCLUSIONS

The exhaust flow in three-dimensional model configuration, used in this work, is successfully simulated. The major flow features are numerically produced. The satisfying results demonstrated that the present implicit finite volume approach is useful to this problem. Further efforts to improve accuracy and robustness of the approach have to be made.

REFERENCES

[1] Gosman, A. D., Johns, R. J. R., and Watkins, A. P., " Development of Prediction Methods for In-Cylinder Processes in Reciprocating Engines " , Combustion Modeling in Reciprocating Engines, Plenum Press, New York, 1980.

[2] Amsden, A.A., Ramshaw, J.D., O'Rourke, P.J. and Dukowicz, J.K., " KIVA, a Computer Program for two and three- Dimensional Fluid Flows with Chemical Reactions and Fuel Spray ", Los Alamos National Laboratory Report No. LA-10245-MS, 1985.

[3] Taghavi, R., Dupont, A., and Dupont, J. F., " Aerodynamic and Thermal Analysis of an Engine Cylinder Head Using Numerical Flow Simulation " , Journal of Engineering for Gas Turbines and Power, pp335-340, Vol. 112, July 1990.

[4] Pulliam, T.H. and Steger J.L., " Implicit Finite- Difference Simulations of three-Dimensional Compressible Flow ", AIAA Joural, Vol. 18, Feb. 1980.

[5] Beam, R. and Warming, R.F., " An Implicit Factored Scheme for the Compressible Navier-Stokes Equations ", AIAA Joural, Vol. 16, Apr. 1978.

[6] Yao, J. and von Lavante, E., " Flow Simulation in a Model Reciprocating Engine " , Proceedings of the Second European Computational Fluid Dynamics Conference, Stuttgart, Sept., 1994.

[7] He, J., Song, C.C.S., " Some Applications of Large Eddy Simulation to Large-Scale Turbulent Flows at Small Mach Number" , FED-Vol. 162, ASME 1993.

[8] El Tahry, S.H. and Haworth, D.C., " Directions in Turbulence Modeling for In-Cylinder Flows in Reciprocating Engines" , Journal of Propulsion and Power, Vol. 8, No. 5, Sept.-Oct. 1992.

[9] Vreman, A.W., Geurts, B.J., Kuerten, J.G.M. and Zandbergen, P.J., " A Finite Volume Approach to Large Eddy Simulation of Compressible, Homogeneous, Isotropic, Decaying Turbulence " , International Journal For Numerical Methods In Fluid, Vol. 15, pp. 799-816, 1992.

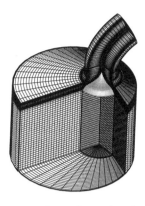

Fig. 1: The three-dimensional computational grid.

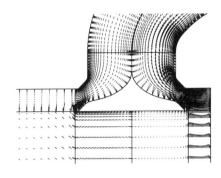

Fig. 2: Velocity vectors around the valve head at $CA = 190^0$.

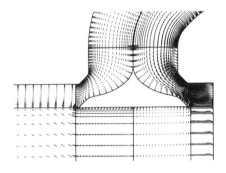

Fig. 3: Velocity vectors around the valve head at $CA = 200^0$.

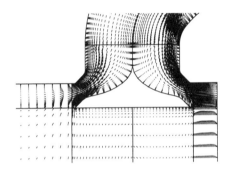

Fig. 4: Velocity vectors around the valve head at $CA = 220^0$.

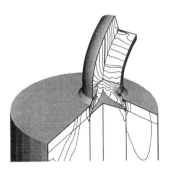

Fig. 5: Pressure contours in the cylinder and manifold.

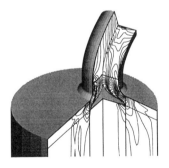

Fig. 6: Mach number contours in the cylinder and manifold.

NUMERICAL SIMULATION OF THREE-DIMENSIONAL COMPLEX FLOWS USING A PRESSURE-BASED NON-STAGGERED GRID METHOD

K.Yakinthos, M.Ballas, P.Tamamidis and A.Goulas

Aristotle University of Thessaloniki
Department of Mechanical Engineering
Lab. of Fluid Mechanics & Turbomachinery
Thessaloniki 54006, Greece.

Summary.
Numerical simulation of complex three-dimensional flows presents a well-known trade off between the accuracy of the solution and the cost of computations. A promising approach for obtaining accurate solutions of complex flows, while keeping the cost of computations at a minimum, is through the use of high-resolution schemes. The objective of this work is to evaluate the efficiency and accuracy of a pressure-based method in complex, three-dimensional, steady, incompressible flows. The method uses the standard pressure-based finite volume procedure and all variables are stored at the centroids of the brick-type elements. Three convective schemes are introduced into a generalized Navier-Stokes solver, the first order HYBRID, the upwind-biased HLPA and the third order MUSCL. Turbulence is modeled using the high and the low Reynolds k-ε models. The results obtained are compared with the experimental data of a jet in a cross-flow and a flow at the junction of a flat surface and a normal-mounted cylindrical wing section.

A. THEORETICAL FORMULATION.

The governing equations for steady, incompressible, elliptic flows can be written in the following general form:

$$\nabla \cdot \vec{J}_\varphi = S_\varphi \tag{1}$$

where φ is a variable, S_φ is the source term associated with this variable and $\vec{J}_\varphi$ is the flux term. $\vec{J}_\varphi$ represents the transport of φ due to convection and diffusion and is given by

$$\vec{J}_\varphi = \varrho \vec{V}\varphi - \Gamma_\varphi \nabla \varphi \tag{2}$$

where $\vec{V}$ is the velocity vector and Γ_φ is the diffusion coefficient of φ. Turbulent flows are calculated by using the standard two-equation k-ε turbulence model. In this model the eddy-viscosity concept is adopted, which assumes local isotropy of turbulence. The turbulent viscosity μ_t is related to k and ε through the following algebraic equation:

$$\mu_t = \frac{\varrho c_\mu k^2}{\varepsilon} \tag{3}$$

In the momentum equations, the molecular viscosity μ is replaced by the effective viscosity μ_{eff} which is taken as the sum of the molecular and the turbulent viscosity, that is, $\mu_{eff} = \mu + \mu_t$. The Reynolds stresses and the generation term in the source terms for k and ε are given respectively by

$$-\varrho\overline{u_i u_j} = \mu_t \left(\frac{\partial u_i}{\partial x_j} + \frac{\partial u_j}{\partial x_i}\right) - \frac{2}{3}\varrho k \delta_{ij} \qquad G = \mu_t \left(\frac{\partial u_i}{\partial x_j} + \frac{\partial u_j}{\partial x_i}\right)\frac{\partial u_i}{\partial x_j} \qquad (4,5)$$

where δ_{ij} is Kronecker's delta. The k-ε model uses several empirical constants which assume the following values: C_1=1.44, C_2=1.92, C_μ=0.09, σ_k=1.0 and σ_ε=1.3 [1].

Equation (1) can be transformed from the physical to the computational space by direct integration over an arbitrary brick-type finite control volume in the physical space. Using Gauss's theorem, [2], we obtain

$$\sum_{cs} \int_{cs} \vec{J}\cdot\vec{n}\, ds = \int_{cV} S\, dV \qquad (6)$$

where the subscripts cs and cV stand for control surface and control volume respectively; V is the volume of the element, s is the surface that encloses V, and $\vec{n}$ is the unit normal on s. The summation is performed over the east, west, north, south, front and rear control surfaces respectively. The total flux through a control surface where ξ_i=const. can be obtained from the following equation [2]:

$$J_{tot}^{\xi i} = \int_{\xi j}\int_{\xi k} \vec{J}\cdot\vec{n}\, ds = \left[\varrho U^{\xi i}\varphi - J a \Gamma(\varphi_{\xi i} g^{ii} + \varphi_{\xi j} g^{ij} + \varphi_{\xi k} g^{ik})\right] \Delta\xi_j \Delta\xi_k \qquad (7)$$

where i=1,2,3, j=1,2,3, k=1,2,3, $\varphi_{\xi i}$ is the partial derivative of φ with respect to ξ_i, and (i,j,k) change cyclically. $U^{\xi i}$ is the contravariant velocity component. The final flux equation [2], is discretized in the computational space with the use of finite differences using the well known procedure presented by Patankar [3]. The coupling of the momentum equations with the continuity equations is done by using the SIMPLEC algorithm [4] extended for grids in nonorthogonal body-fitted system of coordinates. To avoid the wiggles in the pressure field appearing in collocated grids in incompressible flows, the momentum interpolation technique of Rhie and Chow [5] is applied. Incorporating the suggestion of Majumdar [6] for the underrelaxation of the contravariant velocities on faces the contavariant velocity component at a cell face is calculated from

$$(U^{\xi i})_{new} = \alpha_u \overline{U^{\xi i}} + D_{U^{\xi i}}\left(\frac{\partial p}{\partial \xi_j} - \overline{\frac{\partial p}{\partial \xi_j}}\right) + (1-\alpha_u)(U^{\xi i})_{old} \qquad (8)$$

Here, α_u is the underrelaxation factor of velocity, the overbars denote values obtained using linear interpolation and $D_{U^{\xi i}}$ is a coefficient that is a function of the $a^{ui}{}_p$ terms, the Jacobian and the metric terms g^{ii}. With some manipulation the pressure correction is written in a final form [2,7] ready to be solved.

The discretized equation for the three components of velocity and for the pressure correction has the general form

$$a_P^\varphi \varphi_P = a_E^\varphi \varphi_E + a_W^\varphi \varphi_W + a_N^\varphi \varphi_N + a_S^\varphi \varphi_S + a_F^\varphi \varphi_F + a_R^\varphi \varphi_R + b^\varphi \qquad (9)$$

where φ stands either for one of the three velocity components or for the pressure correction and b^φ is the sum of the cartesian source terms and the non-orthogonal diffusive terms. The last terms become less important for high-Reynolds number flows and zero for orthogonal grids. Note that since the two equations of the k-ε model have the sense of a momentum equation, the discretized equations for them have the same format as eq.(9).

A.1 Discretization schemes.

Equation (9) is formed by making some assumptions for the calculations of velocities on faces. The values of unknowns (which are the velocities) on faces are calculated using three schemes: the simple upwind scheme, the HLPA scheme [8] and the third-order MUSCL scheme [9]. In the simple upwind scheme, the value of the velocity at a cell face assumes the value of the velocity of the node in the upwind direction. This scheme, combined with central differences for cell Reynolds numbers less than 2, is well known as HYBRID [3]. This scheme is very simple to incorporate into a code and stable, but has high truncation errors. Despite this latter disadvantage, the HYBRID scheme has been used extensively and still is in use due to its stability and simplicity.

The HLPA scheme is an upwind-biased scheme. This scheme is formulated on the basis of the variable normalization proposed by Leonard [10] and the convection boundedness criterion of Gaskell and Lau [11]. The HLPA scheme for smooth gradients switches to the QUICK scheme. However in steep gradients a limiting procedure is introduced which is simple and efficient. The variable φ is calculated at a cell face using the following expression [9],

$$\hat{\varphi}_{i+1/2} = \begin{cases} \hat{\varphi}_i (2-\hat{\varphi}_i) & \text{if } 0<\hat{\varphi}_i<1 \\ \hat{\varphi}_i & \text{otherwise} \end{cases} \quad (10)$$

where the hatted variables have been normalized based on a special procedure [9,11]. Equation (10) represents a hybrid first order linear / second-order parabolic curve in the normalized plane.

The MUSCL scheme is a third-order scheme with flux limiter, and has been extensively used in the calculations of high speed flows. In this scheme the calculation of the variable φ on a cell interface between west and east direction is performed as following:

$$\varphi^L_{i+1/2} = \varphi_i + \frac{1}{2}[\frac{2}{3}\Delta''_{i-1/2} + \frac{4}{3}\Delta'_{i+1/2}] \quad \Delta'_{i+1/2} = minimod(\Delta_{i+1/2}, 4\Delta_{i-1/2})$$

$$\varphi^R_{i+1/2} = \varphi_i - \frac{1}{4}[\frac{2}{3}\Delta'_{i+3/2} + \frac{4}{3}\Delta''_{i+1/2}] \quad \Delta''_{i+1/2} = minimod(\Delta_{i+1/2}, 4\Delta_{i+3/2}) \quad (11)$$

In these equations, Δ denotes the algebraic differences of velocities, for example $\Delta_{i+1/2}$ is for $u_{i+1} - u_i$ and $minimod(a,b) = sign(a) \max(0, \min(|a|, sign(ab)))$. The use of the left or right value depends on the direction of the contravariant velocity at the cell face.

A.2 Solution procedure.

The discretized equations of momentum and continuity with the two equations for the k-ε model, constitute a system of six non-linear algebraic equations. The system is linearized by resorting to relaxation, where the coefficients in the discretized equations are evaluated using the values from the previous iteration level. The procedure constitutes a predictor-corrector method and is done in one iteration. The solution is considered convergent when the maximum residual becomes smaller than a convergence criterion. For each equation the residual is defined as:

$$R = \sqrt{\sum_{i=1}^{N} Er_i^2} \bigg/ \text{reference flux} \quad (12)$$

where N is the number of nodes, Er_i is the error of an equation at a node i.e. the algebraic difference between the right part and the left part of eq. (9), and the *reference flux* is the flux of the respective variable into the domain. For the residual of the pressure correction equation the error in the discretized equation is equal to the mass imbalance in a control volume and the reference flux is the mass flux entering the domain. Equation (9) is solved using Stone's algorithm [12] where one sweep is performed by solving on a plane and then moving to the next one. Convergence is assumed when the maximum residual of the equations is less than 10^{-4}.

A.3 Boundary conditions.
For the inlet in the computational plane the values of inlet velocity are described. Usually a uniform profile is used and the other two components of velocity have zero values. In cases where there exist experimental data, a measured profile of velocity is used. On solid walls the typical boundary conditions are used, i.e. the non-slip condition for the velocities and the wall functions [13] for the k and ε. The inlet values for k and ε are calculated from the following expressions:

$$k_{in} = \frac{3}{2}(T_u u)^2 \qquad \varepsilon_{in} = \frac{k_{in}^{3/2}}{L_\varepsilon} \qquad (13,14)$$

where T_u is the turbulence level, L_ε is a characteristic length of the domain. For the pressure, a Neumann condition is used on every boundary as the SIMPLE procedure requires.

B. APPLICATIONS.

B.1 Single jet injected into a cross-flow.
The first application is the injection of fluid into a cross-flow at an angle of 45 degrees. The flow domain is shown in Fig. 1. The injection duct is a circular pipe with a radius equal to 7mm. The injection hole formed by the intersection of the injection pipe with the wind tunnel wall is an ellipse with major and minor axes R=10mm and r=7mm respectively. The mean flow velocities in the injection pipe and the wind tunnel are equal to 100 and 33.33 m/sec. Due to the symmetry of the flow field, only half of the domain is considered. The inlet section is located at x=-5R and the exit at x=40R. With this simulation of the experiment, the effects which are important in the development of the flow of coolant in a turbine can be quantified [14]. The flow field generated by injecting a jet into a cross-flow is very complicated, especially for high velocity ratios. The nature of the flow field for the range of velocity ratios from R_v=0.5 to R_v=2 has been well described by Andreopoulos and Rodi [15]. For the very high velocity ratio R_v=3 of the present study the flow in the hole exit should be less affected by the external flow than the R_v=2 case [15], so that the assumption of fully developed velocity profile at the hole exit is valid.

Grid independent results can be obtained on a moderate 105300 cells grid (75X39X36), using the 3rd-order MUSCL scheme for the convective terms. The final run using this moderate grid is performed by using as initial profile from the velocity measurements upstream of the wind tunnel as reported by Ballas and Goulas [14]. Over the hole the axial velocity profile is prescribed and then this is decomposed into u and v profiles while w-profile is zero. The turbulence level is taken equal to 10%. Initial runs show that the best results are obtained by using the 3rd-order MUSCL scheme. Figure 2 shows the vector plot of the u-v velocities in the region near the injection of the jet into the cross-flow. There is a large wake region in the lee of the jet and this region is associated with a low back pressure, which causes significant inflow towards the symmetry plane. Figure 3 shows a vector plot of the v-w velocities at a distance X=35 mm where the vortex is forming by the steep gradients of the v-velocity in the z-direction. Figure 4 shows how the longitudinal mean velocity varies with the distance from the wall, compared with the experimental results of Ballas and Goulas, at various downstream positions and for two planes (symmetry plane Z/D=0 and Z/D=1). From this figure one can see that there is quite good agreement between the computational and the experimental results. The MUSCL scheme has a good behaviour and it is capable in general to predict the experimental velocity distribution. Differences appear when the velocity gradients are strong but one can say that the predictions can give with a good accuracy the variation of mean velocity downstream of the region of the injection of the jet. The v-profiles shown in Fig. 5 suggest a very complex secondary motion. Very close to the wall the v-velocity takes negative values, indicating the presence of a vortex counterrotating the bound vortex, which was not captured by the experiments. The predictions compared with the measurement show that the v-profiles can follow the general trend but at some stations fail to predict correct magnitudes.

B.2 Flow in a wing-body junction.

The second application is the turbulent flow at the junction of a flat surface and a normal-mounted cylindrical wing, Fig. 6, dominated by the presence of a horseshoe vortex. The adverse pressure gradient in the plane of symmetry causes the boundary layer approaching the wing to separate. Upstream of the zone of separation and in the vicinity of the separation point, mean velocity and Reynolds stress profiles are like those in a 2D boundary layer separating in an adverse pressure gradient. Initial values for u-velocity profiles are taken from the experimental data. The turbulence is modelled using the Launder-Sharma [16] low Reynolds number k-ε model. A symmetrical C-type grid is used to simulate the flow field. Initial runs show that a 123840 cells grid (96X43X30) gives grid independent results using the MUSCL scheme while satisfying for the low-Reynolds number k-ε model restrictions for the y^+ values near the wall. The numerical predictions are compared with the experimental results of Fleming et al. [17].

Figure 7 shows the vector plot at the stagnation point on the symmetry plane using the HYBRID scheme (a), the HLPA scheme (b) and MUSCL scheme (c). The differences occured on the shape of the recirculation bubble are caused by the three different levels of accuracy among the schemes. Figure 8 show the u-velocity profiles using the three schemes, compared with the experimental data, on the symmetry plane at a position upstream the wing X/T=-0.1591 and near the stagnation point (T is the maximum wing thickness). The MUSCL scheme behaves better than the two other schemes and that is expected because of the higher accuracy of this scheme. The three schemes though, underpredict the size of the recirculation bubble. In the y-z planes the velocity profiles at X/T=0.75 from the stagnation point are shown in Figures 9 and 10. Figure 9 shows the velocity profiles using the three schemes compared with the experimental data at a distance from the wing surface Z/T=0.925 and Fig. 10 at a distance Z/T=1.00. The HYBRID scheme is not capable of predicting any horseshoe vortex while the other higher order schemes indicate the presence of the vortex although much smaller in size. A reason for this behavior can be the modelling of turbulence by the low Reynolds number k-ε model which is an isotropic one and cannot correctly capture the anisotropic phenomena caused by the presence of the vortex. In the region of nearly uniform velocities, satisfactory results are obtained from all schemes.

C. CONCLUSIONS.

In this work, the performance of three convective schemes on an elliptic Navier-Stokes solver using body-fitted coordinates in a generalized system was tested and compared with experimental data. The aim is to find the best scheme with good accuracy and low cost of computations when moderate grids are used. The third-order MUSCL scheme is found to be the best among them, although it is not as stable as the second order HLPA. The computations were performed on HP 720 and HP 712/60 workstations which use RISC architecture. For both cases presented, the real user-time to obtain convergence was about 24 hours in single user mode providing that the very efficient Stone's algorithm for solving the system of equations of the independent variables was used. The limitations of the k-ε turbulence model were very apparent in the complex case of the wing-body junction where the anisotropic phenomena of the flow due to the appearance of a horseshoe vortex are incompatible with the turbulence modelling.

REFERENCES.

1. B.E.Launder and D.B. Spalding, The numerical computation of turbulent flows, *Comp. Meth. in Appl. Mech. and Eng.*, Vol.3, pp.269-275, 1974.
2. P. Tamamidis and D.N. Assanis, Three dimensional incompressible flow calculations with alternative discretization schemes, *Numer. Heta Transfer, Part B*, Vol. 24, pp.57-76, 1993.
3. S.V. Patankar, *Numerical Heat Transfer and Fluid Flow*, Hemisphere, Washington, D.C., 1980
4. J.P. VanDoormal and G.D. Raithby, Enhancements of the SIMPLE Method for Predicting Incompressible Fluid Flows, *Numer. Heat Transfer*, Vol. 7, pp. 147-163, 1984.
5. C.M. Rhie and W.L. Chow, Numerical study of the turbulent flow past an isolated aifoil with trailing edge separation, *AIAA J.*, Vol. 11, pp. 1525-1532, 1983.
6. S. Majumdar, Role of Underelaxation in Momentum Interpolation for Calculation of Flow with Non-staggered Grids, *Numer. Heat Transfer*, Vol. 13, pp.125-132, 1988.
7. S. Majumdar, Development of a Finite Volume Procedure for Prediction of Fluid Flow Problems with Complex Irregular Boundaries, Technical Report, SFB 210/T/29, Sonderforschungsbereich 210, Univ. of Karlsruhe, December, 1986.
8. J. Zhu, A Low Diffusive and Oscillation-Free Convection Scheme, *Comm. in Applied Numer. Methods*, Vol. 7, 1991.
9. B. VanLeer, Towards the ultimate conservative difference scheme.IV. A new approach to numerical convection, *J. Comput. Phys.*, Vol.23, pp. 276-299, 1977.
10. B.P. Leonard, A stable and accurate convective modelling procedure based on quadratic interpolation, *Comput. Methods Appl. Mech. Eng.*, Vol. 19, pp. 59-98, 1979.
11. P.H. Gaskell and A.K.C. Lau, Curvature-compensated convective transport: SMART, A new boundedness-preserving transport algorithm, *Int. J. Numer. Methods Fluids*, Vol. 8, pp. 617-641, 1988.
12. H.L. Stone, Iterative Solution of Implicit Approximations of Multi-Dimensional Partial Differential Equations, *SIAM J. Num. Anal.*, Vol. 5, 1968.
13. B.E. Launder and D.B. Spalding, The numerical computation of turbulent flows, *Comp. Meth. in Appl. Mech. and Eng.*, Vol. 3, pp. 269-275, 1974.
14. M. Ballas and A. Goulas, Study of the flow from a row of cooling holes, *Lab. of Fluid Mechanics and Turbomachinery report*, prepared for the AER2-CT92-0044 EU project, 1994.
15. J. Andreopoulos and W. Rodi, Experimental inestigation of jets in a crossflow, *J. Fluid Mechanics*, Vol. 138, pp. 101-132, 1961.
16. B.E. Launder and B.I. Sharma, Application of the Energy-Dissipation Model of Turbulence to the Calculation of Flow Near a Spinning Disc, *Letters in Heat and Mass Transfer*, Vol. 1, pp.131-138, 1974
17. J.L. Fleming, R.L. Simpson, J.E. Cowling and W.J. Devenport, An Experimental Study of a Turbulent Wing-Body Junction and Wake Flow, *Exp. in Fluids*, Vol. 14, pp.366-376, 1993.

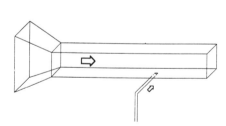

Fig. 1. The schematic of the flow domain

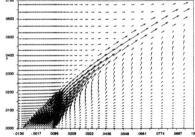

Fig. 2. u-v vector plot at the plane of symmetry (z=0).

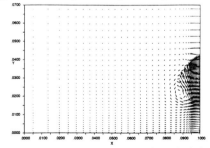

Fig.3. v-w vector plot at x=35mm from the hole.

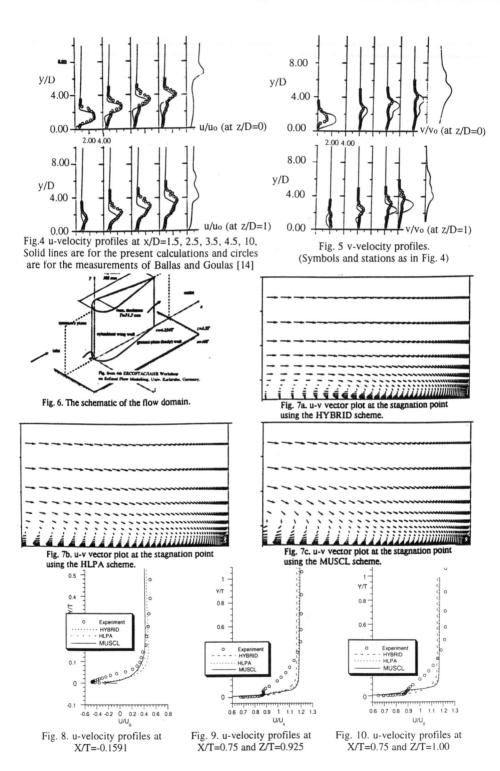

Fig.4 u-velocity profiles at x/D=1.5, 2.5, 3.5, 4.5, 10. Solid lines are for the present calculations and circles are for the measurements of Ballas and Goulas [14]

Fig. 5 v-velocity profiles. (Symbols and stations as in Fig. 4)

Fig. 6. The schematic of the flow domain.

Fig. 7a. u-v vector plot at the stagnation point using the HYBRID scheme.

Fig. 7b. u-v vector plot at the stagnation point using the HLPA scheme.

Fig. 7c. u-v vector plot at the stagnation point using the MUSCL scheme.

Fig. 8. u-velocity profiles at X/T=-0.1591

Fig. 9. u-velocity profiles at X/T=0.75 and Z/T=0.925

Fig. 10. u-velocity profiles at X/T=0.75 and Z/T=1.00

Flows with Interfaces: dealing with surface tension and reconnection

S. Zaleski[1], Jie Li[1], R. Scardovelli[2,3] and G. Zanetti[2]

[1] LMM, CNRS, Univ. P. et M. Curie, Tour 66, 75252 Paris Cedex 05, France
[2] CRS4, Via N. Sauro, 10, I 09123 Cagliari,
[3] Lab. di Ing. Nucleare, Via dei Colli 16, I 40136 Bologna

Abstract

In this lecture, we discuss the numerical simulation of 3D flows with interfaces. We briefly review the Volume of Fluid approach to interface simulation and the conservative modeling of surface tension. Applications to falling droplets are shown.

1 Introduction

How may one simulate an interface in a 3D flow — a thin surface separating two thermodynamic phases — on a computer ? The answers to that question are numerous [1, 2], and the clearest thing that emerges is that it is a difficult problem, especially in the three-dimensional case. The problem is, however, of tremendous applied and technological interest. There are for instance no accurate models that can predict when droplets or bubbles will coalesce or breakup. A precise knowledge of these mechanisms would be useful for a better parametrisation of larger scale computational models.

In this Lecture we will consider the Navier-Stokes equations for viscous fluids. The domain is filled with two fluids, say phase 1 and 2. The position of the two fluids is determined by a characteristic function C equal to 1 in phase 1 and 0 in phase 2. We will consider interfaces with a constant tension σ. We let the mean curvature of the surface be κ. On the surface we define also the unit normal $\mathbf{n}$. With p the pressure and $\mathbf{u}$ the velocity field, the equations read

$$\rho(\partial_t \mathbf{u} + \mathbf{u} \cdot \nabla \mathbf{u}) = -\nabla p + \nabla \cdot \mu \mathbf{S} + \sigma \kappa \delta_S \mathbf{n} \qquad (1.1)$$

where ρ is the density, μ is the shear viscosity, and $\mathbf{S}$ is the rate of strain tensor

$$S_{ij} = \frac{\partial u_j}{\partial x_i} + \frac{\partial u_i}{\partial x_j}. \qquad (1.2)$$

Together with (1.1) we require incompressibility

$$\nabla \cdot \mathbf{u} = 0. \qquad (1.3)$$

Notice that in (1.1) μ and ρ may vary in space. In fact, they are slaved to the volume fraction C

$$\rho = \rho_1 C + \rho_2 (1-C) \qquad (1.4)$$

$$\mu = \mu_1 C + \mu_2 (1-C). \qquad (1.5)$$

The interface itself is located on a manifold where C jumps from 0 to 1. This manifold is assumed to be differentiable almost everywhere, but reconnections may occur when droplets merge or break. The interface motion is given by the condition that the velocity of the interface normal to itself, U_i, equals the fluid velocity $\mathbf{u} \cdot \mathbf{n}$. In a real flow, breakup is a complicated process, involving 3D phenomena [3], [4], interaction forces between surfaces such as Van der Vaals forces, and possibly molecular scale effects. We cannot simulate all these effects with current methods, and it is at any rate difficult to envision 3D simulations going all the way from the molecular scale up to the droplet scale. The pragmatic alternative is to introduce a cutoff scale δ below which we do not attempt to model the physics. Such a cutoff is consistent with the natural behavior of a Volume of Fluid type of code, in which interfaces of thickness smaller than the mesh size h are broken.

If we introduce a velocity scale U (for instance the speed of the surrounding gas) and a length scale in the problem (for instance the droplet size) then the problem has 4 independent dimensionless numbers $Re_i = UL\rho_i/\mu_i$, $We_1 = \rho_1 U^2 L/\sigma$ and ρ_2/ρ_1. Numerical analysis may follow two broad strategies. One is to have a set of Lagrangian markers which follow the interface. The computational mesh may deform to follow these markers or stay fixed while they advance. This creates rather difficult problems for 3D flows, as one has to follow the motion of triangulated surfaces and relate tem to the mesh uised for the solution of the Navier-Stokes equation. Another, simpler strategy is to represent the position of the interface with a volume of fluid function C_{ij} which is the fraction of cell ij filled with phase 1 (see figure 1). This strategy. which we describe here, has two principal advantages. It allows to keep an exact balance of mass on the lattice and may thus avoid large deviation of the total mass in the course of relatively long simulation. It is also simple and robust in the presence of interface reconnection, in which case it introduces automatically a cutoff length as discussed above. These advantages are especially valuable for the simulation of 3D flow.

2 Method

Our version of the Volume of Fluid method includes a kinematic algorithm for tracking the motion of the interface, and a "dynamical" part that integrates equation (1.1).

There are many methods for solving equations (1.1) without the singular surface tension term $F_S = \sigma\kappa\delta_S\mathbf{n}$. In our calculations we choose a simple robust method and perform a first order in time explicit integration on the MAC staggered finite difference grid. The incompressibility condition is accurately satisfied by a projection method [5] with the help of a multigrid algorithm [6]. This yields a velocity field $\mathbf{u}$ which is used to propagate the interface Surface tension is implemented in a new, momentum conserving way, via the introduction of a non-isotropic stress tensor concentrated near the interface. We notice that one may write the capillary force as the divergence of a capillary stress tensor $F_S = \nabla \cdot \mathbf{T}$ where

$$\mathbf{T} = \sigma(\mathbf{I} - \mathbf{n} \otimes \mathbf{n})\delta_S . \qquad (2.6)$$

It is possible to obtain a momentum conserving representation of volume fraction forces from equation (2.6). A crude representation of the tensor $\mathbf{T}$ may be obtained from a smoothed characteristic function $\tilde{C}$. Such a smoothed function may be obtained by a convolution of the true characteristic function C by an appropriate filter, as shown in [7]. This smoothed representation of C, may be used to approximate the normal and the delta function since the unit normal to the interface obeys $\nabla C = \mathbf{n}\delta_S$. This serves to calculate the force F_S using Laplace's law in the CSF method of [7]. In the SURFER method of [8], we calculate $F_S = \nabla^h \cdot \mathbf{T}^h$ where ∇^h is a finite difference gradient and $\mathbf{T}^h$ approximates $\mathbf{T}$:

$$\mathbf{T}^h = \sigma \left(|\nabla^h C| \mathbf{I} - \frac{\nabla^h C \otimes \nabla^h C}{|\nabla^h C|} \right). \qquad (2.7)$$

For a smooth function $\tilde{C}$ the finite difference operators ∇^h converge to an approximation of a true gradient. However the method works even with very little smoothing or no smoothing at all, a fact perhaps related to the conservative nature of the formulation.

This representation of surface tension stresses is especially interesting for the simulation of breakup, since it allows one to avoid the singularity which would occur in the continuum limit when interfaces change topology and the curvature becomes locally infinite. Another interesting property is the conservative nature of the representation in the SURFER method. The total sum of forces added for a finite object such as a bubble then vanishes. The advantage of the method is similar to that of the volume of fluid method for interface kinematics: it allows to keep an exact balance of momentum on the lattice and may thus avoid large deviation of the total momentum in the course of relatively long simulations.

The location of the interface is approximately represented by the volume fraction C_{ij} of fluid 1 in cell i, j (figure 1). We have $0 < C_{ij} < 1$ in cells cut by the interface and $C_{ij} = 0$ or 1 away from it. To reconstruct the interface from the C_{ij}, a simple choice is to use only vertical and horizontal line segments [9], as in figure 1a. We have recently developed a more accurate method, which yields errors of order $\mathcal{O}(h^2)$ [10, 11, 12], [13]. In this method, linear segments of slope approaching that of the interface are constructed as on figure 1b. The interface slope is estimated by inspecting the volume fraction of neighboring cells.

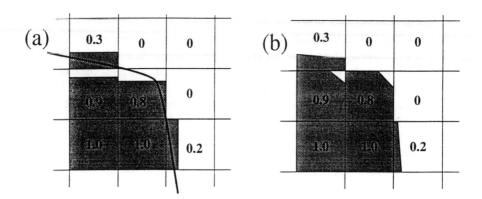

Figure 1 (a) First order interface reconstruction. The interface position is approximately represented by the volume fraction C. At first order, the interface in each cell may be seen as a straight line parallel to one of the cartesian axes (b) Second order reconstruction. In a first step the volume fraction field C is used to estimate the local slope of the interface. The interface is reconstructed as a line segment enclosing the given volume C. The line segments need not join at their ends, although the errors are exaggerated in the figure. In a second step, the reconstructed interface is advected in a Lagrangian manner

The interface motion is then calculated in a Lagrangian manner with velocities obtained by linear interpolation. Finally the volume fractions are recalculated.

3 Results

To illustrate the capabilities of our method, we show simulations of a liquid drop falling through a gas on a layer of the same liquid (Fig. 2). The density ratio is 2, and the Weber number is on the order of 10 in the first simulation. The simulations are performed on a 64^3 grid, in about 8 hours of CPU on an IBM RISC 6000 workstation. In the second simulation the Weber number is on the order of 100. In both simulations the gas and liquid Reynolds numbers are on the order of 10^3. A typical mushroom shaped structure is created after the impact of the droplet on the liquid layer. (Fig. 3). Simulations of colliding droplets have been shown in reference [8]. Several movies of simulation, including the two examples above, may be found on the World Wide Web in the URL

http://www.crs4.it/Animate/Animations.html

4 Conclusion

The simulations presented in this paper show the feasability of our approach. With a simple robust method, it is possible to simulate very complex 3D flow even on relatively modest hardware. However, our ability to represent many flows of interest is limited by the very fast growth of CPU time with grid size. Indeed in 2D we already need 512^2 grids to resolve some problems with sufficient accuracy [14]. Extrapolating to 3D, we face the prospect of 512^3 simulations which, at current computational speeds, would last 8 years on a workstation. Finding practical ways to realize such simulations is at present a formidable challenge.

References

[1] J. M. Hyman. Numerical methods for tracking interfaces. *Physica D*, 12:396–407, 1984.

[2] J. M. Floryan and H. Rasmussen. Numerical methods for viscous flows with moving boundaries. *Appl. Mech. Rev.*, 42:323–340, 1989.

[3] J. Eggers. *Phys. Rev. Lett.*, 71:3458–3461, 1993.

[4] H. A. Stone. Dynamics of drop deformation and breakup in viscous fluids. *Annual Reviews of Fluid Mechanics*, 26:65–102, 1994.

[5] A. J. Chorin. Numerical solution of the Navier–Stokes equation. *Mathematics of Computing*, 22:745–762, 1968.

[6] W. H. Press and S. A. Teukolsky. Multigrid methods for boundary value problems. *Computers in Physics*, SEP/OCT:514–519, 1991.

[7] J.U. Brackbill, D. B. Kothe, and C. Zemach. A continuum method for modeling surface tension. *J. Comp. Phys.*, 100:335–354, 1992.

[8] B. Lafaurie, C. Nardone, R. Scardovelli, S. Zaleski, and G. Zanetti. Modelling merging and fragmentation in multiphase flows with SURFER. *J. Comp. Phys.*, 113:134–147, 1994.

[9] W.F. Noh and P. Woodward. SLIC (simple line interface calculation). In A.I. van de Vooran and P.J. Zandberger, editors, *Proceedings, Fifth International Conference on Fluid Dynamics*, volume 59 of *Lecture Notes in Physics*, Berlin, 1976. Springer.

[10] D.L. Youngs. Time dependent multimaterial fow with large fluid distortion. In K. M. Morton and M. J. Baines, editors, *Numerical methods for fluid dynamics*, pages 27–39, New York, 1982. Academic Press. Institute for Mathematics and its Applications.

[11] N. Ashgriz and J. Y. Poo. FLAIR: Flux line-segment model for advection and interface reconstruction. *J. Comp. Phys.*, 93:449–468, 1991.

[12] E. G. Puckett and J. S. Saltzman. A 3d adaptive mesh refinement algorithm for interfacial gas dynamics. *Physica D*, 60:84–93, 1992.

[13] Jie Li. Calcul d'interface affine par morceaux (piecewise linear interface calculation). *C. R. Acad. Sci. Paris, série IIb, (Paris)*, 320:391–396, 1995.

[14] S. Zaleski, Jie Li, and S. Succi. Two-dimensional Navier-Stokes simulation of deformation and breakup of liquid patches. *Phys. Rev. Lett.*, 75:244–247, 1995.

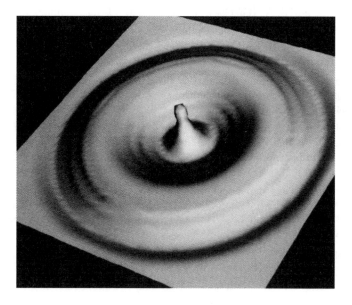

Figure 2 A droplet falling on a liquid surface. The Weber number is of the order of 10, and the simulation box is 64^3. Reconnection occurs easily, and a characteristic jet shoots back up. The figures show two typical instants of the simulation, just before and just after the reconnection of the droplet with the surface.

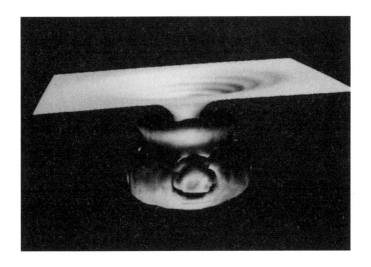

Figure 3 As in figure 3, a droplet falling on a liquid surface is simulated. The Weber number, at $\simeq 100$ is however much larger. The impact of the droplet creates a mushroom-shaped gas cavity (photo(a)) in the liquid layer, reminiscent of the Rayleigh-Taylor instability. A toroidal gas bubble forms in this cavity and climbs back up to the surface, where it breaks (photo (b)).

Numerical Simulation of Three-Dimensional Instabilities and Secondary Regimes in Spherical Couette Flow

Oleg Zikanov
Institut für Mechanik, TH Darmstadt
Hochschulstr.1, Γ-64289 Darmstadt, Germany

Abstract

The solutions of the nonlinear and linearized Navier-Stokes equations are computed to investigate the instabilities and the secondary two- and three-dimensional regimes in the flow of an incompressible viscous fluid in a thin gap between two concentric differentially rotating spheres. The numerical technique is finite-difference in radial direction, spectral in azimuthal direction, and pseudo-spectral in meridional direction. In agreement with the previous experimental results, two different three-dimensional traveling wave flows with spiral vortices are detected.

1 Introduction

The results to be discussed in the present paper concern the behaviour of the flow of incompressible viscous fluid contained between two concentric spheres rotating about a common axis with fixed angular velocities. This spherical Couette flow (abbreviated frequently as SCF) is of basic interest for understanding global astrophysical and geophysical processes which depend crucially on two factors: rotation and spherical geometry.

An important property of the system under consideration is that it displays two types of symmetry – reflection-symmetry with respect to the equatorial plane and rotational symmetry.

Three dimensionless parameters specify the problem – the Reynolds number $Re_1 = R_i^2 \Omega_i/\nu$, the angular velocity ratio $\omega = \Omega_o/\Omega_i$ or, alternatively, the second Reynolds number $Re_2 = R_o^2 \Omega_o/\nu$, and the relative gap size $\delta = (R_o - R_i)/R_i$.

The numerous experimental and theoretical studies have shown that the stability of the basic flow and the type of secondary regime depend heavily on the problem parameters. In thin layers [1], the centrifugal instability of the basic flow results in transition into the steady, axisymmetric flow with two Taylor vortices only in restricted range of Re_2. Outside this range, three-dimensional time-periodic flows with spiral vortices near the equator are developed at the stability limit. The present investigation is believed to be the first extensive study to simulate numerically these secondary regimes. The case of a thin layer $\delta = 0.11$ and both spheres rotating is considered.

2 Method of solution

Any steady or time-periodic flow is found as the limit at $t \to \infty$ of the solution to the initial-boundary value problem consisting of the Navier-Stokes equations for incompressible fluid, no-slip boundary conditions at inner and outer radii and initial condition which is generally taken to be the flow calculated at nearby values of the parameters. A combined method (spectral decomposition in polar and azimuthal directions and finite-difference technique in radial direction) is used.

A solenoidal velocity field is represented as a sum of toroidal and poloidal terms. The toroidal and poloidal scalar fields T and S, and the pressure P are expanded in spherical harmonics $Y_l^m(\theta, \varphi) = P_l^m(\cos\theta)e^{im\varphi}$ (P_l^m is an associated Legendre function). The series are truncated so that $|m| \leq M$, $|m| \leq l \leq L$.

The equations for the complex-valued coefficients are obtained by substituting the expansions into the governing equations and applying the operators $\mathbf{e}_r \cdot rot$, $\mathbf{e}_r$, and $-(\sin\theta)^{-1}(\frac{\partial}{\partial\varphi}\mathbf{e}_\varphi + \frac{\partial}{\partial\theta}\sin\theta\mathbf{e}_\theta)$.

It is shown in [2] that the nonlinear terms of the equations can be represented as finite series of spherical harmonics. The nonlinear coefficients are calculated at each time step using the straightforward evaluation of convolution sums in azimuthal direction and a pseudospectral technique in polar direction. The pseudospectral procedure is produced via the Gaussian integration which can be performed to the machine accuracy since the integrands are finite series of the associate Legendre functions.

The resulting uncoupled systems of second-order in radial coordinate r, first-order in time t differential equations are solved by an explicit-implicit in t, finite-difference in r technique. In the radial direction the equations are discretized by central differences of the second order. The nonlinear terms are computed at the n-th time layer and the other terms are computed at the $(n + 1)$-st one. The boundary-value difference problems are solved by a simple and convenient variant of Gauss elimination known as "double sweep method" [3].

Most of the calculations is performed with the following values of discretization parameters: $M = 6$ ($M = 0$ if axisymmetry is imposed); $L = 60$ or $L = 90$; the number of radial discretization points $K = 10$ or $K = 15$; and the time step $\Delta t = \pi/50 =$ inner sphere rotation period/100.

The linear stability problem consisting of the linearized Navier-Stokes equations, homogeneous boundary conditions, and initial condition is solved in the same manner as the fully nonlinear problem except that the former nonlinear

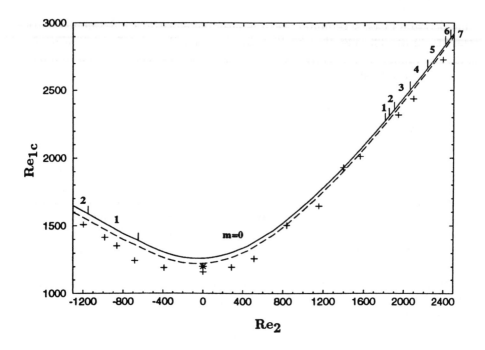

Figure 1 —, critical number Re_{1c} of stability of the basic flow as a function of Re_2; m, the azimuthal wavenumber of the preferred eigenmode; - - - -, experiments [1] with $\delta = 0.1096$; + + +, experiments [4] with $\delta = 0.111$; *, calculations [6] with $\delta = 0.115$.

terms are now the linear functions of unknown variables. The preferred eigenmode is obtained as a time limit of the solution.

3 Linear stability of the basic flow

The linear stability of the basic flow was analyzed at $\delta = 0.11$ and $-1300 \leq Re_2 \leq 2500$. Main results are shown in figure 1.

The solution of the linear stability problem demonstrates the diversity of symmetry-breaking bifurcations of SCF in the range of Re_2 under consideration. This range can be divided into three parts.

When $-646 < Re_2 < 1811$, the first instability of the basic flow is to the eigenmode that is axisymmetric and anti-reflection symmetric. The primary bifurcation breaks the equatorial reflection symmetry, with the rotational symmetry remaining unchanged. The eigenvalue of the preferred eigenmode is always real.

At $1811 < Re_2 < 2500$ the preferred eigenmodes are also anti-reflection symmetric. In addition, they are nonaxisymmetric, that is, the primary bifurcation breaks the rotational symmetry as well as the equatorial one. The nonaxisymmetric modes are azimuthally travelling waves propagating in the sense of rotation of inner sphere.

When the boundary spheres rotate in opposite directions and $-1300 < Re_2 < -646$, a third type of symmetry-breaking bifurcation appears. Again the preferred eigenmodes are azimuthally travelling waves propagating in the sense of rotation of inner sphere. But in this case they are reflection-symmetric.

4 Secondary regimes and transitions among them

The full nonlinear equations were solved to simulate the secondary flows. In agreement with the experimental results [1], steady axisymmetric flows with one and two Taylor vortices per hemisphere and two different nonsteady three-dimensional flows with spiral vortices were detected. We will refer to these flows and to the basic flow as regimes **I**, **II**, S_1, S_2, and **0**, respectively.

Slow rotation of the outer sphere; $-646 < Re_2 < 1811$. At the stability limit $Re_1 = Re_{1c}$, the basic flow equilibrium becomes unstable at an equatorial symmetry-breaking subcritical pitchfork bifurcation and a $0 \to \mathbf{I}$ transition occurs, which was described in [5] and [6].

The transition $\mathbf{I} \to 0$ can be initiated by lowering Re_1 to a value lesser than the turning point of the branch of regime **I**.

By imposing axial and reflection symmetry of the flow on our numerical code, the branch of basic flow solutions can be extended into the supercritical region. The extension reveals the phenomenon previously described in [5]. Linearly unstable regime **0** transforms smoothly into unstable regime **II** which turns stable at higher Re_1.

The three-dimensional flow with spiral vortices (regime S_1 shown in figure 2) was obtained in supercritical region using the solution at $Re_2 > 1811$ as an initial condition. Decreasing Re_1 through the lower boundary of the existence region of regime S_1 leads to a transition $S_1 \to \mathbf{I}$.

Linear stability analysis of axisymmetric regimes **I** and **II** has shown that increase of Re_1 gives rise to the sinusoidal travelling waves on the Taylor vortices.

Co-rotating spheres; $Re_2 > 1811$. The experiments [1] disclosed that at $Re_2 > 1940$ the first instability of the basic flow results in a travelling azimuthal wave flow with equatorially asymmetric spiral vortices located near the equator. The flows with one and two pairs of Taylor vortices were detected in supercritical region.

The bifurcation diagram constructed at $Re_2 = 2050$ and shown in figure 3 explains the laboratory observations. The primary bifurcation of the basic flow breaks the rotational and equatorial symmetry. The transition $0 \to S_1$ into

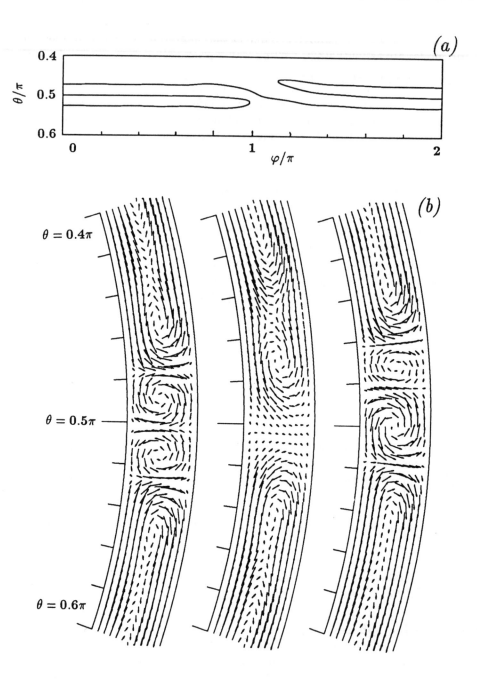

Figure 2 The flow with spiral vortices (regime S_1) (a) Vortex boundaries defined as curves of zero meridional velocity component at the spherical section $r = 1 + 6\delta/11$. The section is mapped into a φ-θ-rectangle and only the part near the equator is plotted. (b) $r - \theta$-projections of the velocity field plotted in the meridional sections crossed at $\varphi = 0.8\pi$, 0.95π, 1.2π (from left to right).

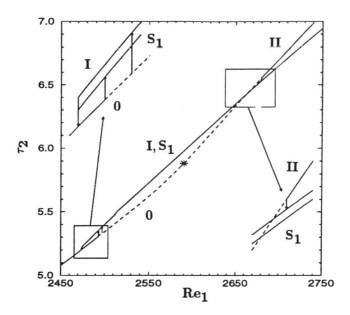

Figure 3 Bifurcation diagram at $Re_2 = 2050$. The torque τ_2 imposed by different flow regimes on outer sphere is shown as a function of Re_1. —— (- - - -), linearly stable (unstable) solutions; ↑, ↓, transitions which are possible under quasistatic change of Re_1; ∗, point of the smooth transformation of regime **0** into regime **II**.

the three-dimensional travelling wave secondary flow with spiral vortices can be produced at the stability limit $Re_1 = Re_{1c}$.

By imposing axial symmetry of the flow on our numerical code, the branch of the basic flow solutions can be extended into the supercritical region $Re_1 > Re_{1c}$. At $Re_1 = Re_{1c} + 4.1$, the axisymmetric, anti-reflection symmetric linear mode is added to the set of unstable modes. When restricting ourselves to axisymmetric solutions and increasing Re_1 quasistatically, we are in a position to obtain at this point the transition $\mathbf{0} \to \mathbf{I}$ which is similar to that calculated with $Re_2 = 0$.

Regimes **I** and $\mathbf{S_1}$ can be obtained in the subcritical region $Re_1 < Re_{1c}$ by quasistatic decrease of Re_1. The backwards transitions $\mathbf{S_1} \to \mathbf{0}$ and $\mathbf{I} \to \mathbf{0}$ occur at the close Reynolds numbers.

Imposing the reflection symmetry condition, we can obtain the smooth transformation of regime **0** into regime **II** which is similar to that at $Re_2 = 0$.

Counter-rotating spheres; $Re_2 < -646$. It was observed in [1] that at $Re_2 < -920$ the first instability of the basic flow results in transition into the travelling wave flow with spiral vortices near the equator. The spirals differ in appearance from those of regime $\mathbf{S_1}$.

The nonlinear simulation was performed with $Re_2 = -1200$. The primary bi-

bifurcation results in transition $0 \to S_2$ into the equatorially reflection-symmetric travelling wave flow with spiral vortices.

An essential property of the instability is that the bifurcation appears to be supercritical. The Reynolds number for transition $0 \to S_2$ was defined by the linear stability analysis as $Re_1 = 1597.3$. The maximum value of Re_1 at which the backwards transition $S_2 \to 0$ was produced by our nonlinear code is $Re_1 = 1597$.

5 Concluding remarks.

The following principal conclusion can be made. The parameter range under consideration ($\delta = 0.11, -1300 \leq Re_2 \leq 2500$) can be divided into three parts depending on the type of primary bifurcation. At $-646 < Re_2 < 1811$, the bifurcation is a subcritical pitchfork breaking equatorial symmetry and resulting in transition into the steady axisymmetric and reflection-symmetric flow with one pair of Taylor vortices. When $Re_2 < -646$ or $Re_2 > 1811$, the bifurcation breaks the rotational symmetry and causes azimuthally travelling wave secondary flows with spiral vortices to appear at the stability limit. The difference between the last two regions is that in the first the equatorial symmetry is preserved and the travelling wave solution branches off supercritically, whereas in the second the bifurcation is subcritical and breaks not only rotational but also equatorial symmetry, resulting in the equatorially asymmetric secondary solution.

We believe that the semi-spectral numerical technique employed here has demonstrated its capability for reproduction of the three-dimensional flows of viscous incompressible fluid in spherical annulus.

Acknowledgments The investigations were began on the initiative of I.M.Yavorskaya. Her help cannot be overestimated. The author thanks N.M.Astaf'eva and Yu.N.Belyaev for many useful discussions. This work was supported in part by the Russian Fund for fundamental research (grants 93(94)-013-2896 and 93(94)-013-17342) and by the Alexander von Humboldt Foundation.

References

[1] YAVORSKAYA, I.M. & BELYAEV, YU.N. 1986, *Acta Astrounautica* **13**, N6/7, 433-440.

[2] ZIKANOV, O.YU. 1995, submitted to *J. Fluid Mech.*.

[3] GODUNOV, S.K & RYABENKI, V.S. 1964, The theory of difference schemes. North-Holland, Amsterdam.

[4] WIMMER, M. 1981, *J. Fluid Mech.* **103**, 117-131.

[5] MARCUS, P.S. & TUCKERMAN, L.S. 1987, *J. Fluid Mech.* **185**, 1-65.

[6] SCHRAUF, G. 1986, *J. Fluid Mech.* **166**, 287-303.

Notes on Numerical Fluid Mechanics (NNFM)   Volume 53

Series Editors: Ernst Heinrich Hirschel, München (General Editor)
                  Kozo Fujii, Tokyo
                  Bram van Leer, Ann Arbor
                  Michael A. Leschziner, Manchester
                  Maurizio Pandolfi, Torino
                  Arthur Rizzi, Stockholm
                  Bernard Roux, Marseille

Volume 35 Proceedings of the Ninth GAMM-Conference on Numerical Methods in Fluid Mechanic
            (J. B. Vos / A. Rizzi / I. L. Ryhming, Eds.)
Volume 34 Numerical Solutions of the Euler Equations for Steady Flow Problems
            (A. Eberle / A. Rizzi / E. H. Hirschel)
Volume 33 Numerical Techniques for Boundary Element Methods (W. Hackbusch, Ed.)
Volume 32 Adaptive Finite Element Solution Algorithm for the Euler Equations (R. A. Shapiro)
Volume 31 Parallel Algorithms for Partial Differential Equations (W. Hackbusch, Ed.)
Volume 30 Numerical Treatment of the Navier-Stokes Equations
            (W. Hackbusch / R. Rannacher, Eds.)

Addresses of the Editors of the Series "Notes on Numerical Fluid Mechanics"

Prof. Dr. Ernst Heinrich Hirschel (General Editor)
Herzog-Heinrich-Weg 6
D-85604 Zorneding
Federal Republic of Germany

Prof. Dr. Kozo Fujii
High-Speed Aerodynamics Div.
The ISAS
Yoshinodai 3-1-1, Sagamihara
Kanagawa 229
Japan

Prof. Dr. Bram van Leer
Department of Aerospace Engineering
The University of Michigan
3025 FXB Building
1320 Beal Avenue
Ann Arbor, Michigan 48109-2118
USA

Prof. Dr. Michael A. Leschziner
UMIST-Department of Mechanical Engineering
P.O. Box 88
Manchester M60 1QD
Great Britain

Prof. Dr. Maurizio Pandolfi
Dipartimento di Ingegneria Aeronautica e Spaziale
Politecnico di Torino
Corso Duca Degli Abruzzi, 24
I-10129 Torino
Italy

Prof. Dr. Arthur Rizzi
Royal Institute of Technology
Aeronautical Engineering
Dept. of Vehicle Engineering
S-10044 Stockholm
Sweden

Dr. Bernard Roux
Institut de Mécanique des Fluides
Laboratoire Associé au C.R.N.S. LA 03
1, Rue Honnorat
F-13003 Marseille
France

Brief Instruction for Authors

Manuscripts should have well over 100 pages. As they will be reproduced photomechanically they should be produced with utmost care according to the guidelines, which will be supplied on request. In print, the size will be reduced linearly to approximately 75 per cent. Figures and diagrams should be lettered accordingly so as to produce letters not smaller than 2 mm in print. The same is valid for handwritten formulae. Manuscripts (in English) or proposals should be sent to the general editor, Prof. Dr. E. H. Hirschel, Herzog-Heinrich-Weg 6, D-85604 Zorneding.